UNWILLING BRIDE

"I warned you before that I would have you any time I desired," Reed growled. "I would rather have you willing, but . . ."

"Never!" she screamed, backing away from him.

"So be it," he said curtly. His hand flashed out, grabbing her nightgown at the throat. With one swift movement, he tore it from neck to hem. She gaped at him, astounded at his savagery. Frightened green eyes met those of frosty blue.

"No, Reed," she whispered past the lump of fear rising in her throat.

Quickly he jerked her toward the bed, flinging her onto it, then threw himself upon her, pinioning her to the mattress. She tried to push him away, but he held her arms beneath her, laughing at her efforts. Her head fell back, baring her throat to his searing kisses.

His lips ground down on hers, bruising her mouth. She bit down hard, and he drew back sharply, wiping the blood from his lip with his free hand. "You little Irish viper! Bite me, will you?"

He kissed her again roughly. She tried to twist her head away, but her hair was caught beneath her, holding her firmly. He drew back, and she glared at him defiantly.

"I'll tame you yet, Kat," he swore hoarsely, his blue eyes blazing.

"I'll see you in hell first!" she spat out.

FIRE
and
ICE

Catherine Hart

LEISURE BOOKS ∞ NEW YORK CITY

For my husband, who coped with and encouraged me; my three children, who ate sandwiches and soup for ages when I had no time to cook; and for Leslie, my sounding board and loyal proofreader for all her help.

A LEISURE BOOK

Published by

Dorchester Publishing Co., Inc.
6 East 39th Street
New York, NY 10016

Printed in the United States of America

Chapter 1

KATHLEEN stood alone on the crest of the hill. A breeze tugged at the skirt of her black silk dress, and her long red-gold hair whipped across her face. That face, lovely enough to start any man's heart racing, was filled with sorrow. Her large emerald eyes reflected the color of the Irish sea into which she tearfully stared. She was an arresting sight even in her mourning dress, with her beautiful young figure outlined in silhouette as the wind molded her dress against her thighs and firm, high breasts. Arms clutched tightly at her waist, she valiantly blinked away the tears that clung to her long, dark eyelashes and traced a path silently down her cheeks. A shudder ran through her slim frame as she tried to quell the sobs which threatened to start anew.

"Mistress Kathleen, 'tis time to go. The carriage is loaded and waiting. Mrs. Dunley sent me to fetch ye."

The girl turned to nod sadly to the elderly man climbing the hill toward her. "Thank you, George. I'm coming, though Lord knows why I have to go anywhere at all when I have a perfectly fine home and friends here." She sighed. "But, no! They are determined to send me half-way across the world to a wilderness full of Indians and crude backwoodsmen, and relatives I've never laid eyes on. It doesn't make a tinker's lot of sense to me!" She stopped her descent to stamp her small booted foot for emphasis, then continued, "And that's another burr under my saddle! Why is it just because I'm seventeen and a female, everyone thinks I haven't the brains to manage this estate by myself? Papa left it to me in his will. He knew I could manage it. For years now I have handled the books for the estate. I'll wager I know more about rents and sales of grains and stocks than most boys my age.''

"I'd bet on that, Miss," George agreed, nodding. "Besides, ye did a fine job of runnin' the house, too, since your poor dear mother took sick of the fever and passed away four years back. She was a fine Irish lady, she was. O' course, yer pa was a fine gentleman, too, even if he was English. No disrespect meant, Miss."

"None taken, George. To be sure, I'll miss him dearly. I'll never understand how his horse stumbled and threw him. Midnight was always so sure footed and Papa was a superb horseman." A fresh sob broke through as she recalled both horse and rider found three days before with their necks broken.

The neighbors had all been very kind. Nearly everyone in the county had attended the funeral the day before. Edward Haley had been well thought of by all who knew him. He had come to Ireland twenty years before, fallen in love with a beautiful red-haired lass named Ann O'Reilly, and married her. Lord Edward did not endorse the harsh way England ruled Ireland and her people, but there was little more he could do than treat his tenant farmers and servants fairly. He fell in love with the country and would never consent to becoming an absentee landlord as many Englishmen did. Thus, he won the admiration and respect of his new countrymen.

Kathleen had been born in Ireland and had only left for two short years to attend finishing school in England. Although she had excelled in her courses, she hated leaving her father alone in Ireland. She worked hard at learning to conduct herself as a lady, not an easy task for such a high-spirited tomboy as she. She quickly learned to speak French, Spanish, and even English without a trace of Irish accent. However, she did not learn to curb her quick temper, nor her razor-sharp tongue. She had a natural grace, never seeming awkward or clumsy as some young girls were. There were lessons in manners, how to walk and talk, dress and dance, eat and sit; how to entertain a gentleman and have him think you were interested in his every word when you were bored to tears. She also learned how to flirt, and to politely reject a gentleman's proposals, whether decent or indecent in nature. Kathleen learned to play the pianoforte tolerably well, and surprised even herself to learn what a beautiful full-throated singing voice she had. But most of this was extremely tiresome to her, so she determined to learn quickly and return home as soon as possible.

Once home, her father taught her more interesting things, such as fencing. As her father had hired the best fencing masters to tutor her, no one handled a rapier better than she. At first the rapier seemed extremely heavy, but gradually she grew used to its weight and length, and it became almost an extension of her right arm. Her reflexes were superb; her responses instantaneous. Soon she was besting each instructor in turn, much to their dismay and embarrassment.

Edward also took her on many of his sailing jaunts to England, checking with buyers for his small shipping line. She became an expert sailor, much of the time captaining the ship herself while her father busied himself with the merchandise or paperwork. The other sailors grudgingly respected her, but eventually grew used to having her command. Many took a great liking to the lass and were proud to sail under her. She seemed to inspire a rare loyalty in her men.

Rarely had Kathleen encountered anything she loved better than sailing, with the wind filling the sails and the sea changing colors beneath an azure sky or brilliant sunset. She gloried in the awesome power of the waves and the freedom of being perched in the rigging and losing herself in the beauty of a perfect dawn. The sights and sounds, smells and feel of the sea made her blood race wildly.

Now she would be sailing on her father's ship again, only this time as a passenger. She was going to Edward's sister's home in America. Aunt Barbara and Uncle William Baker lived in Savannah, Georgia, with their two children, Ted and Amy. Kathleen did not want to go, and determined to try again to persuade her father's solicitor and lifelong friend to let her stay. Somehow, the adventure of sailing unknown seas did not dull the pain of leaving home. "Perhaps," the thought crept in, "forever."

Reaching the bottom of the hill, they continued past the stables where Kathleen had already bid a final farewell to her beautiful black gelding, and circled around the corner of the huge stone mansion to the front entrance where the carriage stood ready. The servants were waiting to wish her well on her journey, all trying to look cheerful and failing miserably. Kathleen stopped before each with a kind word here, a pat there, an embrace for another. There were numerous sniffles and coughs, and nearly all were dabbing at their eyes by the time she climbed into the carriage with her faithful old Nanna and her father's solicitor, Mr. Kirby.

As they started rolling along the curved stone drive, Kathleen could not resist the urge to take a final, long look at the only home she had ever known. The stone mansion stood majestically beyond a rolling green lawn dotted by huge oaks. Steep hills rose behind, their emerald glow intensified against the brilliant blue sky above.

Now they were passing numerous pastures where sheep, cattle, and horses grazed contentedly. Kathleen could see farmfields beyond being readied for planting. All the world was coming to life to greet spring in a profusion of color, yet Kathleen had never felt so numb inside. She felt nothing save her heart-wrenching pain and grief.

"How I shall miss all this," she thought sadly.

Once again she turned to the portly Mr. Kirby and queried, "Why can I not stay here instead and run the estate myself with your help? I know I could do it. I promise not to be a burden. Please! I cannot bear this so soon after my dear papa has gone," she choked.

The kindly Mr. Kirby leaned across and gently laid a

hand on her arm, saying, "I'm sorry, lass, but I have explained all this to you before. It is unwise for you to remain here with political unrest becoming more pronounced each day. You are half English, and though you were raised here it is unhealthy for an unprotected young lady to ever consider staying. It would be very risky indeed. The Irish will resent the English blood you carry, and the English will not trust the Irish part of you. The situation here is fast becoming critical, and you should not be here when the cauldron boils over.

"In the meantime, the estate will be maintained through me, and kept running. The house itself will run with a small staff until you can return. Meanwhile, you must go to America and live with your aunt. A single lass such as yourself can not remain living alone in a huge house with only servants. It would only invite gossip and problems. Being a bachelor myself, it would not be deemed proper for me to take you in. You can not imagine the fortune hunters who would flock to your door. You are quite a wealthy woman with this estate. I know there are young lads aplenty who would sincerely court you for your love and beauty alone, but then there are none you have particularly favored so far. I must warn you there are others who would wed you for your fortune alone, since even now Catholic Irishmen can not own land under English law. An enterprising young scamp could wed a Protestant lass such as yourself, and though he couldn't claim a title, he would own the estates through marriage and his children could inherit the lands and the title of lord or lady. I would hate to see you marry for any reason other than a love such as your parents had. You need time to adjust to your loss so as not to jump hastily into a marriage you may regret. Give yourself this time, girl. You will see the wisdom of it."

"You are right when you say I have not found a man I can love as Mama loved Papa, but I doubt I shall find one in America, either," she said. "I would rather stay here and let the right man find me, but I can see I'll get no place trying to convince you on that matter."

Turning to Mrs. Dunley, she added glumly, "Well,

Nanna, it looks like we are off to America—like it or not. I hope you will have little trouble developing your sea legs. From what I figure it will take six to eight weeks to reach Georgia, depending on the weather.''

"Once you get there I'm sure you'll like it," Mr. Kirby assured. "I'm told Savannah is a port city with a fine ocean breeze in the summer and not too dreadfully cold in winter. At least not as cold as England or the northern American cities.''

"I've written your aunt of your coming, but could not really advise her on a time for your arrival. She is a lovely woman. In our university years, she was always underfoot of Edward and me. Her husband, Mr. Baker, seemed a nice enough sort, but they left soon after they were married, so we never saw either of the children. They are about your age, I suppose. Also," he continued, "your maternal grandmother lives on a plantation just outside Savannah, so you will not lack for people who care for you.''

"Grandmother O'Reilly," Kathleen mused. "I think I remember her and Grandfather from when I was little. When did they move to America?''

"They left when you were just a wee lass of three," contributed Mrs. Dunley. "Your mama used to say although your grandmother loved her new country, she missed her only daughter and granddaughter. You are her namesake, you know, Kathleen. She was called Kate. You look and act exactly like her. She was quite a spirited young lass, too, she was. Always into much mischief so I hear, just as you are most of the time, love.''

"Now, Nanna, don't start on me," Kathleen began. "I truly try to be good. I always have, but boys just have so much more fun. I never could resist climbing trees, swimming in the lake, catching frogs, and all.''

"To be sure," Mrs. Dunley laughed, "but you didn't have to fall out of the trees, swim in the altogether with the lads, and wallow in mud to get the frogs! Oh, you've led your mama and me a merry chase, and I've a feeling you're not done yet. Just now it is the boy's breeches I disapprove of so."

"Nanna, I'm sure you'd not be pleased to see me sail or ride astride a horse or fence in a skirt," Kathleen said. "They are not only uncomfortable and impractical, but also much more revealing than the breeches."

"Now I am wondering," Kathleen said, leaning out through the window of the carriage, trying to catch a glimpse of the bay they were nearing, "if Grandmother lives in Savannah, why can't I live with her? After all, she is probably lonely with Grandfather dead these past seven years."

Mr. Kirby shifted to a more comfortable position on his seat. "No, Kathleen, it is best you live with your aunt. Your grandmother is probably in her late sixties and an active young girl would likely be upsetting to her. Besides, I suppose she is not very active in the social circles any longer, while your aunt and uncle will be attending numerous functions of society and can introduce you to other young people in the proper circles. Your cousins will be much company for you too, I know, especially at first when you know no one else. Of course, you may visit your grandmother whenever you wish, though your Aunt Barbara's family, being English, does not associate with her overly much. In point of fact, they do not acknowledge any family ties to your Irish side of the family at all."

"In that case, I'm surprised they want me to come." Tossing her head, Kathleen thrust her chin out defiantly. Her already slightly uptilted nose tilted a little higher in her anger.

"Don't judge them unfairly, Kathleen. Get to know them first," Mr. Kirby advised. "Ah, we are almost at the docks. It won't be long now. A few minutes at the most. I am glad we started when we did. I would not want to annoy the captain by being late."

Kathleen breathed deeply of the sea and said, "By the way, who is this captain you've hired? It is my frigate we are sailing on and I could have captained it myself with my regular crew."

"I realize what an excellent sailor you are, my dear, but

11

this Captain Taylor I've hired is from Savannah. His family owns a plantation there and he knows the Baker family well. It is better to sail with someone who knows the Atlantic as well as he, having crossed it several times. Besides, many ships have been encountering pirates and privateers lately with England and France being at war. If you don't get stopped by the English or French, a privateer will surely try. Sailing with your father to England was one thing; this is quite another matter," Mr. Kirby stressed.

"Tell me more about Captain Taylor," Kathleen said. "How does he happen to be in Ireland without a ship?"

"Oh, he has a ship, miss. They arrived four nights ago smuggling in contraband arms for Ireland, so the rumor goes. I ran into him in one of the waterfront pubs. Of course, he does not own the other ship either, but sails it for the owners. I hear his family wanted him to stay home to run the plantation, and so would not support him with the price of his own ship. Nevertheless, he still chose a life at sea. He agreed to sail the *Kat-Ann* on the condition he could fill the hold with cargo to take back to America and sell at his own profit. Also, he will bring along half of his own crew, as they are used to sailing under his command, and fill in the other half with men who have sailed with your father. It is my understanding that his quartermaster will sail the other ship—the *Sea Fire* I believe it is—with the other half of his crew and others signed up from the docks. You will be even safer traveling in the company of another ship."

"Why is it we are sailing the *Kat-Ann* instead of the *Starbright?*" Though they sailed the *Kat-Ann* more often, the *Starbright* was Kathleen's favorite.

"Kathleen, you forget. The *Starbright* is in drydock getting her hull scraped and tarred. She'll soon be on the seas again and putting in at Savannah with goods and mail every so often. I'll tell her captain to be sure to let you know when she's in port."

The carriage jolted to a halt, nearly throwing all three passengers from their seats. Not waiting for the coachman to open the door, Kathleen jumped down in a flurry of skirts. She shielded her eyes from the sun with her hand and peered out into the bay. The *Kat-Ann* was riding at anchor about a quarter mile out. Glancing quickly about, Kathleen located the dinghy that would row them out to the ship. Taking off her glove, she put two fingers between pursed lips and gave a shrill whistle. Old Dan Shanahan looked up from the dinghy, gave a broad smile, and waved. Climbing nimbly from the boat, he hastened toward her.

"Cap'n Kathleen, 'tis a pleasure to see ye again. We'll be sailing together this voyage, though I must say 'twill seem queer not having ye in command. Mr. Kirby here warned us not to mention ye being a captain and all, seeing as that Captain Taylor knows yer relatives in America. I guess yer aunt would take a case of the vapors if she ever got wind of some of your unladylike talents, eh?" Dan chuckled.

"To be sure, Dan, that she would. Best keep our secrets to ourselves for a while," Kathleen grinned.

Mr. Kirby assisted a frowning Mrs. Dunley from the carriage, and Dan went to help the men take the bags aboard the dinghy.

"Kathleen, kindly conduct yourself properly," Mrs. Dunley pleaded wearily, "before I lose all these lovely gray hairs you've put on my head."

Grinning impishly at the older woman, Kathleen took her by the arm and together they walked toward the waiting dinghy.

When the dinghy came alongside the *Kat-Ann*, Kathleen was the first to climb aboard. Shrugging off the offer of assistance, she scurried sure-footedly up the ladder. Next came Nanna, very reluctantly and requiring much help and encouragement.

"I knew you could do it, Nanna," Kathleen soothed the older woman. "See, I told you it wasn't so hard." Kathleen patted her lightly on the arm as Nanna stepped wobbily aboard.

As she turned from the rail, Kathleen nearly collided with the captain. The first thing that impressed her was his size. Kathleen stood five foot six, which was tall for a woman, but this man towered at least eight inches above her. He stepped back and she saw that he was powerfully built, his uniform jacket seeming barely able to constrain his broad frame. Though he had wide shoulders, his hips and waist were slim above long muscular legs. She raised her eyes and gazed straight into icy blue eyes fringed with thick black lashes. His dark hair was trimmed to collar length, and an inky wave of it fell rakishly across his forehead. His handsome face was etched with impatience.

Mr. Kirby stepped forward. "Captain, this is Lady Haley, of whom I spoke, and her escort, Mrs. Dunley. Kathleen, may I present Captain Reed Taylor."

The captain gave a curt nod and spoke brusquely, "Ladies, ten more minutes and I would have set sail without you. The bo's'n will see to it that your trunks are taken to your cabins, and you will be assigned a cabin boy to see to your needs."

Through her irritation Kathleen noticed full sensuous lips set above a strong jawline. His cheekbones were high and well defined, and he had straight white teeth that contrasted sharply with his deeply tanned face. His nose looked as if it had been taken directly from a Roman statue. He was by far the most handsome man she had ever met, and she flushed as he caught her studying him.

"Thank you for seeing the ladies aboard, Mr. Kirby," the captain continued, "but as we are about to be underway, perhaps you had better bid them farewell and return to shore. Unless, of course, you are prepared to travel to Savannah." His deep voice carried a slight southern accent that sounded foreign to Kathleen's ears, but in no way concealed his impatience and contempt.

"Ladies, if you will excuse me I'll be about the ship's business." With that Captain Taylor turned and left them.

"A man of few words, I would say," Kathleen commented wryly, wrinkling her nose in distaste. "I do hope he is not so unpleasant during the entire voyage."

"He's probably the type who feels that women aboard

ship are bad luck," Dan whispered over her shoulder. "Little does he know! Hee! Hee!" Dan ambled off chuckling merrily.

Turning to Mr. Kirby, Kathleen inquired peevishly, "Does Captain Taylor know that I own this ship?"

"I did not think it was necessary to reveal family business to him, dear." Kirby secretly squirmed beneath Kathleen's angry gaze.

"I beg to differ with you, and I can assure you that if his present attitude is normal for him, I shall take great pleasure in informing him just whom he is dealing with."

"Kathleen," he warned gently, "don't get your back up and start making problems. Just sit back and enjoy a pleasant voyage and let the captain manage the ship." He cleared his voice nervously. "Now I really must go," he said, hugging her affectionately. "We'll miss having you here to liven up our dull lives. Have a safe journey and convey my greetings to your family, especially your Aunt Barbara," he added.

"Goodbye, Mrs. Dunley. Take good care of our girl."

Kathleen watched the dinghy cast off toward shore. Then with a final wave of her hand, she led Nanna down the passageway leading to the cabins at the sterncastle. A thin red-haired boy of about thirteen was just coming out of the doorway next to the captain's quarters.

"Captain Kathleen—er—I mean Lady Kathleen, you'll be in your usual cabin next to the captain's and the other lady is in the one across from yours. Your trunks are there already." He paused, then added, "Sorry about your father."

"Thank you, Bobby. This," she said, gesturing to Mrs. Dunley, "is Mrs. Dunley. Nanna, meet Bobby, our cabin boy. He will provide us with anything we may need and help you in any way he can."

"Right now all I need is to get to those trunks before our clothes get so wrinkled that I spend the entire voyage pressing them." She shuffled off into her cabin, her grey head bobbing in perfect rhythm with her short, round body.

Kathleen entered her cabin and glanced around. Everything was as she remembered it. She had always occupied this room when she sailed the *Kat-Ann* with her father. It was smaller than the captain's, but larger than the quartermaster's. Definitely, it was not your usual, Spartan cabin. Her trunk sat at the foot of the bunk against the wall which divided her room from the captain's. On the other side of the connecting doorway was a curtained corner with a chamber pot. Along the passageway bulkhead was an anchored washstand with pitcher and bowl. A few towels and washclothes were neatly stacked on shelves below it and there were cupboards above it for storage. To the right of the passageway door stood a roomy oak wardrobe lined with cedar. Near this was an enormous highboy and a low dressing table with hinged mirror. Below the porthole was Kathleen's writing desk. Along each wall lamps were mounted, and a small table was bolted to the floor in the center of the room, circled by four chairs. In the far corner stood a small stove, an orderly stack of cordwood next to it. A rocking chair sat in front of the stove on a colorful little rag rug.

The room had been redesigned by Kathleen for her own comfort. The bunk was extra wide, with a comfortable mattress. Turned down over a blue coverlet was the eiderdown quilt Kathleen's mother had made for her. Curtains to match the coverlet hung over the porthole. Because the cabin was used by other passengers when she wasn't sailing, only two of the desk and dresser drawers could be opened without the key that only Kathleen carried.

Now Kathleen unlocked the drawers and her trunk. She shoved her reticule with her money into a desk drawer along with her jewel case and relocked the drawer. She hung the key on a ribbon about her neck, and tucked it between her breasts. Closing her door, she hurried back on deck.

The frigate had already raised anchor and was heading out of the bay. The wind-filled sails reminded Kathleen of huge clouds. Men were scurrying about setting sails and performing other necessary tasks.

"I feel so useless just standing here," she mused, allowing herself a moment of self-pity. "I should be up there on the bridge with my hands at the wheel." Automatically she raised her eyes to the upper sterncastle.

He was standing there, feet planted firmly apart, hands on the wheel, his head thrown back surveying the sails. As if her look was a physical touch, he lowered his head and saw her standing below him at the rail. Blue eyes caught emerald in a long, curious gaze until she turned away to watch Ireland's coastline grow smaller in the distance, finally fading away altogether. Now, in every direction, all to be seen was the deep blue of the sea melting into the lighter azure of sky.

Reed stood on the bridge and watched Kathleen. Somehow, with just a glance, this girl had the power to unnerve him. Her tall, lithe frame, her emerald eyes, the upturned nose. Or was it her stubborn chin; the tilt of her head, or the way she unconsciously squared her shoulders? Something about her disturbed him. He wasn't sure what it was, but he was positive he didn't like the feeling.

"Lady Haley; Kathleen is what Kirby had called her," he mused. "Probably nothing more than a beautiful spoiled brat used to having men fall at her feet and beg for her favors. Well, Lady, here is one man who won't. What a jolt it will be for you to find your title doesn't mean a thing in Savannah!" Still, he had to admit she was a beauty. "A raving beauty!" Her red-gold hair was glistening in the sunlight and the wind was catching loose tendrils from her coiffure. He wondered how long it was when unbound and flowing down her back.

"And a beautiful ivory back it would be, too." He drew himself up short, scolding himself for letting his mind wander off on such a ridiculous tangent. "Besides," he told himself, "she has brought good old faithful Nanna to keep the wolves at bay." With a terse laugh, he turned the wheel over to the quartermaster and headed for the hold.

Chapter 2

THE sound of Reed's laughter jolted Kathleen out of her reverie. She waited until he descended into the hold, then made her way to her cabin. "I'll have plenty of time to freshen up before lunch," she reasoned to herself. She passed Bobby in the passageway and asked him to bring her some fresh water as soon as he got the chance.

Once in her cabin, she wandered to the door connecting with the captain's quarters. Lifting the latch, she swung the door wide and stepped inside. Instantly a deep feeling of nostalgia swept over her. She sauntered about the room, stopping now and again to touch her father's nautical instruments, his maps, his chair. She opened the wardrobe and suddenly it struck her that her father's personal items were gone. Captain Taylor's clothes hung there. His razor and mug lay on the shelf above the washstand. It was Reed's trunk at the foot of the bunk.

She was standing in the center of the room when the passageway door opened to admit Captain Taylor. Reed stopped in the doorway, his eyes as cold as steel. He was a formidable sight in his anger, his powerful body filling the doorway.

Kathleen, taken aback, stood uncertainly for a moment. Then recovering from her surprise, she stepped toward the connecting doorway.

"Just a moment, Lady Haley," the captain commanded her, his voice deceptively mild. "I would like an explanation as to why you have entered my quarters. No one enters here without my express wishes."

Dumbfounded, Kathleen remained silent.

"Well?" he pressed. "Have you come to steal me blind? Or perhaps," he suggested, "you have thoughts of warming my bed."

At this Kathleen went scarlet and exploded, "On a cold day in hell I'll warm you bed, you arrogant toad! Just who do you think you are to speak to me in such a manner?"

"I am the captain of this vessel, in case you have forgotten, miss, and if that be the case, may I suggest you not forget again in future. You still have not answered my question."

Pushed beyond all restraint, she retorted hotly, "You, sir, may be the captain, but I am sole owner of this ship—a fact which may have slipped your notice. And might I suggest you not forget that little fact in future!" Again Kathleen headed for the doorway.

Quick as a panther he blocked her path, his eyes gleaming with the light of battle. "So, you are the Kat of the *Kat-Ann*. How appropriate that you should have the slanting green eyes of a feline. Still, I am issuing you a warning not to return to this cabin uninvited. For if you do, you shall indeed share my bed, whether you are willing or unwilling."

With one swift move his arms encased her, pinning her arms to he sides. For a long moment he gazed intently into her startled green eyes. Then, lowering his head, he brought his warm, full lips to hers. She struggled to free herself from his hold, and kicked out at him ineffectually. Laughing deep in his throat, he merely increased the pressure on her mouth, and she felt his tongue forcing her lips apart. She gasped and his tongue slid into her mouth, moist and probing. Inside her she felt a warmth start in her stomach and progress downward. Through a daze she realized he was running one hand along her spine, while the other he had slipped into the low bodice of her dress and was gently caressing her breast. The nipple rose at his touch and he laughed again. He pressed her to him and through the thin material of her dress, she felt the bulge of his manhood. The contact sent a warmth coursing through her, and she moaned involuntarily. Never in her life had she been kissed in this manner, and her mind reeled as she fought for control over feelings she had never even imagined existed.

He released her so suddenly she nearly lost her balance. Brought back to her senses, she drew back her arm and slapped him across the face with all her might. Darting swiftly past him, she ran into her cabin and quickly shut the door. Instantly the door flew open again. Reed gave her a long mocking look and said with a low laugh, "This door seems to be sadly missing a lock." Then, as he closed it again he added, "I'll look forward to seeing you at lunch, Kat."

She quickly picked up the empty water pitcher and hurled it crashing at the door. It shattered into pieces on the floor. Her only regret was that it had missed his skull.

Shaken and enraged, she threw herself onto the bunk and cried out furiously, "That big, overbearing, no-good arrogant Yankee!"

A roar of laughter from the next cabin told her he had heard her through the wall. Then she heard the door slam and he was off down the passageway.

When Kathleen had finally calmed down, she began to wonder where Bobby was with her water. Nor had she seen Nanna since they first came aboard. Her clothes were still in her trunk. "Nanna was so adamant about the dresses not getting wrinkled. I wonder what is keeping her?" she pondered.

Rising from her bed, Kathleen straightened her gown and crossed to her dressing table. She smoothed her hair and studied her reflection in the mirror. "Good grief! I look like an old crow in black! I don't care what anyone thinks, I refuse to wear these weeds any longer. Nanna will pitch a fit, but Papa would understand. He was proud of me, and would not want to see me looking like a withered old hag. Tomorrow I go back to colors again. Besides, mourning is a matter of the heart. It is inside of you, not worn on your sleeve or your back. My grief will not be less because I dress in lavender or white."

Deciding she had better check on Nanna, she crossed the hall and tapped lightly on the door. She was surprised when Bobby answered her knock. "Come in, Lady Kathleen," he said. "I'm sorry I haven't gotten your water yet, but Mrs. Dunley here is feeling poorly and I've been seeing to her."

Kathleen crossed to the bunk where a slightly green Nanna lay groaning. "Oh, Nanna, you're seasick!" she exclaimed. "Poor dear, I was hoping you wouldn't have to go through this, but you'll be fine in a couple of days, I promise."

"I feel like I'm dying," Nanna moaned. "In fact, I'm hoping I will and afraid I won't. At least I'd be out of this misery."

"Nanna, you didn't hear a word I just said. Some people don't get seasick at all, and others just need a few days to get their sea legs. Don't worry. I'll stay with you. The best thing is to keep something light on your stomach. Perhaps some bisquits and some strongly laced tea. Dan swears that a shot of whiskey fixes the problem straight away."

The very thought had Nanna reaching for her bucket again.

"Bobby, the captain is expecting me for lunch. Please convey my regrets to him and explain that I must take care of Nanna. Then bring the water and my lunch on a tray to my cabin. If you can sit with Nanna while I eat, then I'll relieve you afterward." As an afterthought she added, "Oh, and Bobby, I've broken my water pitcher so you'll have to find me a new one."

"Straight away, Lady Kathleen." Bobby turned to leave.

"One more thing," she called to him. "Since you can't call me captain, just call me Miss Kathleen. I cannot adjust to being called Lady aboard ship for some reason. And bring the whiskey and tea for Nanna. Good Irish whiskey." She winked.

"Yes, Miss Kathleen," Bobby grinned, closing the door quietly behind him.

Nanna again grabbed for the bucket.

After she had freshened up and eaten, Kathleen was about to take over sitting with Nanna when there was a tap on her door. Thinking it must be Bobby, she called, "Come in, please."

The door opened to admit Reed. He had changed from his captain's uniform into a pair of snug black breeches that emphasized his long, muscled legs and slim hips. He wore high black boots and a black lawn shirt with long, billowing sleeves. The shirt was open to his waist, revealing a wide strip of hairy chest and flat stomach.

Kathleen turned from her desk. "I was just com—oh, it's you!" she amended with a frown. She didn't know if it was because of her earlier encounter with him, or because he looked so dashing, but her heart suddenly began to pound faster.

"That is a rather unusual way for a captain to dress. Have you spilled something on your only uniform, or perhaps you are in mourning over your recent behavior," she quipped.

"If I were in mourning over anything it would be that I didn't finish what I started earlier," he remarked, blue eyes twinkling. "Perhaps I have merely decided to join you in wearing black."

"Well I hate to disappoint you, sir, but tomorrow shall find me once again in gowns of colored fabric. It may be I'll decide to wear pink, and it strikes me that you would look a trifle silly if you decided to follow suit."

"No, pink is definitely not my color," he agreed amiably. "Actually, this is my usual uniform."

"I thought the devil normally wore red," she retorted.

"Only on holidays and special occasions." He gave her a sly grin. "I see you were unwilling to lunch with me, Kat. Could it be you were afraid to face me?"

"If you knew me better you would know that is nonsense," she said heatedly, her face flushed.

"I intend to know you, Kat; very well indeed."

"Stop calling me Kat. My name is Kathleen. However, you may call me Lady Haley." She eyed him coldly.

"No real lady responds to a man's kisses the way you did, Kat," he baited.

Kathleen decided to ignore this last comment. "If you have no other matters to discuss with me, would you please leave my cabin, sir?" She crossed to the door and opened it for him.

Instead of leaving, Reed looked around in interest. "A cozy little nook you have here. All the comforts of home. You must sail aboard this ship often."

"As often as father would have me. We were very close," she explained. "Now, will you please leave, Captain? As I am sure Bobby informed you, Nanna is ill and I must look after her." Going to her dresser, Kathleen removed a lace handkerchief. She locked the drawer, slipped the beribboned key over her head, and let it drop into her bodice.

Reed's rude gaze followed the key. "Ah," he commented lazily, his voice husky, "so that is what else I felt down there."

Blushing to the roots of her hair, Kathleen gestured toward the door and commanded angrily, "Get out! Get out of my cabin this instant!"

Reed gave her a mocking bow and exited, chuckling softly at her discomfiture.

"You, sir," she fumed to herself, "are asking for trouble, and I shall be only too glad to give you more than you bargain for."

As it turned out, Nanna did not respond favorably to the whiskey and tea remedy. Nothing stayed in her stomach for very long. Her seasickness would just have to run its course. After two days and nights of nursing her, Kathleen gratefully relegated her to Bobby's care and that of Dr. Long, the ship's physician.

While she was caring for Nanna, Kathleen had done

little more than eat, sleep and sit with her. Occasionally, she strolled on deck for a few minutes at a time. Once, she had been walking about the deck and wandered to the bow, where she stood leaning on the rail watching the sunset. The air was fresh and cool; the water calm and streaked with orange and red reflections of the sky. The captain's other ship, a brig, was sailing on the port side and slightly aft of the *Kat-Ann,* and offered no hindrence to a perfect view.

Kathleen was completely engrossed in the glorious sunset when Reed barked at her, causing her to jump.

"What are you doing forward? Passengers are not allowed up here as you probably well know."

She swung on him, eyes flashing. "I am not a passenger, Captain. I am the owner."

"For this voyage you are a passenger, and one who should know better than to disturb the crew."

"I have been standing here quietly and am in no one's way. How could I have disturbed the crew?" Contempt laced her voice.

"Don't be dense, Kat. Look around you. There isn't a pair of eyes near that aren't trained on you. They can't keep their minds on their work."

"You are saying it is because I am a woman," she accused.

"Are you, sugar, or are you still just a little girl?" he sneered.

Whirling away from him, she snarled, "You have to be obnoxious, don't you Captain Taylor?"

He tried to take hold of her arm, but she pulled away haughtily.

"That is not the issue," he said. "You will limit yourself to the quarterdeck hereafter. The officers aboard are supposedly gentlemen, which is more than I can guarantee with the rest of the crew. I do not need you inciting a mutiny," he said curtly. "And if you refuse to obey my direct orders, I shall be forced to have you confined to quarters."

Reed grabbed her roughly by the arm and marched her

back to the quarterdeck. "Do not disobey me, Kat. I warn you."

She'd had no more conversation with Captain Taylor after that. She took her meals in her room, and when she had seen him, he had either ignored her or glared. Kathleen was very perplexed by her feelings toward him. She had never met anyone who disturbed her like he did. Most of the boys she'd known in Ireland were just that—boys. This was a man. A very powerful, dominating man. A tingling heat ran through her as she recalled the feel of his lips on hers, his brawny arms about her.

She judged him to be around twenty-six years of age, and very self-assured. From what she observed, he ran a tight ship, and the men seemed to respect him. Only a fool wouldn't! No one in his right mind would purposely tangle with a man of Reed's size and temperament. Undoubtedly he was arrogant, overbearing, hateful, and thoroughly maddening. Still, she had to admit he was extremely handsome with his sleek, dark good looks and piercing blue eyes. He was very agile for such a big man. His movements reminded her of a jungle cat; graceful and quiet, yet full of restrained power and danger.

By the third night out of Ireland, Kathleen felt a pressing need for solitude and freedom. She needed space and time to sort out her feelings. She waited until the crew quieted down for the night and only those men absolutely necessary to sail the frigate were on duty. Then she donned her boy's breeches, boots, and a peasant blouse and slipped out on deck. Keeping to the shadows, she crept to the mainmast and quickly climbed the rigging. She perched herself among the shrouds about halfway up and relaxed.

The breeze was cool, refreshing. The sky was bright with a million stars. They were mirrored by the ocean, as was the moon. Kathleen took a deep breath, relishing the saltiness of the air, and sighed deeply. This was where she belonged. This was her heart's home. "How will I ever survive it if I must stay on land for months at a time in

Georgia? I must devise a way to escape to the sea every so often or I'll start raving."

The *Seafire* was still riding slightly behind the *Kat-Ann*. It was beautiful as the moon turned the white sails to silver, and the sea sparkled all around. Kathleen thought her ship was even more beautiful. The other was a brig, and had only two masts and fewer sails. The *Kat-Ann* was sleek and low and fast with her three masts. She was well made, with only the best timbers, and built for speed and maneuverability, though she had ample space in her hold for cargo. To Kathleen the brig was clumsy by comparison. She appreciated the sleek lines of a good ship.

Caught up in the beauty of the sea under star-filled skies, Kathleen lost all track of time. She let her mind seek release in the serenity of the ship slicing silently through the water. Soon the stars faded one by one and the sky began to lighten in the east. The darkness faded to a silver-gray, then to a golden hue as the sun sought dominance in the heavens. Dawn was a beautiful experience at sea as the waters around the ship took on the golden highlights of the new day.

Kathleen knew she should return to her cabin before the captain came on deck. "Just a few more minutes. It is so exquisite. The sea is like a golden jewel, so alive and sparkling. It is so serene up here; a balm to the spirit."

Finally, knowing she could stay no longer, she climbed nimbly down the rigging. Suddenly, about three feet from the deck, she was grabbed from behind. A strong, hairy arm clamped firmly around her waist and lowered her to the deck with a thud that nearly rocked her teeth loose. She was spun about to face a scowling Captain Taylor.

"Oh, no," she groaned wearily. "Why do you always turn up when I least want to encounter you?"

He said nothing, but gave her a disconcerting look and none too gently steered her toward the quarterdeck.

"Let go of me, you boor!" she cried, struggling to release his grip on her arm. She kicked out at him, her boot connecting with his shin.

Still silent, he bent swiftly and swung her over his

shoulder, face down. Mortified, she beat on his back with her fists and screamed at him, "Put me down!" When he ignored her, she bit him on the back.

Reed promptly slapped her forcefully on her backside, causing her to yelp. "Oh! You cur! You filthy swine! You blackhearted devil, you!"

Opening the door of her cabin, he walked to the bunk and dumped her unceremoniously onto it. Standing over her, hands on hips, his piercing gaze raked slowly over her body, causing her to shiver under the scrutiny. After a long minute, he stated flatly, "I told you not to create any more problems on deck. Not only do you disobey a direct order by leaving the quarterdeck, but you parade about in breeches that show every curve of your body and a blouse cut low enough to reveal most of your breasts. You have caused quite a stir among the crew. I warned you that those men are only human. For your own safety I am restricting you to your quarters. There will be a guard posted outside your door. I have had quite enough of your escapades."

"Surely you can not expect me to pass the remainder of the voyage in here. Even prisoners get fresh air and exercise," she complained loudly, scowling up at him.

"I shall grant you half an hour on deck twice daily in the company of Mrs. Dunley. Dr. Long informed me this morning that she has finally overcome her seasickness. I will send her to you soon."

Tears welled up in her eyes, making them shimmer like cut emeralds. She blinked them back, willing herself not to cry in front of him. "You are being very hateful and unjust, Captain," she accused in a shaky voice.

"I am sorry, Kat, but I warned you what would happen. I try to be fair. We will see how you behave, and perhaps in a week or so I can lift the restrictions."

"Stop talking to me as though I were a delinquent child!" she spat. "You are a spiteful, odious cad. I thoroughly despise you! I'd be eternally grateful if you would remove yourself from my cabin this instant."

Giving her a long, cold stare, he bowed stiffly and

stalked out, closing the door after him.

Kathleen glared at the door and stuck out her tongue childishly. Then she dissolved into a fit of tears.

For an active young woman like Kathleen, the next week was unbearable. Nanna was not particularly sympathetic. "You brought this on yourself with your wild ways, Kathleen. You cannot blame anyone else, and certainly not poor Captain Taylor."

"Poor Captain Taylor!" Kathleen shrieked. "That ogre? Nanna, what has that man done to you? Has he cast a spell over you or simply fed you something in your tea?"

"Really, Kathleen. You are just being spiteful. He is a very nice young gentleman when you get to know him, and so handsome, too."

Kathleen stopped pacing and stared at the older woman. "Are you really talking about the same person? I have gotten to know that snake as well as I ever hope to, and he is certainly far from being a gentleman."

Each morning after Kathleen had washed and dressed, Bobby brought her breakfast tray. She and Nanna would eat together and then take a half hour of fresh air on deck. Whenever they chanced to meet Reed, Nanna would engage him in conversation. Kathleen would stand at the rail and pointedly ignore them both.

After their morning excursions, Nanna would retire to her cabin, and Kathleen would read and pace alternately until lunch was served. Again the two women would dine together. Then Nanna usually retired to her afternoon nap.

Time dragged for Kathleen. The sun shone brightly most afternoons, and the cabin was stifling. She would stand by the open porthole trying to catch a chance breath of fresh air. Sometimes she would sit at her desk and watch the waves until she began to nod, and she too would lie down for a nap, though it was not customary for her to do so. Midafternoon brought tea time, and afterward she and Nanna would play checkers or chat. Occasionally they would play a game of piquet, but Nanna, not being very adept at cards, tired of Kathleen winning so often and easily.

Both ladies would freshen up and change their dresses for dinner, and afterward again traverse the quarterdeck. By lamplight they would sew, or Kathleen would softly strum her guitar while Nanna stitched, and they would retire early. At least Nanna retired early. Kathleen merely tossed and turned in her bunk, often giving up entirely to rise and pace the floor until she was exhausted enough to sleep. All too often what little sleep she managed was broken by dreams of frosty blue eyes, warm, demanding lips, and searching hands that turned her body to fire. She'd awaken trembling until she reminded herself that she loathed Captain Taylor to distraction.

Reed was also having his difficulties. He was glad to have his ship running smoothly once more, without Kathleen's disrupting influence. Still, he regretted being so harsh with her. Each time he saw her he was tempted to apologize, but she would turn away, obviously not wishing to speak with him.

"She hates me," he thought. "I don't know why that should bother me, but it does. That green-eyed vixen is starting to get to me. I'll be glad when I can deposit her in Savannah and forget her." Forget her? Not entirely. Savannah was not that large and he would undoubtedly come into contact with her socially whenever he was in port. His family was very good friends with her aunt's.

"She will become acquainted with many other people, make friends, and perhaps our chance encounters will not be too great a problem. Some other poor man will have to put up with her temper then," he mused. Somehow that thought brought him no comfort. He pictured her dancing and dining with the young men of the town, all of them mooning over her and vying for her attentions. He felt a hard knot of jealousy form in his stomach. He told himself, "Forget her. That kind of trouble you don't need." But at odd times of the day he would catch himself thinking of softly waving red-gold hair, moist pink lips, and emerald eyes behind long, lush lashes. And each evening he strained to hear the low, clear tones of her voice as she sang and played her guitar.

At the end of a week, Reed sent Bobby to inform her that as long as she limited her activities to the quarterdeck and dressed properly, she was free to leave her cabin again at any time, with or without Mrs. Dunley. Kathleen immediately abandoned her quarters, and if Nanna wished her company, she had to seek her out on deck. Kathleen soaked up the sunshine and fresh air as a thirsty person would water. Her face and arms turned a glowing tan, and her hair took on even more golden highlights, resembling burnished copper.

On the fourth morning of her reclaimed freedom, Kathleen awoke to an angry red sunshine. The wind was brisk and the frigate was rocking more than usual. They'd had two weeks of glorious weather, which was rather unusual so early in the spring, and now they were in for a storm.

She rose and dressed quickly. She was stowing everything loose or breakable away in cabinets and drawers when Bobby arrived with breakfast.

"We're in for some rough weather ahead, Captain says," he informed her.

"What direction is the storm approaching from?" she asked him brusquely.

"From the northeast is what Dan said."

She grimaced, "Terrific! A nor'easter, and probably a mean one this time of year. I hope Captain Taylor knows as much as he thinks he does. I'd feel more comfortable if I were in command now, though. Anyone can sail in fair seas, but I know how the *Kat-Ann* handles in the wind."

"Captain Taylor's had a lot of sailing experience, Miss Kathleen. Don't worry. He'll handle her fine."

"I'm sure you are right, Bobby. At any rate, we'll find out soon enough," she said as the ship rolled again, nearly tipping her breakfast tray from the table.

Nanna had declined breakfast, the tossing ship making her queasy. Kathleen ventured on deck by herself. She loved battling a storm, especially when she was in command. It was exhilerating to overcome the elements, to ride the waves and wind. The danger of it made her blood surge faster and her spirits soar. As she stood, feet apart on the rolling deck, she felt the familiar thrill course through her. She fought the urge to dash to the bridge and take the wheel from Reed's hands.

She sighted the captain at the helm, shouting orders to the men. The rain began pelting down in torrents. Kathleen flew down the passangeway to her cabin and dragged on her boots and her breeches under her dress. She grabbed her slicker from a hook near the door and raced back to the deck. She climbed carefully up the slippery ladder to the bridge, fighting her way toward the helm.

Reed didn't see her until she appeared at his elbow. "Get below!" he roared over the wind.

"No!" she yelled back.

"I haven't the time to argue with you about it. Don't be stubborn!"

"I'm staying! It is my ship!"

"Fine! Have it your way! Just stay out of the way and do your best not to fall overboard. I'll not be responsible for your safety if you remain on deck," he snapped.

Turning from him, she scouted around for a length of rope, and with it she lashed herself to the rail near the wheel. Now, even if the force of the gale loosed her grip or she lost her footing, she would not be swept overboard. As long as the rail and rope held, she would be fine, and could observe the storm's progress. From where she stood, she could also observe Reed's actions.

He was busy instructing the men to reef the sails before they were torn to shreds. Close hauling, he tacked the frigate into the wind. The gale was battering the ship back and forth, and the waves rose dangerously, spraying the decks with showers of foam.

Reed threw a glance at Kathleen. She smiled her approval. He was indeed a very capable captain. He was handling the situation exactly as she would have and she felt confident of his abilities, and strangely proud. He flashed her a gleaming smile in return. Unexpectedly they shared a moment of closeness, an unspoken comradship born of the sea and the storm.

Kathleen turned back to her post. The storm built in strength, and her spirit soared with it. Then, through the rain, she glimpsed a flash of black skirts. Nanna! As she shouted a warning, the next wave hit the deck. Nanna, caught unprepared, slid across the deck. She could find nothing to grab hold of, and as the wave receded, it pulled her with it. She caught at the bottom of the rail and held on.

Kathleen frantically untied herself from her post. "Hold tight, Nanna! Someone help her!" she screamed over the wind. She had just loosed the last knot when the next wave struck. Kathleen lunged for the ladder and felt arms of steel encircle her, holding her back. She looked toward the rail to see Nanna being drawn into the seething ocean. Desperately she tried to break loose, but was held firmly.

"Nanna!" she cried.

Nanna's head bobbed twice and was lost in the boiling sea.

Kathleen ceased struggling and stood disbelieving for a moment, and then whispered hoarsely, "No! O God, no! Not Nanna! Not my dear, sweet Nanna!" She had not the strength to resist as Reed pulled her back with him to the helm. Tears coursed down her face and her shoulders shook as he placed her between himself and the wheel, enclosing her with his arms. She grabbed the wheel for support and leaned onto him as the tempest raged on.

It seemed eons had passed before the storm began to abate; the wind to lessen. Gently Reed pried her stiff fingers from their hold and turned her toward him. "Kathleen?" he asked softly.

She looked up at him dumbly, not really seeing him at all. He motioned for the quartermaster to take over. Picking her up in his arms, he carried her to her cabin and laid her carefully on the bunk. Entering his own quarters, he returned with a bottle of whiskey and a glass. He poured a generous amount in the glass, and propping her up, held the glass to her lips. When she failed to respond, he sloshed some of the liquid down her throat. As she sputtered and coughed, reality forced its way back into her mind, and she began to cry again. Huge sobs escaped her, and her enormous green eyes glittered as the tears flowed freely.

"It's all my fault! All my fault!" she gasped repeatedly.

"Ssh, little one. It's not your fault. Hush, kitten."

Reed held her gently, rocking her to and fro as she wept. She clung to him, burying her face in his chest, until gradually her sobs subsided. Still she rested against him, drawing strength from his nearness. He stroked her hair with his large, rough hand, and murmured to her while her diminishing shudders shook them both.

Finally he released her and tenderly removed her wet slicker. He pulled off her boots and slid her under the coverlet. When he attempted to leave, she grasped his hand. "Don't leave me. Please, Reed," she pleaded with upturned eyes.

Seating himself on the edge of the bed, he whispered, "I'll stay until you fall asleep, Kat. Now rest, darling. Close your eyes and rest." It did not register with him that she had called him Reed for the first time.

Vaguely Kathleen wondered if he'd really called her darling or if she was dreaming.

Chapter 3

KATHLEEN awoke slowly the next morning. Her brain felt foggy, and her body like lead. She knew there was something she should remember, but she was having trouble recalling it. Suddenly her mind registered the previous day's disaster. Vividly she recalled the storm, and again she saw Nanna dragged into the thrashing sea.

She leaned back on her pillow, letting her tears flow unchecked, and thought sadly of Nanna. Nanna was always so good, so kind and patient. Most of the gray hairs she'd carried so proudly, Kathleen had helped to put there. She had always been there, through good times and bad, always ready to praise, or scold, or comfort. Now the sweet old woman was gone. Kathleen's conscious mind refused to suggest that Nanna's bones might rest on the ocean bottom, but that her body would feed the fish. The thought was too horrid to entertain.

Bringing her thoughts back to the present, Kathleen threw back the coverlet, preparing to climb out of bed. She sat up in shock! She was completely nude! She tried to concentrate on what had happened after Nanna's death, but remembered very little, and that foggily. She could not recall returning to the cabin, nor undressing and getting into bed. Vaguely she had the impression of Reed carrying her, holding her in his strong arms, comforting her with tender words.

Surely he hadn't undressed her! "Even he would not

take advantage of an unconscious woman," she brooded. She rose and walked to her mirror. Twisting and turning, this way and that, she inspected her body. She could see nothing to prove or disprove her suspicions. Everything looked normal. She would just have to ask him.

"How do you ask a man if he has defiled you while you were in a state of shock?" she mused. "The very thought is indecent! Surely I would feel different in some way if he had. Maybe he left and I undressed myself." She pressed shaky fingers to her throbbing temples. "If only I could remember! Perhaps he undressed me with his eyes closed. No, that rake would gawk until his eyes bulged!" she steamed.

She was still worrying over her predicament when Bobby knocked. He carried her breakfast tray. She scurried to her curtained corner and emerged in a dressing gown.

"How you feeling this mornin', Miss Kathleen?" he inquired, regarding her curiously.

"Why is he staring at me so?" she wondered. Aloud she said, "I am as well as can be expected, I suppose, thank you, Bobby."

Picking at her breakfast, she contemplated her problem. "I wish I knew how long the captain stayed, but I can't ask Bobby, or even Dan. It would be too embarrassing. Besides, they would not know what went on behind closed doors. Blame it all!" she fumed. "I will have to find out from him, somehow."

She had just finished her meal when Captain Taylor appeared at her door. She ushered him in.

"I see you have recovered from your ordeal fairly well," he observed.

"Yes," she said simply, but worried to herself if he was referring to Nanna's drowning or something more.

"There is a matter we should discuss if you are up to it. May I sit down?"

Mutely she indicated a chair. "Oh, heavens!" she thought. "Here it comes!" Hands folded to control their shaking, she stood opposite him.

"First I should tell you that we will hold services in remembrance of Mrs. Dunley this morning on the quarter-deck." His matter-of-fact tone told her nothing. "If you feel capable of dressing, the ceremony will be in an hour."

"Sir," she said coldly, blinking back tears, "I am quite capable of dressing, and also of undressing myself."

His eyes widened at her remark, and suddenly it dawned on him what she might be thinking. A slow grin spread across his face.

"Pray, tell me what is so amusing," she inquired hotly, hands on her hips.

"Not a thing, Kat, but your comment has served to remind me of the topic I wished to discuss with you." He leaned forward, gazing at her intently, and said, "I think we should be married immediately. Today if possible."

Kathleen stood rooted to the floor. She was stunned. Thoughts raced through her tormented brain. "So he did ruin me! My God, please help me!" she prayed silently. "What am I to do now?" She tried to swallow and nearly choked.

"Kat, are you all right? Sit down. You've turned a ghastly shade of white. Did you hear what I said about being married?" Reed was half concerned and half amused. Now he was sure she believed he had bedded her.

Anger loosened Kathleen's tongue. "I heard you, you devil! You wolf in sheep's clothing! You—you Yankee!" she stuttered.

"Now, Kat, is that any way to speak to your future husband; the man who is willing to give up his freedom in order to protect your reputation?"

"What shreds of it I have left, you mean!" she countered.

"Would you feel better if I told you I hardly noticed that intriguing little mole on your backside?"

Kathleen gasped, then screamed, "Just turn this ship around and take me straight back to Ireland. I'll get someone else to escort me to America. Someone I can trust." She was near tears, and closer yet to murdering him.

Reed shook his dark head determinedly. "No, Kat. If I turned back now, we'd lose four weeks sailing time, which

I can't afford. You and the cargo will arrive on schedule, like it or not. In the meantime, it is best you marry me with haste. You are now the only woman aboard and you cannot sail an entire ocean and still retain your respectability unless you do so.''

Kathleen's face was crimson with embarrassment and rage. "I refuse to marry you, Captain. You are the last person on this earth I would want for a husband," she declared loudly.

"The feeling is mutual, I assure you, but there is no other way." He stood and walked calmly to the door. "I'll leave you to think about it. You have one hour. I want your decision before the services for Mrs. Dunley. You have few alternatives, Kat," he reminded her.

"I'd rather kill myself," she muttered after the closed door.

Reed had spent a restless night after he'd left Kat sleeping. When she'd fallen asleep, he'd eased her out of her clothing. He had stood for a moment admiring the beauty of her perfect form, then bent to kiss a rose-tipped breast. Reluctantly he'd covered her and slipped into his own cabin.

For a while he tried to sleep and finally gave up entirely, spending the night alone on the bridge with his thoughts. He'd considered returning her to Ireland, but used the excuse of the cargo to discard that notion. Next he thought of locking her in her cabin for the remainder of the voyage, but that was not realistic. He even speculated on letting matters stand as they were and chance the results. Finally he had come up with the idea of marriage.

He'd wrestled long and hard with himself for hours before deciding to speak to her. He could think of dozens of reasons why he shouldn't marry her. She was spiteful and had a sharp tongue. She had a fiery temper and was very self-willed and independent. She was bull-headed, and they would be forever fighting. She was still a spoiled child. Still, she was so soft and feminine; so beautiful. He recalled the perfection of her naked body. He saw again the lovely smile she had flashed at him earlier during the storm. Those captivating green eyes; those sweet lips made him want her as he had never wanted anothe woman. The thought of losing his freedom was not nearly as bad as the thought of another man having her.

"I must be out of my mind," he told himself. But in the end, he'd decided to make her his wife and suffer the consequences.

Kathleen battled with herself, too. Alone in her cabin, she paced, her thoughts flying furiously. He refused to go back to Ireland because of his precious cargo. She could not lock him up and sail the frigate there herself. Half of the men were loyal to him. Besides, at this point it mattered little to her in which direction they sailed. With Nanna gone she did need his protection as captain, however. A locked door would not save her, and she certainly couldn't go around with a rapier in her hand every moment of the day, or sleep with one eye open. Her men she could trust, but his? And who was to protect her from him?

Of course, what was left to protect, actually? Her virtue was obviously gone. Perhaps she ought to marry him. After all, he was responsible for her ruined condition.

"Yes," she thought. "I'll marry him and make him rue the day he stole my virginity." The thought of a lifetime with that arrogant, mule-headed, conceited, smart-mouthed snake-in-the-grass made her reconsider again.

Then she thought of Nanna. Nanna would not want her reputation tarnished. She would not approve of Kathleen sailing without protection with a ship full of lonely men. Nanna had liked Reed. She'd thought him a handsome, well-mannered gentleman. His family was socially acceptable, owning property and slaves and a grand home in Savannah.

"For you, Nanna," Kathleen sighed resignedly. "For you I'll marry the Yankee."

Her decision made, Kathleen slipped on a black dress and hurried on deck, there locating Reed on the bridge. A few feet from him, she hesitated briefly and said, "I've decided to accept your proposal after all, Captain Taylor. I will marry you.''

He regarded her silently for a time, then stated authoritatively, ''I will not marry a woman dressed like a crow. Change into something suitable, Kat.''

''This is perfectly suitable for a funeral, Captain,'' she argued defiantly. ''I will change immediately after the service for Nanna is concluded, and no sooner.'' She gave him an icy glare and threw her chin up.

''See that you do,'' he answered bluntly. ''We are tying up to the *Seafire* for the ceremonies,'' he want on to explain. ''I will conduct the services for Mrs. Dunley. I shall make arrangements with my former quartermaster, Captain Venley, for him to marry us. Don't worry,'' he interjected. ''He is a licensed captain, so our union will be perfectly legal. As you can probably understand, we cannot keep the two ships tied up indefinitely, so please try to hurry in changing your clothing.''

He spoke stiffly, as if he regretted offering her his name. ''By the way, do you require a Catholic ceremony?''

''No. Father was Protestant and I have been raised as such,'' she told him. To herself she thought sullenly, ''Don't put yourself out, you swine.''

The services for Nanna over, Kathleen hastened to her cabin. She rifled through her wardrobe until she located a white silk evening gown. The bodice was low and caught up under the breasts in the empire style that was currently fashionable. It sported capped sleeves and was studded with hundreds of tiny seed pearls. The skirt fell in smooth folds to the floor. Swiftly she donned the shimmering gown with slippers to match. She caught up the sides of her hair and tied them back with a white ribbon. leaving two small curls near her temples, and left the remaining hair hang straight down to her hips in softly waving curls. In her dresser she found a delicate white lace mantilla which she draped over her head and shoulders. Drawing a deep breath and squaring her shoulders, she walked sedately to the quarterdeck.

At Kathleen's approach, a hush fell over the crew. All eyes turned on her with appreciation and awe as she walked toward the captain. Reed stepped forward and tucked her arm beneath his own. His steel-blue eyes were warm with admiration as he whispered softly, "You make a very beautiful bride, Kat."

They were a stunning couple as they stood before Captain Venley and exchanged their vows; he in his uniform once more, and she all in white. Vows exchanged, Captain Venley pronounced them man and wife. Reed's dark head bent to hers of copper as he kissed her sweetly on the lips.

The men congratulated them heartily. There were shouted wishes of long life and happiness and many children, which promptly made Kathleen blush. Within minutes the crew from the *Seafire* cast off from the *Kat-Ann,* and they were sailing toward Savannah once more.

Feeling suddenly shy, Kathleen left Reed on the bridge and returned to her cabin. Sitting on the edge of her bunk, she stared at her reflection in the mirror across the room. Her cheeks were flushed, and her emerald eyes were large and luminous. She raised a shaky hand to touch trembling lips that had so recently repeated vows of matrimony. His warm kiss still lingered on her lips. Shaking her head in wonderment, she repeated, "to honor and obey till death does us part. O God, what have I done? I have married this man in Your sight and recited vows I don't know if I can keep. I am Kathleen Haley no longer. No longer my own person, but half of a pair. I am Mrs. Reed Taylor, and I have no idea who she is or what she is supposed to be like. Is she me? Am I still to be the same individual, or will I be absorbed into Reed's personality? I am so scared! If only I knew what to expect!"

Her worrying was interrupted when Bobby delivered her lunch. She picked at her food, and most of it returned to the galley untouched. Afterward, she sat in the rocker and stared out the porthole, lost in remembrances of happier times at home with Mama and Papa and Nanna. When

Bobby reappeared with tea, Kathleen was startled to realize it was midafternoon already. She sipped slowly, letting the warm liquid soothe her shattered nerves. Finally she lay down across the bunk, resting her arm on her forehead, and tried to sort through her confusion.

She must have dozed for quite a while, for when she awoke the shadows in the cabin told her it must be near the dinner hour. She turned her head and saw Reed standing in the connecting doorway, again dressed in a clean black outfit. He stood regarding her with an unreadable expression. He arched a dark eyebrow at her. "You have time to freshen up before dinner is served. We will dine in my quarters," he told her and returned to his cabin, leaving the door ajar.

Kathleen rose and washed quickly. She smoothed her gown and brushed her hair lightly. She waited until she heard dinner arrive, and only then did she venture into Reed's quarters.

The cook had outdone himself. There was succulent pink pork roast, sliced thin, mashed potatoes and gravy, fluffy golden biscuits with butter, and cauliflower with cheese sauce.

Reed seated her opposite himself at the small table and filled both their goblets with wine. Raising his glass to hers, he said, "I propose a toast to us. May this marriage be more successful than either of us now supposes it will be."

She smiled at this, realizing that he too had doubts similar to hers. Too nervous to eat much, Kathleen nibbled sporadically at her food. During dinner she nervously plied him with questions about Savannah. What was the weather like? What was the town like? How big was it?

He answered her courteously, explaining patiently

Too soon, Bobby came to clear away the meal. When they were alone once more, Reed walked to his dresser, took something out of the top drawer, and turned to her. "Kat, I have something for you." He took her left hand

and slipped onto her finger a large square-cut emerald surrounded by diamonds.

She was dazzled by the beauty of the ring. "Oh, Reed, it is breathtaking," she whispered, daring a look up at him.

"So are you," he stated huskily, pulling her gently into his arms. Before she could protest, he lowered his lips to hers in a long slow kiss that set her heart pounding. His tongue sought to part her lips, and she opened them to him. With her own tongue she tentatively touched his, tasting sweet wine. A shock wave of warmth spread through her. His fingers nimbly unfastened the buttons of her gown and he pushed it from her shoulders to fall to the floor in a rustle of silk. Her shift and camisole and pantalettes joined it in a heap.

Releasing her lips, he picked her up in his arms and carried her to the bed. Bending over her, he kissed her again, this time running his hands lightly over her bare skin. She trembled beneath his touch. His mouth sought her breast, and she caught her breath at the thrill she felt. Adeptly he removed her slippers and stockings, creating even more delight by his touch. Then his hand slipped up the inside of her leg, and she gasped when he touched the flesh between. He caressed her gently, his mouth devouring hers. She opened eyes wide with wonder to gaze into the sky-blue depths of his.

"Reed?" she questions shakily.

"Trust me, kitten," he reassured her in a whisper. Bending his head, he gently tugged at her hardened nipples with his teeth. The heat growing between her thighs intensified, and she unconsciously spread them wider, allowing him better access.

Then she was aware of him rising from the bed. He discarded his clothing quickly and rejoined her. Kathleen felt the warmth of him against her. Her fears transfered themselves to him through her trembling body, but soon he had her senses drowning once more as he caressed and stroked her. His kisses became more demanding, and his touch set her on fire as he explored her most intimate places.

He entered her carefully, moving tentatively a few times and suddenly plunging deeply into her. Silencing her cry with his lips, he continued an easy rhythm. When he lunged into her the pain startled her, but as he continued his steady movement, the hurt became a tingling feeling, and now he was arousing sensations she had never dreamed of. She raised her hips to his and began to naturally match his movements. Low moans of pleasure escaped her lips. She ran her hands along his back, feeling the muscles move beneath her fingertips.

Burying his face in her hair, he whispered, "Kat, my sweet, my darling." His cadence increased, and with it their passion. Suddenly they were exploding into a world of ecstasy. The universe seemed to spin, flinging stars in a spectrum of color.

Kathleen cried out at the wonder of it, astonished by the force of her emotions. Reed groaned and held her tightly as they slowly seemed to float back to earth. He rolled to her side and held her tenderly as she lay quivering in the aftermath of their lovemaking. He sighed deeply. She turned to him with eyes and lashes moist. "Did I hurt you, darling?" he asked, brushing away her tears.

"No," she said softly. "I'm not really crying. It is just that it was so beautiful, so wondrous. I never imagined."

"That is just the beginning, Kat. It gets better, and there is so much more to learn. The first time is not generally so enjoyable for a woman. You really are full of surprises." She snuggled close, her head on his shoulder, and he held her tightly until she drifted off to sleep.

Early the next morning, Kathleen awoke still wrapped in Reed's strong arms. She studied him as he slept. He was a magnificent figure of a man, well built and ruggedly handsome. "He must have many women chasing after him in Savannah," she thought. "He is extremely good-looking and very nice when he wants to be. Maybe we do have a chance in this marriage after all."

She ran her fingers across his chest, playing with the curly black hairs growing thickly there, and followed the line they traced down his stomach. Unexpectedly his hand covered hers, and she raised her eyes shyly to meet his.

"Good morning, Mrs. Taylor," he drawled.

"I didn't mean to wake you," Kathleen said softly.

"I'm glad you did. I'm very hungry this morning," he answered with an easy grin.

"Shall I ask Bobby to bring breakfast?"

"It's not food I'm wanting, Kat. It's you." He brought his lips to hers in a long, tantalizing kiss, running his hands along the length of her slim body. Her trembling became desire, and soon they were caressing each other, both becoming aroused to a fever pitch. Then he mounted her and once more took her hurtling with him toward the stars and beyond. A rainbow of colors burst within her as she felt her tensions released in an explosion of such magnitude that she was left trembling in its wake.

"Reed. Oh, my darling! You make me feel so strange. I'm absolutely giddy!" she exclaimed.

He laughed, "See, I told you the second time would be better. Now, let me up before this ragtag crew of mine has us heading for China."

"Second time," she echoed with a frown as he rose. She eyed him suspiciously. Scooting aside, she discovered her dried blood staining the sheets. "Oh, God! I'm right!" she raged. "Last night was the first time for me! That is why there was pain last night and none this morning. You deceived me, you blackhearted beast! You let me believe you'd taken my virginity the night before last. You horrid, low-down liar!" She flung herself from the bed. Grabbing her dress from the floor, she covered herself with it and backed toward the door to her cabin.

"I don't know why you tricked me into marrying you, but I'm through playing your games, Captain Taylor. Don't you ever touch me again, I warn you!"

"And don't you make the mistake of thinking you can

44

stop me if I so desire, Mrs. Taylor," he stressed in a low, tightly controlled voice. He fastened his belt, gave her a mocking look, and left.

Kathleen scrubbed as if to rid herself of his very touch, then dressed and went on deck. She picked a spot at the rail out of his sight. Dan tramped up and leaned on the rail next to her. He watched her silently, then spat out a stream of tobacco juice and stated flatly, "Well, Cap'n, I guess congratulations are in order. Though, personally, I think it was a stupid move on your part. Thought ye had more sense."

"Let's just say he pulled the wool over my eyes before I realized what was happening," she answered glumly.

"Yea, well, he's a handsome one all right, and a good captain, too. Guess he's right happy to own a frigate like the *Kat-Ann*."

"What are you talking about, Dan?"

"It's his now, Cap'n, by law. You really messed up good this time. Everything ye owned is his now, ye know."

"I never thought of that!" she exclaimed, then added, "but I'll bet that snake did."

"O'course he did. Any man would unless he was daft in love," Dan agreed, nodding. "Didn't think ye would do somethin' that stupid."

"Well, there's nothing I can do about it right this minute, but I'll certainly have a few words for that stinking Yankee pirate! I'll make him pay for this treachery!" she promised, her eyes flashing green sparks. "And a neat piece of thievery it was, too. How could I have been so blind?" She paced a section of the deck, thinking furiously. "Dan, I need a favor."

"Sure thing, Cap'n. Name it."

"Half of this crew is Irish. Most know me and have sailed under me. I want you to tell them that Captain Taylor has stolen the *Kat-Ann* from me by marrying me under false pretenses. Don't let on to Taylor's men at all, but tell ours not to let it slip out that Papa owned more ships. So far Reed assumes the *Kat-Ann* is the only one, and I suppose he realized about the estate in Ireland. As long as I can hide the fact of Papa's shipping line from him, I'll have something I can call my own at least."

"Consider it done." Dan grinned, spat again, and ambled away.

A little while later Reed stalked quietly up, appearing at her elbow. "Out for some air, Mrs. Taylor?"

"Drop dead, Reed," she said tersely.

"Now why would I want to do that when I've just acquired such a charming wife?" he taunted.

"Acquired," she repeated angrily, "like you so recently acquired the *Kat-Ann* and an estate in Ireland by marrying the owner, you conniving cur?"

"By George," he mused. "Thank you for bringing it to mind, Kat. To tell the truth, it hadn't dawned on me until you just mentioned it. It is nice to know there is some compensation for being married to an ill-tempered Irish wildcat with a nasty mouth!"

"You wouldn't know the truth if it stood up and bit you! Don't try to tell me you didn't have the *Kat-Ann* in your sights from the outset. You are no better than a pirate, Reed Taylor, and I'll not be forgetting it," Kathleen stormed. She turned heel and headed for the passageway.

Reed strode after her, caught her arm, and swung her around to face him. "Kat, I swear by all that's holy I never gave the ship a thought. I was concerned about your reputation and your safety. If we'd been closer to port, and I didn't have the cargo to worry about, I'd have turned back and the devil take you and your ship. Marriage just seemed to be the best solution to a sticky problem."

"Even so, you decided the best way to convince me was to let me think you had defiled me! That was a rotten, lousy trick to pull, Reed. For that alone I could tear your eyes out."

"I admit my guilt on that count, Kat, but I was at a loss as to how to convince you, and when you made your suspicions so obvious, I just took advantage of the situation.

I'm sorry," he said earnestly.

"And I played right into your hands," she groaned.

"Look at all the fun you'd have missed otherwise." He grinned at her rakishly.

Kathleen blushed. "You're incorrigible, an arrogant, smooth-talking Yankee devil, and I want nothing more to do with you!" She brushed his hand aside and ran to her cabin.

Upon entering her cabin, Kathleen stopped short. Not only had the door which connected her cabin with the captain's been taken down, but her bunk had been dismantled and removed. Reed's voice sounded amused behind her, "I decided a little redecorating was in order in celebration of our wedding. I hope you like it."

She whirled to face him. "You scoundrel! You know darned well I don't! Get my bunk and that door put back up immediately!" she ordered, stamping her foot angrily.

"Don't order me around, kitten," Reed cautioned with a low growl. "You forget. I am not only the captain of this frigate, but its owner as well. If anyone takes orders from now on it will be you, my sweet. It seems you could use some practice in how to be a meek, soft-spoken, obediant wife."

She reacted exactly as he expected. "Have you been out in the sun too long?" she demanded. "You really should not delude yourself with such fantasies! Face it. You married the wrong woman, if that is the kind of wife you are looking for. Furthermore, I'll sleep on the deck tonight if you refuse to return my bunk."

"And what do you propose to use for blankets and pillow?" he questioned calmly. "Unfortunately for you we are not into warmer climes, yet."

When she said nothing he added, "I'll not be having lunch with you today, but be ready at eight for dinner, Kat. We'll dine alone. I'm sure my officers will understand since we're newly wed."

"I'll see if I can arrange to have arsenic added to your portions," she snarled, "but with your disposition you'd probably thrive on it.''

"Such honeyed words from my sweet young bride," he taunted.

She slammed the door behind him as he swaggered out.

Chapter 4

KATHLEEN fumed for the rest of the day. Finally evening arrived and she dressed for dinner. She selected a sleeveless green silk gown, high waisted, with a darker green sash under the breasts. The bodice was cut from the shoulders in a very low vee that revealed much of her breasts. The pale green of the gown enhanced the emerald of her eyes, and the cut of the fabric made it cling smoothly to her figure. She swept her hair high on her head in an artfully careless array of curls.

Exactly at eight she heard Reed enter his cabin. He walked directly to her doorway, surveyed her from head to toe, and said congenially, "Good evening, Kat. Give me a few moments to clean up and I'll call for dinner. Be a sweetheart and pour us both some wine, will you please?"

He returned to his cabin and chuckled softly as he heard her muttering oaths under her breath while she poured the wine. He shook his head slightly. It would take all his wits to tame this little Irish spitfire.

Dinner went smoothly. Reed entertained her with talk of Savannah, and even had her smiling reluctantly a few times during his discourse. After the meal, he suggested, "Let's take a turn around the deck, Kat. Maybe some fresh air will remove that sour look from your face."

"I doubt that seriously and I really don't relish your company, but if you insist." She prepared to rise.

"I do insist," he replied, holding her chair for her. He

placed her hand in the crook of his arm. When she attempted to remove it, he clamped it tightly, silently daring her to defy him.

They walked slowly, saying little. The evening was cool and Reed drew her close to him. Kat grimaced, but refrained from commenting. She was at a loss as to how to deal with him. He was certainly not the type she could control or manipulate. Ordering him about only made him more difficult.

Kathleen turned her attention to the night. The sky was brilliant with stars. The air was crisp and clean, smelling of salt and wet wood. It filled her senses and buoyed her spirits. She revelled in life at sea. Nature at her best and worst. The sea could be calm, soothing, like an old friend; or churning, angry, challenging, a dangerous enemy. Kathleen loved it either way, feeling a kinship similar to that of a daughter for her mother. The sea seemed to reach out to her, envelop her being, whisper secrets. Kathleen had developed a strange sense of communication with the sea and learned to read the signs of its varying moods. Kat was truly a daughter of the deep, green sea, temperament, and all.

Without realizing it she had stopped walking. She stood peering into the dark depths as though mesmerized. Her lips parted in a slight smile, her arms were extended as if reaching out in supplication. An eerie calm pervaded her, giving her an aura of shining serenity that transformed her face with an ethereal beauty.

Reed stood quietly at her side. He sensed her withdrawl and was awe-struck at the intensity of her mood. He studied her intently, beholding the beauty of the metamorphosis taking place before his eyes. The wonder of it left him with a feeling of reverence, as though he were witnessing something holy. The spiritual contact Kathleen was encountering communicated itself to him almost intangibly, like the light brush of a feather.

Love of the sea was not new to Reed. He had long revelled in the feel of a deck rolling beneath his feet, the power of the waves and the wind, and the sense of freedom it brought. He, too, enjoyed the sea and held a deep respect for it, but he had never felt the overpowering sensations he realized Kathleen was experiencing. He marveled at the depth of her concentration, and wondered to himself.

"It is almost as though she is part of the sea itself; as if she has sprung from it, like some strange sea nymph sailors speak of. Will I ever fully understand this complex, mysterious wife of mine with her changing moods and hidden facets? She is a magnificent, fiery creature, a masterpiece of nature's work in all her varying forms and states. If only I could find the means to tear down that great wall she is trying to build up between us. If I could just reach out and touch her; know how to make her respond to me honestly and without reserve, without hatred. I must make her understand that I am not the thieving beast she mistakes me for. To melt her anger; to feel her soft and tender beside me instead of bristling at my every move or twisting my deeds and words. God, what a magnificent woman she could be then!"

Kathleen slipped out of her trancelike state. She sighed tremulously and leaned into Reed for support. He held her close, breathing in the clean lemon scent of her hair. As they strolled back to the cabin she allowed herself to enjoy the secure feeling of his arms about her.

"Why do I feel comfort in his arms?" she pondered. "Why do I feel drawn to him as I do to the sea? He's not to be trusted. He is a liar, a cheat, a thief, and a pirate! Yet I yearn for his embrace, I melt beneath his kisses, and am moved by his lovemaking. I tingle at his touch and delight in the sight, the sound, and the smell of him. He's the perfect man; at least he would be if he weren't so despicable. I find the man I can finally give my heart and soul to totally, and he turns out to be my enemy. Dear Lord, it is like being given a glimpse of heaven and having the pearly gates slammed in your face!

"I will have my revenge. I must have it for my peace of mind, my sense of justice. But in order to have my satisfaction I must destroy my heart's desire, so can I really win either way? No matter. He would never truly love me anyway. But I can still take my joy now while I may, and later I will wreak my vengeance when the time is right. He has what he wants, the *Kat-Ann*, and soon I will repay him for my loss. But for now I'll taste of heaven, for I may never find it again.''

They entered the cabin and Kathleen turned to him. She curled her arms about his neck, and standing on tiptoe, offered him her lips. Reed was astonished, but pleasantly so. He kissed her sweetly, savoring the taste of her. Her tongue darted into his mouth, and he drew her closer. Slipping her hands inside his shirt, she ran them lightly across his bare chest and around to his back, clutching his shoulders, drawing him nearer. He untied her sash and slipped the dress from her shoulders. It slid to the floor silently. Her fingers worked to loosen his belt and buttons as he deftly removed her remaining clothing. He released her and stepped out of his breeches and shrugged off his shirt. He gathered her into his arms and led her to the bunk.

Sparkling emerald eyes gazed deeply into fathomless eyes of blue. He took her lips to his in an eager, demanding kiss. His lips traced their way across her face to her ear, sending chills through her. His mouth traveled to the base of her throat, his tongue sending flashes of flame along her body as he moved to her breasts. He nipped at the pink tips and they rose to his touch. Every nerve in her body came alive as he licked a path down her stomach, stopping briefly at her naval, then continuing downward. He urged her legs apart, and she raised her hips to meet his touch. One exquisite thrill after another raced through her as his tongue probed and prodded. She held his head tightly as she tangled her fingers in his dark hair. Her world exploded into a thousand dazzling fragments, and she pleaded, ''Oh, Reed, stop. Please! I am dying from the thrill of it!''

He mounted her, entering her swiftly. She met his thrusts ardently, as together they climbed to the heights of heaven. Rapture was theirs as their passion hurled them beyond the stars into endless space. They descended slowly, as if drifting on a silken cloud, and lay entwined, hearts beating as one; needing no words, no explanations. Each needed only the feel of the other's soft caresses as sleep overtook them.

When Kathleen awoke the next morning, Reed had already gone from the cabin. Ordinarily she never slept late, but periodically through the night she and Reed had resumed their lovemaking until the sky began to lighten, inviting the new day. She smiled happily and hopped from the bed. Her muscles rebelled, twinging.

"It's obvious I am not used to this type of activity." She winced. "I walk like I've been riding a horse for a solid week. I do hope it is not evident to the crew, for I would hate to spend the entire day in two rooms."

She dressed, breakfasted, and went on deck. Spotting Reed on the bridge, she smiled tentatively, blushing as she recalled the heights of passion he had revealed to her. He threw her a roguish grin and motioned for her to join him. Drawing her close beside him at the wheel, he kissed her lightly.

"Good morning, kitten. Did you sleep well?"

"Finally," she quipped, and immediately could have bitten her tongue.

Reed laughed good-humoredly.

"Have you had any sleep at all?" she asked.

"I require very little," he answered lightly.

"I should have known!"

They stood for a while enjoying the quiet of the morning. A seagull flew over and perched on the rail nearby. Kathleen asked, "Are we nearing the Azores?"

"We'll anchor early this afternoon. The men will be glad for a brief shore leave. It is grating them that I am getting special treatment with you on board." He grinned again.

"How long will we stay?"

"We'll sail again tomorrow evening with the tide. We'll be adding tea to the cargo here, and we need fresh water, fruit, and enough stores for the rest of the voyage."

"Aren't we putting in at Bermuda? It's on our way."

"For heaven's sake, Kat!" he said with sudden impatience. "Do you want our entire crew impressed? Bermuda is British controlled."

"Then we sail straight for Savannah," she surmised.

"No, I'm setting a course for an island called Grande Terre in Barataria."

Kathleen twisted away from him, eyeing him suspiciously. "Why not Savannah, Reed?" she inquired testily.

"Because all American ports are closed to foreign trade. The Embargo Act last December allows for only coastal trade between states."

"Then you had planned on stopping at this island—"

"Grand Terre," he supplied.

"Whatever," she waved his words aside. "You had planned this all along."

"I had considered it. Just long enough to drop off the cargo. Otherwise, we would have had to slip through the blockade at night and run the risk of getting caught and having the cargo confiscated."

"And just what is this precious cargo you carry? I've never asked you." She surveyed him through narrowed green eyes, hands on her hips.

"Nothing spectacular, Kat. Just some lace and linen, wool, Irish whiskey, hand-painted china, a few fashion plates for the ladies, and some fine cloth for their gowns. With the blockade in effect, American ports are suffocating. Planters cannot get their cotton to English mills, and imported merchandise is extremely hard to come by."

"So you smuggle and probably quadruple your profits," she concluded.

"Smuggle is a strong term," he frowned. "You either run the blockade or find another way to get your goods to market without getting caught. The public is happy, and you can command good prices. Everyone is satisfied except the government that created the turmoil to start with."

"Where is Grande Terre?"

"It's off the coast of Louisiana just west of the mouth of

the Mississippi River, about sixty miles overland from New Orleans," he explained. "Grande Terre sits at the entrance of Barataria Bay and is Jean Lafitte's base of operation."

Kathleen's delicate eyebrows raised in surprise. Everyone had heard of Jean Lafitte! "Jean Lafitte! You deal with pirates?" she exlaimed. "So! I had you pegged from the outset!"

"Jean is not a pirate. He is a privateer, an accomplished businessman, and a gentleman," Reed declared indignantly. "I admire him greatly. We are the best of friends, and I will hear no ill of him from you or anyone else. Do I make myself clear, Kathleen?" His look was cold as an iceflow.

Kathleen stood at attention, eyes flashing, and saluted smartly. "Quite clear, Captain, sir," she replied in a clipped Irish brogue. Marching briskly away from him, she descended to the quarterdeck, deciding her own company was better than putting up with his.

The two ships dropped anchor at the island of Terceira. They had reached the Azores just after noon. Reed set up work details, instructing the men that the supplies must be brought aboard before anyone entertained thoughts of pleasure on shore.

"We are fortunate to arrive and find no British ships in port. How long our luck will hold only God knows. We must get our stores laid in first, in the event we have to run for it."

He supervised the loading, and at last, the work finished, he let most of the crew go ashore. He assured the others that they would be relieved by midnight, and set up lookouts to sound the alarm if enemy ships were sighted.

Once ashore, Reed veered off onto a footpath leading away from the town, pulling a reluctant Kathleen along behind.

"Where are we going?" she pouted.

"You'll see when we get there," was all he would tell her.

Soon the trees began to thin out, and they entered a

small, sheltered clearing. A clear stream flowed over the high rocks in a majestic waterfall and formed a steaming pool beneath. Birds flew in and out among the trees and rocks, and a myriad of butterflies dotted the air with color.

"Oh, Reed! It's magnificent!" Kathleen clapped her hands in glee. "It's a miniature Garden of Eden!"

"Without the forbidden fruit," he added. "Do you like it?"

"Oh, I do! I do indeed! I've never seen such a beautiful sight."

Reed stooped and untied the bundle he had carried under his arm. Handing her a bar of soap, he started undressing. Kathleen gave him an incredulous look.

"Can you swim, Kat?" he asked as he pulled her around and began unbuttoning her dress.

"Yes, I can swim, and what do you think you are doing?" She swatted his hand away.

"You can't swim with your clothes on, and frankly you could use a bath," he teased, slapping her playfully on her now bare posterior. He bent quickly, picked her up, and threw her shrieking into the pool. He dived in and surfaced next to her.

"Reed, this is so warm! It's like bath water," she marveled.

"This is a hot spring. There are many of them on these islands. They are caused by the volcanoes," he explained. "Now I suppose you've lost the soap," he added flatly.

"You want soap? I'll give you the soap," she said mirthfully. She reached out and swiftly dunked him, rubbing the soap in his hair. As he surfaced, she tossed him the bar and nimbly eluded his grasp, swimming a safe distance from him.

After Reed had scrubbed himself thoroughly he advanced on Kathleen, a resolute expression on his face. Excellent swimmer though she was, Kat could not escape him in the small pool. Reed caught her. Instead of giving her the soap, he began lathering her body himself. His hands slid smoothly along her limbs, creating exotic sensations.

She quivered under his intimate stroking and clung to him, murmuring his name in low, honeyed tones. "Reed. Oh, my sweet, my darling Reed."

She brushed back his hair from his forehead and met his searching blue eyes, deepened to a sapphire by his passions. Her own had darkened to a forest green, soft with love. He led her to shore, where they flung themselves on lush green grass. There they made wild, frenzied love to one another until both lay languidly basking in the sun.

"You'll get a sunburn, Kat," he commented huskily some time later.

"I won't tell if you won't. Besides, you'll have one to match," she giggled. She kissed him lightly on the neck and waded into the clear water. After a few dives, she located the soap and washed her hip-length hair. It spread about her in the water, forming a fiery fan. Like a lovely mermaid, Kathleen romped in the water until she tired, and returned to sit next to Reed. The sun and warm breeze soon dried her hair. While she dressed, Reed took a final swim. She sat on a large, flat rock, watching him; admiring his sleek, powerful body.

"Ready to go?" she questioned as he dried off.

"Not quite yet." He fished out of his bundle a hair brush. Seated behind her, he gently brushed the tangles from her long tresses. The ends curled around his fingers in a feathery caress, and he buried his face in the coppery mass, nuzzling her neck.

"You must wear your hair loose like this for me more often, Kat. It is very sensuous."

"Only for you, Reed, ever," she promised in a whisper as his lips claimed hers once again.

Early the next morning most of the men had already staggered aboard and were sleeping off the effects of too much rum. A few stragglers still arrived from time to time, climbing laboriously up the ladder and weaving their way to their bunks. Gallons of steaming hot coffee waited to be served in the galley.

Kathleen awoke and stretched lazily.

"Good morning, kitten." Reed looked up from his

desk where he was reviewing his charts. Kathleen gave him a slow smile. He poured her a cup of coffee and brought it to her. Sitting on the edge of the bunk, he ran his fingers through her shining, luxurious hair.

"You are beautiful in the morning with those slanting green eyes and your hair streaming down. Would you like a last visit to our pond today?"

Before she could answer, there came a furious pounding on their cabin door. Bobby called excitedly, "Captain Taylor! It's the British!"

Reed flung open the door. "Have they spotted us? Does Venley know?"

"I don't know, sir. Dan just said to hurry!" Bobby exclaimed.

"Stay below, Kat." With that Reed was racing to the bridge.

Kathleen heard the commotion on deck as men ran for their posts. Reed's voice carried to her as he issued orders. She peered out the porthole, deciding the tide was still with them. Reed shouted orders to weigh anchor and give him full sails. He would need all the canvas possible to outrun the British warship.

Kathleen dressed hurriedly, praying fervently that they could avoid capture. Unable to stand the suspense, she scurried down the passageway. At the entrance onto the quarterdeck, she stopped and glanced upward. The *Kat-Ann* was flying the British flag, trying to outwit the enemy. Stepping out onto the deck, she noted that Venley was doing likewise. Hopefully they would sail right by and not be stopped. Just in case the ruse didn't work, the men were readying the guns.

"Thank God Papa believed in being prepared," Kathleen thought. The *Kat-Ann* carried twenty guns—two twelve-pounders, eight eighteen-pounders, and an equal number of twenty-four pounders, plus two thirty-two pounders. She carried a crew of fifty men. The warship, Kathleen knew, would carry twice the men and guns, but the *Kat-Ann* was sleeker, faster, and easier to maneuver; responding instantly in her captain's hands. Her size, speed, and craftsmanship could make the difference, depending on the captain's experience and judgement.

Dan spotted her and she waved him over to her.

"Captain," he said, "what's it to be?"

"Dan, you must tell our men to obey Captain Taylor's every command, and in case we cannot avoid a confrontation, to fight like madmen. I have no desire to spend the rest of my days in Newgate."

"Aye, Cap'n." Giving her a parting glance, he added, "Better arm yerself if ye're staying on deck."

"Reed has ordered me to stay in the cabin, so I'd best get back. Tell the lads not to worry that they don't see me, and to give the British one for good old Ireland," she grinned.

Kathleen hurried back to her cabin and dug through her trunk until she found her rapier. She drew it from the scabbard to inspect it. It was razor-sharp, of the finest blue-gray Toledo steel. The weight of it felt familiar, the balance perfection. She sliced the air experimentally, wondering how out of practice she was. It had been over a month since she had last practiced with her father. She limbered up a bit; lunging, parrying with her invisible opponent. Satisfied, she sat in a chair facing the door, rapier in hand. She waited tensely, straining her ears to hear what was happening on deck.

The wind and tides were in their favor. All sails were trimmed and billowed out in the morning air. The *Kat-Ann* skimmed along the top of the water, the *Seafire* at her starboard. The brig was larger, older than the *Kat-Ann*, but well made, fast, and quick to respond.

As they neared the sloop-of-war, Reed signaled to Venley, and the two ships veered apart. *The Seafire* approached the British sloop from a forty-five degree angle, passing her stern without incident. The *Kat-Ann* tacked on the same angle to the sloop's port side, heading away from it; presenting only a limited portion of her stern as a target. By the time the British captain grew wise to their deceptive tactics, it was too late. The sloop fired at the two receding ships, but the shots fell short. The sloop swung about, preparing to give chase. The race was short-lived as the two faster ships soon outdistanced the wide, lumbering sloop, and her captain turned back to the islands.

Kathleen heard the roar of the sloop's cannon and the ensuing splash, and knew the shot had fallen short. She leaped to the porthole, estimated the distance between the ships, calculated the *Kat-Ann's* speed, and relaxed. The crew was cheering wildly. She smiled. Reed had really outfoxed them this time!

Laughing aloud, she was sheathing her rapier when Reed entered the cabin. He glanced from her face to the sword and back again. "What did you intend to do with that?" he gestured.

"Defend myself, naturally, if need be," she replied calmly. "I certainly wasn't peeling carrots for supper!"

Reed's laughter filled the cabin. He stood shaking his head at her unexpected wit. He'd thought to find her in tears and all atremble. "Someday when we have time, I'll show you how to use your father's sword if you wish. Otherwise you might seriously injure yourself," he offered.

Kathleen gave him a strange look, then shrugged her shoulders. "We may try that someday," she said quietly, smiling to herself.

"Did your father build the *Kat-Ann?*" Reed asked, pouring himself a fresh cup of coffee.

"He had her built according to his specifications, yes," she answered hesitantly.

"Well, he sure knew what he was about. I'll say that. This frigate handles beautifully. I've never sailed one better. Her lines, her craftsmanship, and her response are all magnificent. She's one of the best of the seas, Kat. A ship to be proud of."

"Papa would agree with you I'm sure," she said non-committally.

Two weeks later, nearly two-thirds of the way to Grande Terre, everything was going smoothly—too smoothly.

They had good trade winds; smooth sailing. Although they'd had some days of choppy sea and rainstorms, there were no dramatic storms of consequence. They were making good headway.

Then the wind stopped. The sails slackened and the ship rocked to a standstill. All eyes turned upward to the lagging sails. All on board held their breath, hoping they were mistaken. Silence permeated the ship. Reed glanced quickly to Venley's direction a quarter mile to their starboard. The *Seafire* was in the same fix.

Kathleen had been wandering about the quarterdeck for some afternoon exercise. She caught Reed's eye immediately and gained no comfort in what she found there. They had encountered a calm, one of the most dreaded phenomena of the sea.

A low buzz of voices started near the bow and worked its way to the sterncastle. Reed stepped to the rail of the bridge. "Men!" His voice boomed across the frigate. "You all must realize we've hit a calm. Some of you have experienced this before, and some have not. There is no need for panic. We do, however need to plan and be prepared. We have no idea how long this will last, perhaps not long, at least we all hope not. In the meantime, I am cutting all water rations in half. Use it for drinking purposes only. Your life and that of your mate may depend on it. Anyone caught taking more than his share or attempting to steal water from the barrels will be hanged on the yardarm at the first offense. There is no room for leniency under these circumstances. Obey orders, try to stay cool, do not overexert yourself or tax your strength. Stay calm and we will all be better off.

"Tempers may grow short, so all weapons except short knives will be collected today, to be reissued when we are again underway. Try to find quiet, easy ways to occupy yourselves, and if you know how, pray."

Reed turned from the men to the quartermaster and boatswain. "See that awnings are constructed today extending from the forecastle and sterncastle. They will provide some respite from the sun. We are far enough south now that we'll bake, even in the shade. Also, select a few trustworthy men to see to the water rationing and guarding of the barrels. And be sure to collect those weapons right away. We have enough problems without a mutiny, though God knows how they would sail anyway. If I die, I do not want it to be at the hand of some madman slitting my gullet.''

The first day went fairly well. Everyone was anxious over the situation, but carried on pretending normalcy. The evening cooled things off nicely. The second day was slightly worse. On the third day, some of the men were starting to grumble irritably. The sun was a blazing white ball in the cloudless sky, and the glare off the clear water pierced their eyes. The heat was stifling; the dead air suffocating, searing the lungs and drying the mouth.

There were no attempts to steal water. All feared the captain too much. By the fifth day, Kathleen was hard put not to tear her hair out at some of the gruesome tales the men were telling of calms they had heard of. Horrifying stories of ships found foundering with nothing but the skeletal or half-decomposed remains of the crew. There were appalling accounts of cannibalism, all too vivid descriptions of deaths, mutinies, and suicides, narratives of how one ship could be stuck in a calm while other ships passed her, too far away to hear a cry for help or helpless to do so for fear of also becoming becalmed.

On this day, Reed ordered that any livestock that died or was killed be served up cooked on the rare side so as to preserve as much of its juices as possible and provide more fluid and nourishment to the crew. He also cut water rations again.

The seventh day again dawned clear and bright, offering no relief. The men now lay around listlessly, too tired and thirsty to talk; to irritable to play checkers or cards, or to whittle. Too depressed to do anything but pray for wind, clouds, and rain. That evening Reed suggested to Kathleen that she bring her guitar on deck and play for the crew, to distract them from their agony.

Kathleen started out strumming soft and low until men started milling about, gathering in groups to sit and listen. Once she had their attention, she played a few lively Irish ditties and even managed to pick out Yankee Doodle after one sailor hummed the tune for her. Some of the crew managed to sing along hoarsely through swollen lips. As her final number, Kathleen had saved something special she had learned at school from her Spanish roommate— the fandango. She began strumming softly, slowly and gradually increasing the tempo. Clear, sweet tones issued miraculously from her parched throat, blending in perfection with the vibrant chords of her guitar. Her audience beheld her spellbound, as her fingers flew over the strings, building ever faster, louder, to a crashing crescendo and sudden silence. The men were quiet, almost reverently so, for quite a few seconds. Then they began to disperse, sauntering away to sleep, or think, or reflect on their lives; their turbulent spirits soothed somewhat by the music they had shared.

Kathleen felt drained. She lay her instrument aside and walked to the rail, peering at the still water beneath. She drew in her breath and let it out slowly. Bit by bit, she slipped into her unique reverie with the sea. Inwardly, almost unaware, she begged the sea to hear her plea and move her waters; create waves where there were none now; to communicate in nature's way with the wind and urge the filling of the sails once more. Tears flowed down her cheeks from swimming emerald eyes as she recalled her love for the sea. Tremors swept her body and huge sobs shook her as her emotions surfaced and broke.

Reed had again noticed her queer behavior and stood nearby, observing her carefully. He knew her tears were not from fear alone, but from the strange rapport she shared with the gods and goddesses of the deep. He imagined her imploring them for help. "Daughter of Neptune, if you can't make them listen, nobody can," he decided thoughtfully.

Late that night as Kathleen slept dreamlessly, the sky clouded over. A slight breeze ruffled the sails. Steadily it built, waking Reed from his fitful slumbers. He raced on deck, throwing his muscular arms wide to receive the wind. "Bless you, Kat," he breathed inwardly. "Somehow I feel you are largely responsible for this saving grace. I don't know how for sure, but thank you, love."

Chapter 5

THANKS to Reed's prudent measures and Kathleen's private petition, the two ships sailed into Barataria Bay on the last few drops of their water supply. They dropped anchor in the bay and were rowed ashore to be greeted by a curious throng. Reed pulled Kathleen close to his side, sheltering her from the pushing crowd. Over the tops of heads, he scanned the faces, finding the one he searched for. His features split in a wide grin, exposing his white teeth, startling against his deeply tanned skin.

"Jean, my friend!" he shouted, dragging Kathleen forward.

When he stopped, Kathleen was surprised at her first impression of the famed privateer, Jean Lafitte. Before her stood an immaculately groomed man approximately Reed's age. He was slim, about three inches shorter than her husband, but he seemed taller than he was because of his proud bearing, his dignified air. He was light complected beneath a slight tan, with curly chestnut hair and lively hazel eyes. He sported a spotless white shirt, with dark trousers neatly tucked into shining black boots. His facial features were those of a born aristocrat. Altogether, he presented a handsome picture of an elegant gentleman.

He took Reed's outstretched hand, embracing him in a brotherly show of affection. "I was beginning to think something had happened to you, Reed."

"Something did." Turning toward Kathleen, Reed added, "Kat, this is my good friend Jean Lafitte. Jean, meet my very Irish wife, Kathleen. We were married seven weeks ago."

Surprise registered momentarily in the hazel eyes, to be instantly replaced by warmth and open admiration as he studied Kathleen. Reaching for her slim hand, he pressed it to his lips, his eyes steadily on hers. "Welcome to Grande Terre, Kathleen. May I call you that since you are the wife of one of my closest friends?"

"Of course, Monsieur Lafitte."

"Jean," he corrected. "Come. Let us go to the house where we can share refreshment and conversation in comfort. It is uncommonly hot for the middle of June." He threw his arm across Reed's broad shoulders and they walked toward Lafitte's house, Reed still clasping Kathleen's small waist.

Grande Terre was a long, thin island; one of two blocking the entrance to Barataria Bay. The shoals around it were tricky to navigate, and only those familiar with them or an exceptionally skilled captain dared enter there. A fort had been erected on the southwestern tip, bordering the Gulf of Mexico, making an advantageous observation post. Men were constantly on alert for approaching vessels.

Jean's home sat some distance from the fort, set apart from the other houses. It overlooked the Gulf, but was set high enough so that he also had an excellent view of the bay. It was a sprawling two story mansion with huge white pillars supporting a wide double gallery, which completely encircled the house.

Across from the fort were the prisoner's barracks and a small hospital. Nearby there were about twenty small homes for officers and guests. Numerous ships, docks, and storehouses crowded the bay's waterfront. Farther down were the auction block, and the slave quarters. Across from the docks, on the Gulf side, the rest of the residences of Lafitte's men were built. At the far northern end of the island, where it widened out, vegetables and produce and a few farm animals were raised by a number of trustworthy, loyal slaves.

The island foliage was lush; semi-tropical, with palm trees and Spanish moss, and brightly colored flowers of every description. Parrots and other colorful birds delighted the eye. North of the bay, the swamp was also beautiful to behold, but deadly to enter. Many men had entered its maze, never to be seen again. Alligators, venomous snakes, and bogs of quicksand lay silently lurking, awaiting the foolhardy.

Jean had first stopped here in 1804, to bury his wife Rachel. She had died bearing their third child on their way to New Orleans. He had befriended the native Baratarians, and later chose this site for his base. The Baratarians had shown the Lafittes the twisting waterways, and they were wary of the dangers. They had learned routes through the swamp to New Orleans and other towns, and some that circled back to the Gulf. His men knew the routes well, traveling them safely almost daily carrying merchandise by flatboat from Grande Terre to Jean's private warehouses in New Orleans. His island base was nearly impregnable. Jean had selected his site well.

As they neared Jean's house, a young woman ran out of the door toward them. She had light brown skin and black hair flowing down her back. White teeth flashed in a smile. As she ran, her full red skirt flew up, revealing long, brown legs. Her voluptuous breasts bounced dangerously, nearly popping from her low-laced blouse. Black eyes full of joy, she flung herself at Reed, nearly knocking Kathleen down in her urgency.

"Reed, lover, what took you so long this time? I thought I would die for missing you!" She planted full red lips firmly against his in a passionate kiss.

Kathleen stiffened visibly.

Reed pried the woman's arms loose from his neck and stepped back. He chanced a glance at Kathleen and winced inwardly. Jean came to his rescue.

"Rosita! Calm yourself!" he commanded. "I would like you to meet Reed's new wife."

"His wife!" she shrieked. She gave Reed a searching look, then turned a hate-filled face toward Kathleen. Hands on hips, she walked slowly around Kathleen, scrutinizing her from every angle, her lips pursed in distaste. Kathleen eyed the girl cooly, but stood still, saying nothing.

"Bah!" Rosita spat. "Such a skinny thing! How can she warm a man's bed on a cold night? Such a prim and proper lady cannot know how to please a man as a warm-blooded Latin such as I!"

Kathleen continued her silence, but a catty grin turned up the corners of her mouth.

"What is so funny, skinny one?" Rosita demanded.

"If you must know, I was thinking you have probably had much practice and should indeed know how to please a man," Kathleen replied bitingly.

The wicked smile on Rosita's face melted as her anger rose even higher. "I ought to scratch your ugly green eyes out!" she screeched.

"Don't let anything but fear and good common sense stop you," Kathleen taunted.

Rosita lunged, claws extended toward Kathleen's face.

In a flash of movement, Kathleen grabbed Rosita's wrists firmly, swung her right leg behind Rosita's knees, and shoved. Rosita landed on her rump in the sand.

Before she could rise, Jean stepped forward. "Rosita! Enough of this jealous display! Reed and his wife are my guests, and I demand that you show respect or suffer the consequences. Now go—and don't return until you can behave decently." Lafitte's voice, though quiet, demanded obedience. Rosita picked herself up and flounced off, swaying her ample hips as she went.

Jean turned to Kathleen, arms extended in entreaty. "Kathleen, I am so sorry. Please forgive Rosita's rude behavior. I trust it will not happen again."

"It wasn't your fault. I'm not really sure it is completely Rosita's fault either. She is obviously smitten with Reed, and it came as a shock to her to find he had married. I can understand that, but I will not forgive the attack on my person."

"Your point is well taken. Shall we go into the house?" Jean offered her his arm, which she accepted, leaving a bewildered Reed to bring up the rear.

Their drinks were served by a house slave in the gallery facing the Gulf, where they enjoyed the first privacy since docking. The gallery sat on the edge of Lafitte's flower gardens, which were beautifully laid out and tended. Kathleen let the men talk as she enjoyed her lemonade and admired the gardens. She noticed several statues that intrigued her. At the first opportunity, she interjected, "Monsieur Lafitte."

"Jean," he insisted.

"Jean, may I presume on your hospitality and inspect your lovely gardens?"

Jean started to rise.

"No, please," she added. "Continue your conversation with my husband. I'll just wander about by myself, if you don't mind."

"Indeed, enjoy yourself," Jean said.

Kathleen gave Reed a questioning look. He nodded his assent.

Those intriguing statues were even more interesting up close. They were figures of sea gods and goddesses. Neptune, god of the sea, was the most prominent, with trident in his right hand and dolphin his his left. He was majestic looking, with his flowing hair and beard.

Kathleen stood before him for a long time, green eyes shining, and as she studied his stony face, he almost seemed to come alive.

Reed and Jean sat talking of Reed's voyage and the *Kat-Ann*. Reed explained to Jean how Kathleen believed he married her for the frigate and the estate. "She refuses to believe otherwise, Jean. She's very warm and tender at times, then she turns either cool or angry. I've never professed to understand a woman's mind, but Kat really has me baffled."

He glanced up, catching sight of Kathleen still standing before the image of Neptune. "There is another puzzlement," he said, gesturing toward her with his head.

Jean followed his friend's look, frowning to show he did not understand. "What are you referring to, Reed?"

"Jean, tell me if I'm wrong, but watch Kat for a while. Those statues in your garden are of sea gods and goddesses,

69

and unless I miss my guess, Kat is completely entranced already, and she's not past Neptune yet.''

"Still? I know those things are fascinating, but she's got to have been standing in one spot for ten minutes,'' he reasoned.

"Exactly! Now watch her,'' Reed said as she snapped her head about and strolled to the next figure.

This one was of Nereus, the old man of the sea: aged, bearded, holding his scepter. She stopped for a moment, appeared to nod, and went on to the next, Nereus's daughter. Nereides was a sea nymph. Her hair hung over her shoulders, shielding her breasts as she sat proudly astride her dolphin. Again Kathleen stood for a long while. Finally she reached out and stroked the dolphin. Walking on, she came to Apollo, who controlled the winds of the seas. He was a beautiful sight to behold. Handsome beyond belief, his face was a study in perfection, his body well muscled.

"He reminds me of Reed,'' Kathleen mused. Touching his thigh, she murmured, "Thank you for guiding us so well.''

Moving along, she passed statues of Pontus, god of the Black Sea; Doris, mother of Nereides; and Rana, another sea nymph with seaweed tangled in her tresses. Last, she stopped before Venus, the goddess of love, most beautiful of all goddesses; Venus, who mythology says was born of the foam of the sea waves. She stood in the center of an open sea shell, her hair flowing down around her slim hips. On her delicate face there was a unique look of innocence and sensuality. Her eyes were wide set over fine cheekbones; her nose small, her lips full and inviting. She was long of limb and beautifully proportioned, with high breasts, a small waist, and nicely rounded hips. She looked every inch a goddess of love. Her look seemed to beckon Kathleen, as if she would divulge her timeless secrets and share her wisdom.

Kathleen was mesmerized before her, as if she waited for her to come alive and speak. Indeed, it seemed she had, for as Reed and Jean looked on, Kathleen closed her eyes. When she opened them again she stood before the goddess, arms outstretched in unspoken plea, eyes trained on the lovely face. Kathleen's lips moved in silent communication, a look of reverence on her face. Silently she poured out her heart to the love goddess, speaking of her love for Reed; the passions he awoke in her; his treachery; her turmoil over wanting revenge, and yet not wanting to destroy her heart's only desire; her confusion and feelings that Reed would never truly love her, but seek only to win her, then break her heart. She pleaded to Venus for guidance and wisdom; to plant in Reed's heart seeds of love.

"You are so beautiful, so wise in the ways of love. Surely you can help me win his love. Yet I still must have my revenge, and I don't know how to gain one without losing the other. Please help me, for if you cannot, there is none who can. Only you, Venus, born of my wondrous, mysterious sea, the supreme creation of the sea, love goddess of all time, can solve my dilemma."

As she finished speaking, a calm swept over Kathleen, as though Venus had heard her and would take care of everything. Kathleen whispered aloud, "I leave it in your hands." Backing away, she strolled to a far corner of the garden where she sat on a bench until she had composed herself sufficiently to rejoin the men.

As Kathleen left the statue of Venus, Reed turned to Jean. "Well, what do you make of what we have just witnessed?"

"It is one of the strangest yet most beautiful experiences I've ever seen. Was she actually praying to that statue?"

"I don't believe praying is the right word, Jean. Kat believes in God, I know, but she seems to have a unique affinity with the sea. It is almost as if she can communicate with it in some way that we cannot fathom. Tell me, of all the figures in your garden, why did Venus affect her so instead of a sea god or goddess?"

"It is not well known, but Venus actually is a sea goddess, Reed. Most people think of her only as the goddess of love, but actually it is said she was born of the sea, rising

71

from the foam. In fact, she is the most beautiful of all the goddesses and most important of the sea goddesses. Perhaps that explains your wife's reaction to her, although I still don't fully understand what you are trying to tell me, my friend.''

Reed related to Jean the two times aboard ship that he had witnessed Kathleen's behavior. ''So help me, Jean, I don't know exactly how or why it happened, but that night the wind began to blow and the waves to roll once more. It sounds crazy, but I honestly think Kat may have had a hand in it.''

''It does sound crazy, but I cannot discount what I saw here today, and I do not doubt your word,'' Jean stated matter-of-factly.

''She confounds me. On one hand she is a fiery tempered Irish wildcat, or a cuddly, purring kitten in turn. At other times she is a tempting, untouchable sea nymph in all her glory. Do you think Kat is deranged, or possibly possessed somehow?'' Reed asked hesitantly.

''From what you tell me of Kathleen, I would say no. She does indeed have a unique ability beyond ours, but that does not deem her unstable.''

Reed sighed in relief. ''I am glad to hear you say it. I am of the same opinion, but I wanted your view on the matter. I value your judgement as much as your friendship. This thing with Kat is almost as you described with Venus. It is almost as if she becomes a part of the sea itself, as if her roots and her being spring from its sources. It's as if sea water runs in her veins and she draws her strength from it. It is strange, and yet marvelous to behold, this fascinating change that occurs in her. Her face glows and she seems wrapped in an ethereal web of serenity. She is at once eerie and magnificent, my copper-haired sea goddess.''

''She is beautiful enough as it is,'' Jean complimented, ''even without her mantle of mystery.''

When Kathleen again joined them, Jean escorted them to the guest house. He invited them to dine with him that evening and left. As soon as Reed left to see to matters of the ship, Kathleen, feeling drained of energy, collapsed onto the bed. Her head had barely touched the pillow before she was asleep.

Dinner that evening was a sumptuous affair. After weeks aboard ship, it was luxurious to view a table laden with fresh food. Hickory smoked ham, fried chicken, and succulent shrimp were piled high on huge platters next to sugared yams, mashed potatoes, gravy and rice. There were fresh sweet garden peas with baby onions, green beans in almond sauce, carrots swimming in butter, and much more. The aroma alone made Kathleen's mouth water.

Kathleen had dressed carefully that evening in a copper colored taffeta gown that matched her hair and set off her golden tan perfectly. Her skin seemed to pick up the luster of the gown. It had a modestly cut bodice that molded itself to her shapely bosom, and the skirt followed smoothly the slim lines of her body.

A young slave girl sent by Jean had done her hair for her in shining braids twisted into a crown atop her head. Around her neck hung a single rare black pearl on a thin gold chain, with smaller matching pearl earrings as her only adornments other than her wedding ring. The look in Reed's eyes told her she had succeeded in her efforts.

Jean sat at the head of the table with Kathleen on his right and Reed next to her. Across the table from her sat a lovely young woman with dark brown hair and lively dark eyes. She was very petite, with a delicate heart-shaped face. Next to her sat a man who resembled her. Jean introduced them as Dr. Charles de Beaumont of New Orleans, and his sister Eleanore. Captain Venley was to Reed's right, and across from him sat a tall, dark man known as Dominique You. He was Jean's best gunner, an expert in his own right, and if anyone looked like a pirate, he did. He had black hair and eyes with bushy brows and a long scar the length of his left cheek, and a large hoop earring in one ear. Although he was clean and neatly dressed, he presented a formidable image with his brace of pistols, his saber, and his long knife, all of which he wore to the table.

"All he needs is a sash and an eyepatch," Kathleen speculated. He must have felt her staring at him, for he looked up from his plate, grinned broadly, and winked at her. Kathleen turned red, smiled weakly, and returned her attention to her plate.

Reed chuckled softly and whispered, "That ought to dampen your curiosity a bit. I do believe Dominique is charmed by you, Kat."

"Not half as charmed as your little Spanish-speaking friend is by you," she countered.

"I take it you are referring to Rosita." Reed shrugged. "She isn't so bad once you get to know her."

"I'll take your word for that since I doubt I'll ever know her as well as you do, dear," she said snidely.

"*Touché,*" he said lightly with a rakish smile, and returned to conversing with Jean.

Eleanore leaned forward, saying, "Your gown is exquisite, Mrs. Taylor. It is delightful to have another woman of good breeding and style grace this island."

"Do you live here?" Kathleen inquired.

"Oh, no, my dear. My brother and I have a townhouse in New Orleans, but we visit here often so Charles can tend to the medical needs of the people here. I enjoy coming with him to see Jean." Here Eleanore's eyes turned to Jean with a soft, loving look that passed between them. "It is tiresome not to have another woman to converse with when the men are busy, though. Will you be staying here long?"

"I really don't know, Miss de Beaumont."

"I really would be pleased if you would call me Eleanore, and I shall call you Kathleen. It is silly to be otherwise when we are the only ladies on the island." She stressed the word 'ladies.' Catching Reed's eye, she asked, "I was wondering how long you and your lovely wife will be staying, Reed."

Jean interjected, "I was just discussing the matter with him. Pierre will be returning soon from the West Indies with another cargo of slaves, and I would like Reed to look

74

them over.''

Kathleen gave Reed a questioning look, and he replied, ''The crew needs a rest anyway, so I agreed.''

Kathleen gave a slight shrug. ''You're the captain,'' she said cooly, and from her tone, Reed knew they would be debating the issue later.

Kathleen turned her attention to Jean. ''Who is Pierre, one of your captains?''

''No, *mon petit,* he is my brother,'' Jean answered with a devilish grin. ''He reminds one of a banty rooster at times,'' he chuckled, ''but you will meet him for yourself soon enough.''

The meal over, the men retired to Jean's library for brandy and cigars, and Eleanore led Kathleen to the parlor where they enjoyed coffee and talked of fashions and got better acquainted. Kathleen wished she could ask Eleanore about Rosita, but resisted the urge. A lady simply did not ask about her husband's lovers. ''I'm sure Rosita will delight in giving me full details, unsolicited.'' She grimaced thoughtfully.

As the evening was quite late when the men again joined them, Reed suggested that they retire. Eleanore, who was staying in Jean's house, invited Kathleen for tea the next day. ''And don't forget that you can call on me for anything you may need. The guest house has been empty for some time and undoubtedly needs some attention,'' she added.

Once back in their house, Kathleen slammed and banged until she located a brown lawn nightgown in the bottom of her trunk. She tugged it over her head, and was hanging her dress in the armoire when Reed spoke sharply. ''Enough, Kat! You've had your little fit, now settle down. Get that ugly rag off your back and come to bed.''

She whirled on him, eyes spitting green flame. ''So! We are staying, just like that!'' She snapped her fingers. ''No matter that I have an aunt waiting who may worry about me, or that I may not wish to stay!'' She flounced to the dressing table, shook the pins from her hair, and unbraided the plaits.

"Be reasonable, woman. This way your aunt will merely receive news of your coming in time to prepare for your arrival. It won't be for long."

"No, just long enough for you to renew acquaintances with your Puerto Rican slut!"

"Now we come to the crux of the matter. You've stewed about that all day, haven't you?" he shouted. He leaped from the bed, scattering the covers. "Well, we both know this was no love match, and I don't have to explain my actions to you—past, present, or future!"

As he advanced on her, Kathleen swung about. "Stay away from me, Reed. There may be no love in our marriage, but I will not be played for a fool!" She aimed the hairbrush at his head and let it fly. He ducked it and grabbed her by the shoulders, shaking her roughly.

"Let me go, you fiend!" she hissed. "Go to your precious Rosita. She will welcome your touch I'm sure, but leave me alone!" She wrenched away from him.

"No, Kat," he growled. "I warned you before that I would have you any time I desired. I would rather have you willing, but—"

"Never!" she screamed, backing away from him.

"So be it," he said curtly. His hand flashed out, grabbing her nightgown at the neck. With one swift movement, he tore it from neck to hem. She gaped at him, astounded at his savagery. Frightened green eyes met those of frosty blue.

"No, Reed," she whispered past the lump of fear rising in her throat. She clutched at the shreds of her gown. He reached out again, knocking her hands aside, and ripped the gown completely away. Quickly he jerked her toward the bed, flinging her onto it. Desperately she tried to escape him, but he threw himself upon her, pinioning her to the mattress. She tried to push him away, but he held her arms beneath her, laughing at her efforts. Her head fell back, baring her throat to his searing kisses; her breasts arching upward invitingly.

His lips ground down on hers, bruising her mouth. She bit down hard, and he drew back sharply, wiping the blood from his lip with his free hand. "You little Irish viper! Bite me, will you?"

He kissed her again roughly. She tried to twist her head away, but her hair was caught beneath her, holding her firmly. He drew back, and she glared at him defiantly.

"I'll tame you yet, Kat," he swore hoarsely, his blue eyes blazing.

"I'll see you in hell first!" she spat out.

He laughed deep in his throat and clamped his mouth around the rosy tip of one breast, pulling hard on it, making her cry out in pain and surprise.

He pried her thighs apart with his knee, and with his hand still binding her wrists, he arched her to him. He entered her with a mighty thrust, filling her entirely, searing the very core of her being. She cried out again, and bit down on her own lip to squelch any further outbursts. As he continued his attack, tears rolled down her cheeks, wetting her hair. "Please, Reed," she sobbed brokenly. "You are hurting me."

He eased up, his face softening at her liquid emerald eyes. "Then give it to me willingly. You'll find it easier than making me use force, which I will do when I find it necessary." He released her wrists. Seconds ticked by until she slowly raised her arms, locking them behind his back, pulling him to her in defeat. Now she offered her lips to him, and he took them in a tender kiss, igniting the familiar fire deep within her. Her passions roared out of control as he took her with him to the heights of the heavens, and the stars burst like a billion brightly colored suns around them.

Afterward, he held her tightly when she would have turned from him. "Don't ever fight me, kitten. You can't win," he assured her.

"Don't bet on that," she reflected bitterly to herself. "Someday you will be in for the surprise of your life, Reed Taylor, when your kitten suddenly becomes a raging tigress."

The next morning, before Reed left the house, he picked up the torn nightgown and tossed it on the end of the bed. "Get rid of that and any more you have like it," he directed sternly. "I want you in my bed as God created you."

Kathleen turned her back to him, defiantly pulling the covers over her ears. It was a useless gesture, for she still heard his wry chuckle as he trod down the steps. "Insufferable cur!" she muttered into her pillow.

After a leisurely breakfast on the veranda overlooking the bay, Kathleen spent the morning investigating the house. It was small, but comfortable. A parlor, library, dining room, kitchen and pantry occupied the ground floor with two bedrooms upstairs. There was a veranda off the rear of the house upstairs, and a large front porch off the downstairs hallway, and a patio to the rear off the dining room. Someone had planted a few flowers around the house, and a small vegetable garden in the back. Beyond the garden the ground sloped gently to the beach.

Jean had sent over a cook, a valet for Reed (who doubled as a butler), a lady's maid for Kathleen, and a housemaid. Lally, the pleasant girl who had coiffed Kathleen's hair the previous evening, was her lady's maid. The cook, Mae, was a jolly old woman whose husband Joe was the valet. The sour-faced, but efficient house maid, was Tess.

Kathleen set Joe and Lally busy with the trunks, unpacking and hanging and pressing her clothes and Reed's. Mae put some order to the kitchen. Donning an old dress, Kathleen pulled back her hair in a bandanna, rolled up her sleeves, and joined Tess in cleaning, airing the dining room, parlor, and library. So absorbed was she in her labors, that she failed to hear anyone enter until Rosita's accented voice echoed from the library door.

"Not so fancy today, eh, skinny one?"

"The name is Mrs. Taylor," Kathleen corrected, wiping the dirt from her cheek.

Rosita shrugged and sashayed into the room, hips swinging. "Reed, he put you to work where you do the most good, I see."

"Reed did not put me to work. The house needs a good cleaning, and we prefer not to breath dust, Rosita. A little work never hurt anyone. You might try it sometime,'' Kathleen retorted.

"He would not want you to bother myself with such things as cleaning. We spend our time in other ways,'' Rosita taunted.

"I'll bet. Of course, you probably can't work very well standing up, can you?''

"Who do you think you are to talk to me this way, you Irish baggage!''

"For your information, I am Lady Kathleen Taylor,'' Kathleen maintained, using her full title. "Furthermore, if you do not leave this house immediately, I will have you forceably ejected; and don't come back again unless you are invited.''

"Reed will hear of this!'' Rosita screeched as she huffed out the door.

"I'm sure he will,'' Kathleen grumbled to no one in particular, and went back to dusting books.

Kathleen spent a pleasant afternoon with Eleanore, returning in time to bathe for dinner. Joe lugged up several buckets of hot water, pouring them into the wooden tub in the corner of the bedroom. After Lally had shampooed her hair with lemon-scented soap, Kathleen lay back, luxuriating in her first hot bath in months.

She bolted upright as the bedroom door was thrust open, admitting Reed. Quickly she slid back into the water. He strode to the dresser, removing a clean white shirt. Throwing her a glance in the mirror, he said, "Don't let me interrupt you, Kat.'' He pulled a pair of dark blue trousers from his wardrobe, and laying them aside, perched on the edge of the bed to watch her.

"Must you stare, Reed?'' she snapped.

"Just enjoying the view,'' he smirked. "I heard you had another encounter with Rosita today.''

"Yes. She invited herself in, and I invited her to leave.''

"That wasn't very neighborly of you. Perhaps she wanted to borrow a cup of sugar,'' he teased.

"The kind of sugar she wants to borrow doesn't come in a cup.''

His rumbling laughter filled the room. "Hurry out of that tub, will you?"

"Why? I'm enjoying it."

"Because you are what I intend to enjoy, my copper-haired vixen," he declared huskily, his blue eyes smoky with desire.

Chapter 6

THE next week went by without incident. Rosita kept her distance. The house got a thorough scrubbing, and all their clothes were cleaned and neatly pressed. Kathleen spent most of her afternoons with Eleanore. They lunched, chatted, and sometimes toured the island. Reed kept her evenings well occupied.

At the end of the week, Eleanore and her brother returned to New Orleans, and Kathleen took to wandering about Grande Terre by herself. Whenever she did, she got the eerie feeling someone was following her, watching her. Each time she whirled about, she saw no one, but the feeling persisted. Soon she noticed that wherever she went, be it the warehouse, fort, beach, or supply store, Dominique You appeared. He would nod politely, inquire about her day, and go on.

One day she was strolling in Jean's gardens when she glanced up and saw him standing in the doorway. She approached him and boldly asked, "Dominique, are you shadowing me by any chance?"

White teeth gleaming, he grinned. "You finally caught me."

"Why? Did Reed ask you to?"

"No, *petit,* I took it on myself to see to your safety. There are those on the island who do not know how to treat a lady, and wouldn't recognize one if he saw her."

They walked together down the garden paths. "The

only person I've had a problem with so far has been Rosita,'' she confided quietly.

He nodded, and after a few paces said simply, ''She is a little tramp. I will try to keep her occupied so she will have little time to cause trouble.''

''She loves him,'' Kathleen offered.

''Reed is out of her class. She must learn to content herself with those who are on her level.''

''You are not on her level, either, Dominique,'' Kathleen assessed. ''I sense you are a gentleman born and bred.''

''Those green eyes are too sharp,'' he stated quietly.

''I won't divulge your secret, sir.'' She smiled up at him. ''You are Jean's brother aren't you?''

A startled look crossed his dark face and disappeared as quickly as it came, to be replaced by a scowl. ''No one knows this but Jean, Pierre, and Reed.'' Then he added, ''And now you.''

She touched his arm. ''Have no fear, for I'll tell no one. What is your given name?''

''Alexandre,'' he admitted, and he told her how he had killed a Spanish prison guard and changed his name when he came to Louisiana. He was believed dead in a shipwreck, or he would be a hunted man now. He told her of his childhood on St. Dominique with Jean and Pierre and the five other children. He spoke of the slave uprisings on the island, and how they had eventually fled to New Orleans. He talked quietly, weighing his words carefully, studying her face.

She in turn told him of Ireland and her childhood. She related her father's death, and that of Nanna, and her subsequent marriage to Reed. But Kathleen said nothing of her father's shipping firm, nor of Reed's deceit, letting him think Reed had indeed married her to save her reputation.

When she had finished her tale, she said softly, ''I wish you were my brother instead of Jean's.''

''And I wish I were Reed instead of Dominique You,'' he laughed.

"But he does not love me," she sighed.

"I think you are wrong, little one," Dominique counseled. "He just doesn't realize it or won't admit it. Either way, the man is a fool." At her frown, he added, "And you love him, too."

She opened her mouth to protest, but he gently laid a calloused finger against her lips. "I see it in your eyes."

"I like you, Dominique, my handsome pirate friend."

He stopped short, drew out his knife, and reached for her hand. She gave it trustingly. "If you still wish to have me for a brother, I can make it so," he offered. At her nod, he pricked her finger with his knife, and his also. When the blood appeared, he joined hers with his, and said quietly, "Now I am your blood brother, and you are my sister."

"It's not often you get to chose your relatives," she observed with a smile.

"If you are in trouble; if ever you need help, turn to me, little sister, and I will be there. Remember this."

Eyes shining with unshed tears, she stood on tiptoes, kissed him on his scarred cheek, and whispered, "I'll remember." Quickly she fled for home before the tears could fall.

From then on, whenever Kathleen roamed the island, Dominique was along. With his big bulk, he silently dared anyone to interfere with her. She enjoyed his company, and since he treated her as a gentleman should, showing only respect, Reed did not seem to mind too much at first.

In order to give her escort some time to himself, Kathleen took up painting. Reed sent to New Orleans for her art supplies, and Jean opened his garden to her. There she set up her easel, and drew the statues one by one. Not knowing how long they would stay on Grande Terre, she penciled each first, adding detail as time passed. Later she would add the oils that would bring these stone gods and goddesses to life in vibrant color.

She was sitting before her easel one early afternoon when a sentry shouted that Pierre's ship was sighted. Pandemonium broke loose as everyone ran down to the docks to greet the returning brother and his crew. Curious as she was, when Dominique poked his head into the garden to ask her to accompany him, she declined.

"You go ahead, Dom. You don't need me there while you try to talk to him. Besides, it will be hot and crowded on the docks, and I would probably run into Rosita. I will meet him soon, I'm sure."

She was still sitting there drawing an hour later when the men burst into the gallery. Reed, Jean, and Dominique were all eagerly questioning Pierre about his voyage. Pierre broke off in mid-sentence, sighting Kathleen.

"Ho! Brother Jean! Have you added another goddess to your collection? I like this one better than the others. She looks warmer, and has more color."

Kathleen looked up to see a shorter, sloppier version of Jean coming toward her. He had Jean's coloring, and similar features, but had an unkempt appearance that Jean never did, and an impudent attitude where Jean was more reserved.

She looked hurriedly toward Reed, noting the dark scowl on his handsome face. Dominique stepped quickly ahead of Pierre. "This is Reed's wife, Pierre, and a fine lady. Show respect or I'll throttle you if Reed doesn't first."

Pierre, promptly changing tactics, bent low over her head, and grazing his lips across it, commented, "My congratulations to you both on your nuptials." The spark in his hazel eyes belied his words.

Pulling her hand away, Kathleen replied primly, "I am pleased to meet you, Monsieur Lafitte. Jean has mentioned you often."

"Can it be that all the fire is in the hair?" Pierre inquired in jest, raising his eyebrows at Reed.

"Not quite all," Reed said stiffly.

"Reed," Kathleen interjected, "could you or Dominique spare a few minutes to help me carry my sketches home? I am quite finished here for today."

"Dominique will, I'm sure, Kat. I have a few things to discuss here. I'll be along soon."

"May I offer my assistance?" Pierre volunteered.

"No, thank you. We will manage fine," she informed him cooly as she and Dominique gathered up her supplies and departed.

Reed was still upset when he arrived at the guest house. He thundered up the stairs like a wounded bull. As he burst into the bedroom where Kathleen was resting, she grumbled sleepily, "Reed, for heaven's sake, you are making enough noise to wake the dead."

He stood at the end of the bed glowering at her.

"What is it?" she asked testily. "What have I done now?"

"Nothing, I suppose, but do you have to be so eternally beautiful? Every man you meet falls at your feet! Jean contends you are one of the most ravishing women he's met, Venley is star-struck, Dominique is at your beck and call. And that I can put up with, but that rutting banty rooster, Pierre, is too much!"

Kathleen stared at him in amazement. "Well, it seems you are the only one immune to my charms," she retorted. "Perhaps you can invent a cure for all the other unfortunates. If I try really hard, perhaps I can make myself ugly. Would a homely wife better suit you?"

"Don't be silly!" he exclaimed.

"Silly? I'm not the one tearing around the room like a dog with a burr under his tail."

"All right. I'm overreacting, but I'll warn you now," he said gruffly as he sat on the bed and yanked her to him. "Don't let me catch you batting those big, green cat eyes of yours at him, or I'll thrash you soundly. What is mine, I keep!" That said, he bent his head and crushed his lips to hers possessively, commanding her response to his blazing kisses.

As they dressed for dinner, Kathleen suggested to Reed that he go alone. "I will be the only woman there."

"You have been all week, and it hasn't bothered you."

"Yes, but with Venley gone again, there were only four of us, and I like Jean and Dominique. It was interesting to hear you talk of your adventures at sea and your daring exploits.''

"And now?" he asked.

"Now Pierre will be there, too, and the four of you could talk business if I took dinner alone here," she ventured.

He paused to mull it over in his mind. "No," he decided. "You can not hide away. Pierre simply must learn to control his urges, or pay the consequences."

"And you to control your beastly temper," she supplied. "Personally, I am getting very tired of having you rail at me everytime something does not go your way."

He came to her, folding her tenderly in his strong arms. "I'm sorry, kitten. I have been acting like a bear lately, haven't I?"

"No. More like an angry jungle cat," she murmured against his broad chest. Lifting her lips to his, she said softly, "Don't growl so much. Purr for me instead, my sleek, black panther." She ran her slim fingers through his ebony hair, her green eyes sparkling.

"We'll be late for dinner," he predicted, already carrying her toward the bed.

"Better a cold dinner than an unsatisfied wife," she joked.

"Woman, you are insatiable!"

"Stop complaining, and start doing something about it," she giggled, and then shrieked as he smacked her on the rump.

By scurrying about, they arrived at Jean's only a few minutes late. Kathleen looked flustered, and at Jean's wry smirk, her face turned beet red. Reed laughed heartily and leaned to her ear. "I told you so."

She darted him a sly look out of the corner of her eyes and wrinkled her nose at him saucily.

Pierre entered the dining room a short time later, his arm clasped around Rosita's waist. At Jean's questioning look, he plopped her into a chair next to his. "I invited Rosita to dine with us. I thought she might liven things up somewhat."

Kathleen groaned inwardly, trying desperately to keep her face in placid lines. Reed seemed about to explode in mirth, and the slight frown on Jean's brow was mild compared to the dark look on Dominique's face.

Rosita bestowed a flashing smile on Reed, and a smug look toward Kathleen. Pierre bellowed, "Jude! Get your lazy carcass in here! I could use some service! And bring another plate for my guest!"

Kathleen jumped when he shouted. Dominique gave her a look that clearly showed his embarrassment for his younger brother. She smiled sweetly at him, letting him know she understood.

Pierre caught only her smile. "Ah, the goddess smiles after all, and she bestows such sweet looks on old Dominique here. Could it be the honeymoon is over so soon, Reed?" he teased. "Perhaps your young bride has a roving eye?"

"Don't push me too far, Pierre," Reed warned lightly.

"I wouldn't be one to talk of roving eyes, Pierre. Not with a loving wife and children waiting in New Orleans," Jean admonished. "Everyone, including dear Francoise knows of the way you behave when you are gone from her. I often wonder why she puts up with you, brother."

"It is because I am so charming, I suppose," Pierre replied, undaunted.

At this, Kathleen promptly choked on her food, hiding her mouth behind her napkin as she coughed delicately. Tears gathered at the corners of her eyes, making them shimmer even more merrily.

"Oh, Reed," she gasped. "Help me up, please. I need some air."

Reed led out out into the garden as Dominique grinned broadly into his plate. Jean was pretending to dab his mouth with his napkin, while hiding a smile, and Rosita and Pierre wore confused looks. Reed had his head bent toward Kathleen's, faking concern.

"Oh dear, thank you for rescuing me," Kathleen giggled into his shirt. "I was about to die in there. It struck me so funny!"

"I know," he chuckled back, "and the funniest thing is that Pierre was so sincere."

They returned to the table, their composure restored. Rosita was shoveling her food down as if there were to be no tomorrow. Even Pierre began to realize his mistake in bringing her. Kathleen kept her eyes demurely on her plate, hoping she could keep a straight face for the rest of the meal. When the meal was finished, Jean, realizing that they could not leave Kathleen to entertain Rosita, suggested that they all adjourn to the gallery for brandy, provided the women would allow the gentlemen their cigars.

Once outside, Kathleen left the men and Rosita to their conversation and wandered about the gardens in the moonlight. She sat on a shadowed bench near Venus, watching the stars twinkling above her. Some inner voice urged her to look back toward the gallery. There she saw Rosita sidle up next to Reed, who was lounging against one of the pillars. Kathleen watched as the woman ran her hand up his arm and reached inside his open shirt to stroke his chest. He brushed her aside as one would a pesky mosquito, but she persisted. She rubbed the full length of her body against his side, then reached up to nibble at his ear.

Kathleen did not wait to see Reed's reaction. It was more than she could bear. On slippered feet, she ran toward the far corner of the garden, deep in shadows. Strong arms reached out for her, and before she could scream, one hard hand clamped across her mouth.

"Now, *cherie*," Pierre's sneering voice sounded in her ear, "while Reed is occupied with Rosita, we will find a little recreation of our own. I sense a fire beneath that cold exterior you put on."

As she struggled to free herself, he pulled her deeper into the darkness of the trees. One hand still on her mouth, and pinning her arms against her sides, he pushed his other hand down inside her bodice. The delicate fabric ripped away, baring her breasts to his large, abusing paw.

While he was ogling her, she worked her arm free.

Giving him a vicious jab in the ribs, she pushed herself free. Then, before he could recover, she spun about to face him and delivered a mighty kick to his groin. As he grunted and doubled over in pain, she sped down the path toward the gallery, pulling the torn bodice together with shaking hands. Weeping through tear-filled eyes, she saw Rosita still openly fondling her husband.

Jean and Dominique sighted her as she neared the gallery, and were instantly on their feet. Reed, whose back was to the garden, spun around to see what had alarmed them. He pushed Rosita aside and started toward Kathleen, his eyes narrowing as he took in her disheveled appearance.

Rosita's voice resounded clearly in the night air. "Well, look at who's been for a romp in the garden!"

Kathleen stopped abruptly. She glared long and hard at the two of them. Knocking Reed's arm aside, she snarled, "Don't let me interrupt your fun, Reed! Go back to your games with your slut! She's welcome to you!"

She turned pleading eyes on Dominique. "Please! Take me home!"

As Dominique moved to her side, Reed tried to intervene. Jean reached out, firmly clamping his hands on Reed's arms, holding him back as Dominique gently led Kathleen away, her shoulders shaking with suppressed sobs.

Late as it was, Kathleen requested a bath, and proceeded to scrub as if the devil himself had touched her. She wept until she had no more tears, and then sobbed dryly. She lay awake for hours, listening for Reed's return, dreading the scene that would follow. As the downstairs clock marked off the wee hours of the morning, she began to wonder where he was, and what he was doing. She waited anxiously, longing to hear his footsteps on the stairs. Finally, as dawn tinged the sky, she gave up her lonely vigil, angrily imagining him locked in Rosita's embrace.

"Oh, God, how I love him! He could never hurt me this way if I didn't. Why?" she sobbed. "Why must he always hurt me so? Coldy, deliberately wounding me. He's taught me to need him; to long for him, to love him, and now he turns to her.''

She rose from the bed and went to the washbowl. Soaking a washcloth in cool water, she dabbed at her swollen eyes. Studying herself in the mirror, she asked her image, ''Why do you cry so over him? You've known since your wedding night how deceitful he is. He stole the *Kat-Ann* didn't he?

''But he can be so charming when he wants to be,'' she argued with trembling lips. ''He is so strong; so tall and sleek, with the body and face of Apollo. He is so handsome with his icy blue eyes and dark hair. And he is such a marvelous, masterful lover, at once tender and demanding. How could I help falling in love with him?

''Ha!'' her second self expounded. ''Strong arms, a broad chest, slim hips, and other ample endowments are all you care about, huh? What about true love, honesty, fidelity? So he awakens your passions, sets your very soul on fire! It is nothing without love, you fool! Stop being such a sniveling ninny! You've made a mistake, but learn from it. A broken heart is not fatal. Weren't you the one who swore vengeance on his head? Aren't you interested any longer in paying him back for his treachery?

''You're right,'' she said to the girl in the mirror. ''I'll give that jackal what he so justly deserves, and when he cries out for mercy, I'll give him no quarter! He may have won a battle, but we'll see who wins the war! I only wish it didn't tear me up so inside. Dying must be far less painful!'' she choked.

Kathleen felt as if she had just gotten to sleep when Reed stumbled up the stairs and into the room. She shut her eyes tightly, feigning sleep.

''I know you're not asleep, Kat,'' he muttered. He plopped on the bed beside her. ''Order me up a bath.''

''Order your own bath,'' she snapped, rolling over with her back to him.

''No need to scream,'' he groaned, holding both hands to his throbbing head. Easing himself off the bed, he wobbled over and tugged the bell-pull, sinking into the chair beside it. When Joe arrived to answer his ring, he was told to prepare a bath.

"Yes, suh." He bobbed his grey head. "I'll get Mae to he'p me empty dis ole tub right way," he said and scurried off.

It was then that Reed noticed the yellow dress Kathleen had worn the night before. It was lying in a heap next to the tub. Leaning forward in his chair, he retrieved it from the floor. He puzzled a moment over the torn bodice before his mind registered the events of the previous evening. A dark look crossed his face, drawing his brows together.

"Kat!" he barked, making her jump. "How did your gown get torn?"

"We'll discuss it later, if you don't mind. I'm too tired to argue with you about it," she grumbled sleepily.

Striding to her side of the bed, hangover forgotten, he yanked the covers back. "We'll discuss it now, Kat!" His eyes widened as he observed the bruises showing clearly on her arms. He scanned further and saw that her wrists were bruised also. As she rolled to look up at him, he saw the slight reddening around her delicate mouth, and drew in a sharp breath at the ugly bluish marks on her breasts. She followed his gaze, wincing as he touched her tender breast.

"What went on last night? Who is responsible for this?" he demanded angrily.

Brushing his hand aside, she commented dryly, "I could ask the same of you, Reed. What happened to you last night, and with whom?"

"Don't sidestep the issue, Kat. It was Pierre, wasn't it? Wasn't it?" he roared as she didn't reply.

She flinched, but wouldn't answer, opting to glare at him defiantly.

"How far did it go?" he ground out through clenched teeth, then exploded as she remained silent. "Did he rape you, damn it? Or maybe after some persuasion you gave in freely!"

"No!" she screamed, unable to believe she'd heard correctly.

"Did you enjoy it, Kat? Did he satisfy you as well as I?" he persisted with a sneer. Holding her chin firmly, he glared at her, his eyes as cold as blue diamonds.

Tears of hurt and anger coursed down her face onto his fingers. "Nothing happened, Reed!" she shouted at him.

"Are you sure?" he snarled. "I know how easily you can be won over; how willing you can become! I can even imagine you encouraging him." His hand moved to circle her slim throat.

"Reed, I swear to you, nothing happened!" she croaked. She looked him square in the eye and added, "No thanks to you! You were too pleasantly occupied with Rosita to notice anything else."

His hand tightened on her throat. She held his gaze with her own, and forced herself to relax in his grip. "Go ahead, Reed. Kill me," she gasped. "The *Kat-Ann* will still be yours, and you'll be free of me. Frankly, I don't care anymore."

After what seemed a lifetime, he released her, his big hand shaking. "It's not my intention to be free of you, Kat. I told you before, what I own I keep, but I'll share you with no one."

Joe arrived with Mae, preventing further discussion. When they had lugged the tub away, Kathleen left the bed, pulling on a light robe. She searched through her armoire, selecting an aqua gown and slippers, the first ones her hands found. Her eyes lit on the hilt of her rapier, and for a brief moment she considered using it on Reed. She grabbed a handful of pins and her hairbrush, and headed for the door.

"Where are you going?" Reed asked gruffly.

"Where you aren't, you loathsome Yankee!" she answered in a trembling voice. She slammed the door and made her way to the guest room, slamming that door also.

Reed bathed, shaved, and dressed in fresh clothes. He regretted the harsh words he'd said to Kathleen, but still his anger held, and he wondered if she had told the truth. He knew she would not have given in to Pierre willingly, but perhaps she was ashamed or afraid to admit he had violated her. When he was ready to leave the room, Reed walked down the hall to the guest room. He tried the door, finding it locked. He pounded on it with his fist.

"Kat, unlock this door!"

"Go away!"

"If you don't let me in, I'll break the door down. Now!"

"Leave me alone! I don't want anything to do with you!"

Turning his shoulder to the door, he heaved. Under his weight the door gave easily, breaking the lock.

Kathleen sat in a chair, staring out the window.

"I want to talk to you," he said evenly.

Wearily she turned to face him, eyes puffed from crying. "All right, talk," she sighed resignedly.

"Tell me exactly what happened last night," he instructed, pulling up a chair to face her.

"Ask Pierre," she said stubbornly.

"I'm asking you, Kat. Now quit hedging."

"All right!" she screamed. "He grabbed me from behind and pulled me into the trees. His hand was across my mouth, and I couldn't scream. He pushed his hand into my bodice, tearing my gown. He was so involved in groping at my breasts that I struggled free. Just to insure he wouldn't be bothering me further, I kicked him in the groin."

Her voice rose as she spoke, and she trembled at the memory of Pierre's hands upon her. "Then, when I would have run to you for help and comfort, I found you in Rosita's capable arms! Now are you satisfied?" she spat.

"Not until I settle a score with Pierre," he swore.

"Why? Because he attempted to have his way with your wife while you were likewise involved with another woman?" she jeered. "I'd say I have more to settle with Rosita since you spent the night with her!'"

"I got drunk and passed out at Jean's. They bedded me down there," he admitted sheepishly. "Alone," he said.

"I'll bet!" she said hotly.

"Ask Jean."

"He's your friend, Reed. I'm sure you two have cooked up some fantastic lies. By the way, what do you intend to do about Pierre?"

"We'll duel, of course," Reed stated calmly.

"How good are you, or should I ask, 'how good is he'?" she queried.

"He's good, but Jean is better and so am I. Were you hoping to play the merry widow?" he asked snidely.

"No, Reed." Kathleen turned serious leaf-green eyes on him. "Of all the things I've wished on you, I would not want to see you dead or seriously injured," she told him sincerely.

"Thank you for that much, Kat."

"You can't kill him either, you know."

"Why not?"

"Because he is Jean's and Alex's brother, and they could not remain your friends after you'd killed Pierre."

"So! Dominique even tells you his most guarded secrets! Maybe I should keep a closer eye on that sly dog, too."

"We have become very close friends, nothing more. We respect each other. He treats me like his sister and I like him very much. You will not harm him, Reed," she warned.

Reed leaned close, tangling his fingers in her silky hair. "Are you sure you do not feel more strongly about him?"

"I do not love him, except perhaps as one would a brother."

Reed nodded, satisfied. "I'll talk to Jean today. If Pierre agrees, we will fight until first blood is drawn, and I will be satisfied with that, since you say you have come to no great harm."

"My word is good, Reed. Better than yours, I'd wager. You must believe me. I would want his head on a platter and his body feeding the fish regardless of Jean or Dominique if it were otherwise."

Reed watched her face closely, then sighed in relief. "I believe you."

Chapter 7

THE duel was set for dawn. Kathleen had lain awake throughout the endless night, alternately hating Reed and praying for his safety. True, he was a scoundrel of the first degree, and she swore vengeance on his head, but she loved him. Any revenge would be at her hands, and she could not bear the thought of him wounded, maimed, or dead.

She'd tossed and turned, finally choosing to sit in a chair so Reed could get some rest. He had refrained from touching her, speaking only when necessity demanded it, both of them still angry and upset. His voice came to her now out of the dark.

"Why can't you sleep, Kat?"

"I'm worried for your safety," she replied softly. "You should be resting."

"I've never been good at resting before a duel at dawn," he chuckled dryly.

"You should try. You'll be dead for lack of sleep in the morning."

"Good heavens, you certainly have a way with words, woman! Never say 'dead' to a man just before he enters combat!"

"I'm sorry," she muttered meekly. "Are you sure you can beat him?"

"Nothing is ever a certainty, kitten, but the odds are in my favor. Pierre has grown slovenly lately, not so quick on his feet.''

They were silent for a time, but she knew he wasn't sleeping.

"Kat?"

"Yes?"

"If anything should go wrong, will you go to my mother in Savannah?"

"Yes."

"Jean will take you, and Dominique will protect you from Pierre."

"Reed?"

"Yes?"

"Please be quiet," she whispered softly through her tears.

Finally the hour had arrived. The contest was to take place on the beach below Jean's gardens. The morning sun had not yet burned off the mist, and the breeze from the Gulf was cool. Kathleen had insisted on attending the match, snatching down her rapier from its place in the armoire.

"What are you going to do with that rapier, Kat?" he asked hesitantly.

"I shall kill Pierre with it if he does not stop when first blood is drawn. I do not trust him."

"Don't be silly. Jean will mediate the duel and see that the rules are followed. Put that thing back before you hurt yourself."

"No," she insisted adamantly.

The men faced each other in the wet sand. Both had discarded their shirts, the better to judge who scored a cut first. Kathleen proudly noted that while rolls of fat hung over Pierre's belt, Reed's stomach was flat and hard; his chest and arms muscular and fit.

Jean announced the rules and both men agreed to adhere to them. The two adversaries saluted one another, then with a touch of rapiers and a shout of *en garde*, the duel was on. Kathleen held her breath as Pierre lunged for Reed's broad chest. Reed parried the blow with ease and followed through with a counter attack toward Pierre's midriff. Pierre thrust again, and this time Reed side-stepped and thrust at Pierre's sword arm, missing it by a hair's breath, to Pierre's relief. The men circled each other warily, alternately thrusting and parrying, lunging and recovering; each trying to judge the other's skill; each watching for an opening to present itself. Their rapiers flashed in the morning sun as they continued to fight.

Kathleen's experienced eye told her that Pierre was wanting more than just to wound, and as she glanced at Jean's worried face, she could tell he knew it, too. The only sound was the clash of steel on steel, no one willing to distract either opponent. Pierre was tiring, sweat glistening on his face and chest. Reed was breathing easily, his hairy chest only slightly damp.

"Reed is almost toying with him," Kathleen thought. She watched as they clashed swords again and again, each seeking to wound the other. She bit her lip to keep from crying out as Pierre landed a lucky stroke, his sword glancing off Reed's belt buckle. Reed counterparried, but Pierre recovered too quickly. Pierre thrust again, and Reed leaped aside with the grace of a cat. A flash of steel, and Reed's blade slashed across Pierre's left cheek, creating a deep gash from just below his eye to his chin.

Jean jumped in immediately, declaring Reed the victor, as he pressed a handkerchief to Pierre's face. He congratulated Reed on his skill, smiled kindly at a very pale Kathleen, and returned to say a few words to Dominique on the edge of the crowd.

Reed walked toward Kathleen, his back to Pierre. Pierre bent, seemingly to pick up his shirt, and suddenly lunged toward Reed's retreating back, rapier extended. Faster than the eye could see it, Kathleen swung her rapier up, stepped forward, and brought the tip of it to Pierre's throat. A gasp went through the crowd as all eyes turned toward them.

"One more step and you are a dead man, Pierre," Kathleen snarled through her teeth.

Pierre gaped at her in amazement.

"Enough, Kathleen. Step aside and let me deal with

97

this back-stabbing coward," Reed ordered.

"Stay out of this, Reed," she snapped, her eyes never leaving Pierre's. "He's mine, now!"

Pierre, feeling confident of her inabilities, brought his rapier up quickly, trying to jab for her left breast. She jumped neatly aside, her green eyes flashing, meeting him with a lightning parry. With one swift slice, she opened his sword arm to the bone from shoulder to elbow. He bellowed in pain, his rapier falling from his hand; his arm dangling.

A murmur of astonishment issued from those watching. A look of disbelief was carved on Reed's face. Kathleen backed away from Pierre, slowly lowering her blade. *"Touché,"* she said softly, then fled along the beach away from the curious stares of the onlookers.

"My God, I've really done it now," she thought. "I've tipped my hand and thrown away my trump card. Now Reed knows I can handle this rapier." She looked at the bloody weapon in her hand. "How do I bluff my way out of this?"

She didn't have long to wonder, for Reed came loping up behind her. "Kat, stop a minute." He reached out, clutching her shoulders, and spun her about. "I think you owe me an explanation," he panted.

"About what?" she asked innocently.

"Start with where you learned to fence," he suggested dryly.

"I used to watch my father," she said. "He was very good."

"I suppose donkeys fly, too. No, Kat, you couldn't have learned to be so quick and light on your feet just from watching," he countered wryly.

"I admit I practiced with him at times, but my agility probably comes from my ballet and dance lessons at school," she told him a half-truth, not willing to tell him an outright lie. "Call it beginner's luck."

"In a pig's eye."

"Look, Reed, this is silly. I was infuriated at Pierre back there and reacted before I had time to think," she told him truthfully. "Good grief! One lucky stroke and you think I'm ready for the world championship fencing tournament! I was mad, damn it, and Pierre was too tired to move quickly, and I was damned lucky! When I think what might have happened, I start to shake."

"Hmm," he grunted non-committally, still eyeing her dubiously.

"Perhaps you had better teach me how to use this sword after all," she suggested innocently, hoping he was falling for her ruse. "Papa always said there was nothing more useless than an inept amateur."

"All right," he agreed. He stepped away from her, taking her stance. "Present your weapon," he instructed her.

"Now?" She stared at him, eyes widening.

"It's as good a time as any."

"But aren't you tired?"

"No. Now quit stalling. I want to see how good you really are."

"You'll be surprised at how bad I can be," she predicted hopefully.

She copied his stance with studied formality, presenting her weapon stiffly.

"Loosen up," he told her.

"Easy for you to say," she grumbled back.

They touched tips and he called out, *"En garde."* He thrust straight toward her midsection. She parried slowly.

"Too slow," he cautioned her, drawing back. "Try again." He lunged forward, the tip of his rapier slicing a button from her bodice before she turned his blade away. She jabbed with deliberate clumsiness at his chest. He parried her attack and sliced her sleeve, neatly avoiding cutting her skin. She had to admire his skill.

"Reed!" she shouted, feigning fright.

"I won't cut you, Kat. Keep trying."

She attacked again, this time barely nicking his chest. She gasped in pretended concern.

"Better," he encouraged, ignoring the slight cut. He thrust again and this time she jumped lightly out of his way. They continued in like manner for several minutes, Kathleen trying to show him that she was agile but unskilled at the art, clumsy enough to be convincing—she hoped.

Finally she threw up her arms and sank to her knees, breathing heavily. "I'm worn out," she pretended. "This thing is so heavy!" she huffed, letting the rapier fall.

Reed sat down beside her on the warm sand. "You'll get used to the weight in time." He lay back and studied her delicate profile. "You could be good with lessons, Kat. You seem to have a natural aptitude for the footwork and the rhythm. But," he added, "you are definitely not ready for any duels, not even with beginners. You are much too slow. You are light on your feet, but awkward with your blade. You try to lunge from too far back, your recovery is hideous, and your parries carry no weight."

"Perhaps we'd better just forget the whole thing," she suggested, trying to hold back a smile of delighted triumph. "I'd much rather let you protect me anyway."

"It would probably be best," he agreed. "Still, I haven't thanked you for saving my life. Thank you, kitten."

"You're most welcome, sir," she quipped. "Just don't depend on me too often."

He shook his head in wonder. "I still can't understand how you bested Pierre."

"I was just blind mad that he would attack you from behind," she stated vehemently.

"I never expected that, even from him. Anyway, he must have been caught off guard by the suddenness of your attack, I suppose."

"I'm sure that must be it," she said, forcing back a grin.

Immediately following the duel, Jean had sent to New Orleans for Charles de Beaumont, urging him to hurry. He arrived with Eleanore by midafternoon. Pierre already had a rising fever, his eyes glazed with pain. After the doctor had examined him, Eleanore informed Kathleen that it would take all of Charles's knowledge and a lot of good fortune to save Pierre's arm. Once the arm was sound, only time would tell how much use Pierre would have from it.

The only remorse Kathleen felt was for Jean and Dominique. Pierre had gotten what he deserved, but they were suffering their brother's pain with him. Kathleen talked briefly with Dominique toward evening. "Do you and Jean hate me now, Dom?"

"Why would we, Kathleen? It was a dirty, sneaking trick Pierre pulled. None of us were close enough to stop him in time. Reed was fortunate you were so quick to react, or we would be attending a funeral. Jean and I cannot get over the way you disarmed Pierre so easily. You were magnificent! Where did you learn to fence so beautifully?"

"I was just furious and impulsive, and very, very lucky, Dominique." At his skeptical look, she continued, "Ask Reed to explain it to you sometime."

"Your anger and luck, if that is what it was, served you well, little sister."

"Too well. I would not blame you or Jean for being angry that I may have destroyed Pierre's sword arm."

"We regret it of course, and share his pain, but he broke the code and is paying for it dearly. Better his arm than his throat, which is what I believe you would have liked to have sliced."

She lowered her eyes, not wanting to meet his look.

"It is all right, *cherie,* we understand, and we are grateful that you were able to save Reed's life. He is a dear friend to Jean and I, and so are you, little one."

She raised her head, gazing into his fathomless black eyes. "I love you, Alexandre Lafitte," she said tenderly. "Not the same way I love Reed, but I truly do love you."

Kathleen and Reed dined alone that evening, not yet ready to face the others socially. Reed had talked to Jean earlier in the day. Jean's feelings were similar to Dominique's, the fault totally Pierre's. They were appalled at Pierre's treachery. Still, the atmosphere was strained with their brother lying so ill.

Kathleen picked morosely at her food, doing little more

than pushing it around on her plate.

"Don't take it so hard, kitten. What is done is done. If you are feeling sympathy for him, remember that he tried to kill me. The dog deserved to die!"

"I am not feeling sorry for Pierre so much as I am Jean and Dominique, and I hope I have not destroyed the friendship between you."

"The alliance between us is not so frail as all that. This will pass, and in time will be forgotten."

"Pierre will not forget so easily," she fretted.

"Pierre and I have rarely seen eye to eye anyway. We've had our disagreements before, and always will. He is the least of my worries."

All through dinner they talked, and afterward they walked along the beach in the moonlight, his arm around her waist. A feeling of closeness, contentment, enveloped them. They strolled slowly, chatting easily, feeling no urgency as they enjoyed each other's company. She leaned comfortably into him, and he pulled her close to his side. The warm night air caressed them gently, wafting the sweet fragrance of tropical flowers to them. The waves lapping gently at the shore and the songs of the night-birds provided a symphony seemingly created just for them.

They sat on the sand listening to the soft sounds of the night. He held her close, nestled in the curve of his arm. They sat for a long while without talking. Then, one by one, he plucked the pins from her hair and it tumbled down her back like a shining curtain, the ends resting on the sand. He ran his hand down the length of it, thinking how much it felt like silk. Leaning nearer, he buried his face in its thickness, breathing in the familiar lemony scent of it. Turning her toward him, he kissed her gently, lingering over the sweetness of her lips. His fingers traced a path up and down her arm. She shivered and leaned closer, caressing the nape of his neck with soft fingertips, delighting in the masculine smell of him.

It was only when his kiss grew more demanding, and his hand brushed her bruised breast that she stiffened. The tenderness of her breast reminded her of the ugly scene in the garden. She recalled a long night of waiting for Reed to return, and her thoughts turned to Rosita. She thought of Reed in Rosita's arms, and now in hers, and it sickened her. She dropped her arms from his neck and pulled her lips from his.

"What is it, kitten? Why do you pull away from me?" he whispered.

She shook her head mutely.

He kissed her again, and this time all thought was blocked from her mind as she was swept up in a whirlwind of heavenly ecstasy so pure that it took her breath away. She felt as if her head were spinning and a thousand butterflies had taken wing in the pit of her stomach. She longed to cry out her love for him; felt as if her heart would burst for want of him. Nectar from the heavens could not have held more sweetness than the kiss he gave to her. Her breath came unevenly, and it seemed as if a bird had lodged where her heart should be, rapidly fluttering his wings within her breast. She whimpered softly as his lips released hers.

Reed gazed into her moonlit face and beheld an angel, more beautiful than any creature he had ever seen before, love radiating from every delicate feature. With a feathery touch, she traced the bold lines of his face, admiring the excellence of its form, delighting in the texture of his skin beneath her fingertips. With each petal-soft stroke she silently bespoke her love for him. She sighed tremulously and pulled his head to hers, presenting her lips in sweet surrender to the powerful, magnificent man before her.

With tender caresses and softly spoken words, he removed her clothing and his. There on the warm sand, her hair spread beneath them like a sheet of glowing satin, he made love to her; slow languorous love, exciting every sensitive nerve in her being; eliciting pleasures so intense they seemed almost painful. She clung to him, running her trembling hands along the rippling muscles of his back, drawing him nearer to her quivering body. Soft purring sounds issued from deep within her throat. When it seemed that neither of them could stand more of this exquisite torture, their rapture broke, soaring them to a pinnacle of passion so high that the wings of angels seemed to brush their fevered flesh. All the stars of the universe surrounded them, radiating a light more brilliant than a million diamonds in the sun. They glided gradually earthward on a gossamer veil of joyous fulfillment.

They lay relaxed, their senses soothed by the ecstasy they had shared, until the night air chilled, driving them to seek the comfort of their bed.

With Eleanore back on Grande Terre, Dominique took over some of Pierre's duties while Reed was overseeing the new slaves and helping to catalog cargo. Everyone was busy readying for the slave auctions, and carting other goods upstream to Jean's warehouses in New Orleans for sales he would hold there. The only ones with time on their hands were Kathleen and Eleanore.

Kathleen talked her reluctant friend into visiting the docks one morning. She knew Dan would be there somewhere and longed to see how her men were faring. She said nothing of this to Eleanore, of course, and was not sure how she could escape the woman long enough to speak to Dan alone. Maybe an opportunity would present itself, for she knew she could not approach the docks alone. Both Reed and Dominique would be angry as it was.

The women strolled leisurely along the waterfront, Eleanore protecting her creamy complexion with her parasol, and Kathleen soaking up the morning sun. Luck was with her, for they hadn't gone far when she spotted Dan. He was busy helping to unload cargo from a barkentine that had docked that morning.

"Let's watch awhile," Kathleen suggested to Eleanore.

"Do you think it is wise?" Eleanore questioned.

"Just for a few minutes. I see my husband's bosun, and if I can catch his attention I'd like to ask him something."

"Kathleen Taylor! That is akin to flirting! Reed would surely kill you if he heard of it!"

Kathleen exploded in a tinkling laugh. Dan and several

others turned her way. She motioned for Dan to join her. "There. You see the old man who hobbles toward us?" She pointed Dan out to Eleanore. "Reed could hardly be jealous of him, don't you agree?"

Quickly, before Dan could say anything, Kathleen stepped forward. "Good morning, Mr. Shanahan," she greeted him, giving him silent warning with her eyes.

He gave a curt nod, an answering spark in his own eyes. "Mornin', Lady Taylor." He spewed a stream of tobacco juice onto the sand and looked at Eleanore.

"I would like you to meet Dr. de Beaumont's sister, Eleanore," Kathleen continued. "Eleanore, my husband's bosun, Mr. Shanahan."

"Ma'am." Dan bobbed his head toward Eleanore.

"I see you are busy, but I was wondering if you might do me a slight service," Kathleen explained. "I have left a few personal items aboard the *Kat-Ann,* and I was hoping you could spare the time to row me out to retrieve them."

"But, Kathleen," Eleanore objected. "I cannot stand here on the waterfront alone waiting for you. I feel so conspicuous as it is."

"Then come along with me. It shan't take long."

"There is a dinghy just down here," Dan offered, leading the way.

"Well, come on, Eleanore," Kathleen said, pulling her along.

Once aboard the deserted ship, Kathleen left Eleanore on deck, taking Dan along to bear the needed items. Closing the cabin door, she headed for her desk. "Dan, you must help me find a safer place to stash these old log books. So far Reed has not asked to see them or bothered to investigate my desk, but one day it may enter his mind and my secret will be out. He'll wring my neck if he finds out I can master a ship and said nothing. I prefer he not know anything about it. My hidden talents may prove useful someday if kept secret from him."

"Aye, Cap'n. That's the first mite o' good sense I've heard from ye since we left Ireland. I was beginnin' to think ye'd gone soft on him, 'specially when I saw what ye did to that Pierre the other mornin'. O' course, that didn't seem to be too smart, either, lass." He scratched at his beard. "Sort of let one cat out of the bag there, didn't ye?"

Kathleen's face split in a mischievous grin. "You'd be surprised, Dan," she laughed. "I wish you could have seen what happened after. Reed demanded to know where I'd learned to fence. I told him I watched Papa often and sometimes practiced with him, but he asked about my agility and footwork, which I tried to explain as from my dance classes at school. Needless to say, he wasn't buying that excuse, so I told him Pierre was surprised and tired, and I was angry, and very lucky. He still wasn't sure, so he had me fence with him."

Dan's eyes widened at this.

"Oh, Dan!" she exclaimed through tears of mirth. "You should have witnessed the awful performance I gave! You would have been proud of my acting abilities, old friend. I was worse than any beginner you've seen. In the end, he conceded that I knew little of the art of fencing. In fact. he criticized my abilities royally!"

Dan chuckled merrily. "I should have liked to have been there to see ye dupe him." Then he sobered. "Ye've got to be more careful in the future though, Cap'n, or he's sure to find ye out."

"Aye, Dan. That I will, for he's nobody's fool." She handed the log books to him. "You hide these some place safe while I find a few things to support my lies to Eleanore."

Kathleen rounded up a few books, miniatures of her papa, mama, and Nanna, and her guitar. Dan came back with a hammer and the log books wrapped securely in oilskin. Drawing back the curtain from around her chamberpot, he removed the pot from its seatlike enclosure. Prying loose the boards around the platform of the wooden structure, he slid the package into the base of it and nailed the planks back in place. The log books now rested beneath where the chamberpot sat; between it and the floor, enclosed completely. Dan chuckled at Kathleen's amazed look. "Not too many folks go pokin' around there, I'd guess."

"Dan, you're a genius." She snickered. Handing him her bundle, she led the way from the cabin. Pausing in the passageway, she told him, "You were right when you said I'd gone soft on Reed. I love him, Dan," she said sadly. "But he may never return my love. Besides, I still have a score to settle with that Yankee, and I won't rest easy until he pays dearly." Again she paused. "By the way, I meant to ask how you and the men are faring."

"We be fine, Cap'n, but some of them are homesick. How long do ye suppose before we head for Savannah?"

"I don't know, Dan, but surely not much longer. Jean has put Reed in charge of the slave auction. Probably after that."

"Why don't you just steal the *Kat-Ann* back from him and head for home?" Dan suggested wistfully.

"And have him chasing after us?" She gave him an incredulous look. "No, I will bide my time, and someday he will be the one to back down. Just taking the *Kat-Ann* would not be nearly sweet enough revenge if and when I decide to seek it. I'll leave things be at present. Besides, the ship is legally his, not to mention the fact that we'd have half the pirates this side of the Atlantic on our trail." Then she added, "But if I should need to flee some night, listen for my whistle."

"Aye. Even above all the snoring those tars do, I'd hear that pipe o' yers, Cap'n K."

As she had thought, when Reed heard she had been on the docks, he was hopping mad.

"Good God, Kat! Where are your brains, in your shoes?" he exclaimed.

"Thanks a lot, Reed. I took Eleanore along."

"Fine! That way there were two of you flitting your skirts about!"

"Oh! You are so maddening! Why can't you ever try to understand?" Kathleen stamped her foot at him.

"The only thing I understand is that you need a constant keeper!" he said, angrily stubbing out his cigar.

"And here I thought you only married me for my money!" she retorted, her green eyes flashing.

"So we're back to grinding that old axe, are we? Talk about people who never try to understand! I wish you'd just shut up about that once and for all."

"What's the matter, Reed? Does the truth hurt?"

"I'm tired of trying to convince you otherwise. You are set on seeing things one way, so let's forget it."

"I'm not about to forget it, you thieving cad, not as long as I can draw a breath!" she railed.

"Sometimes, Kat, you tempt me to correct that problem!" he said, gritting his teeth.

She stood before him and flipped her hair back, baring her throat to him. Then she glared at him defiantly.

Reed glared back through stormy blue eyes, hands clenched into tight fists at his sides. "Blast your Irish hide! Just stay off the docks, Kat!" He strode from the room, swearing under his breath.

For once Reed stayed angry with her all night. Both kept to their own sides of the bed, carefully avoiding contact with the other. Neither slept at all well.

The next day Kathleen decided to make amends. The thought of another evening of silence, and sleeping with one eye open held little appeal for her. She got Mae to pack a picnic hamper full of goodies. There was fried chicken, ham sandwiches, cheese, warm bread, fresh fruit, and wine. Eleanore walked with her as she set out full of good intentions for the warehouse where Reed was sorting cargo that day. As they neared the warehouse, Kathleen heard Rosita's voice and then saw her perched seductively atop a barrel just inside the warehouse entrance. Her full skirt was pulled high, revealing most of her thighs and bare legs, and her voluptuous breasts were almost popping out of her blouse as she leaned forward. There was the sound of Reed's voice from somewhere inside the building, followed by Rosita's lusty laugh.

"Somehow that wench gets on my nerves," Eleanore claimed disgustedly.

"Yours too?"

At the sound of female voices, Rosita twisted around on her seat, swinging her brown legs before her. She eyed Kathleen hatefully, looking her up and down. With a sneer, she called in Spanish over her shoulder into the warehouse, "Darling, your skinny wife is here." Turning back to Kathleen with a sly look, she said, "I just told Reed you are here."

"I know," Kathleen said with a slight smile, as she passed Rosita. Abruptly she whirled, catching hold of Rosita's leg, and tipped her off the barrel onto the dirt floor. "Now you can tell him his skinny wife has dumped his fat slut on the floor!"

As she spoke, Kathleen reached into the picnic hamper and withdrew the bottle of wine. She yanked off the cork and poured the contents over Rosita's head. Eleanore dissolved into peals of laughter. Rosita rose sputtering and screeching just as Reed appeared around the corner of a stack of crates.

"What in the name of Hades is going on here?" he demanded, looking from the dripping Rosita to Kathleen, who still held the empty bottle. Eleanore let loose another burst of laughter. As Reed turned his look to her, she tried unsuccessfully to regain her composure, tears streaming from her merry brown eyes.

"Good afternoon, Reed," she choked out.

Rolling his eyes heavenward, he returned his frosty look to Kathleen. "Are you going to answer my question?" he demanded.

"No," Kathleen stated flippantly, handing him the empty bottle. "You are so smart. Figure it out for yourself."

That said, she took Eleanore by the arm and walked out the doorway. "Come, Eleanore. We seem to be minus a bottle of wine for our luncheon."

She glanced over her shoulder at Reed to find him staring after them, a puzzled frown on his face, the bedraggled Rosita at his side.

Reed stayed out late that night, staggering in drunk about midnight. He lit the lamp near his side of the bed and stood glowering down at Kathleen's stiff back. He sat on the edge of the bed and removed his boots, dropping each noisily onto the floor. Blowing out the lamp, he lay down on the bed fully clothed. "Women!" he grumbled grouchily, and promptly fell asleep.

Chapter 8

REED was gone when Kathleen awoke the next morning. She dressed in a pale pink muslin dress trimmed with delicate rose-colored lace. Amazingly, the gown set off the color of her hair instead of clashing with it, and enhanced the rosy glow of her cheeks and lips. She was breakfasting on the patio when she noticed a commotion on the beach below the house. Looking closely, she saw half a dozen men and women circling about an object on the sand. Several dogs were barking and running crazily around, trying to squeeze into the circle. As one of the men stepped back to swat at a pesky mongrel, Kathleen realized what was going on. A dolphin had beached itself, and the men were trying to drag it back into the bay.

Instantly she was on her feet, nearly tipping over the table in her haste as she raced on nimble feet to the beach. Shoving her way through the crowd, she reached the wheezing animal. Her heart was thundering as if she felt the animal's pain as her own. This was a fellow creature of her beloved sea, and all she knew for sure was that she had to help the poor, stranded thing.

She pushed the men aside, shouting, "Get away! Stop! You'll hurt him!" She knelt beside the dolphin, sizing up his condition with a quick examination. She knocked a yapping dog back and ordered, "Keep those dogs away!"

Glancing up, she spied Reed just entering the growing ring of onlookers. "Reed, give me your shirt," she demanded brusquely, holding out her hand. At his astonished look, she grew impatient. "Would you rather I used my dress?" she asked tersely.

Quickly he removed it, handing it to her. With sure hands, she dug a trench in the sand below the stranded animal. Wetting the shirt, she squeezed it out over the dolphin's drying skin. After repeating this several times, she slid the shirt beneath the dolphin's belly. Motioning for Reed to help, together they tugged at the shirt, ever so gently easing the dolphin toward the water. Kathleen crooned soothingly to the frightened beast.

When they had reached the water's edge, Kathleen bade Reed to step aside. Putting her back into it, she gave a mighty tug on the shirt and the dolphin slid into the water as Kathleen splashed in on her backside. Recovering quickly, Kathleen wrapped the shirt around the dolphin's middle, and clucking sympathetically to it, pulled the shock-stricken animal gently around in a huge circle in the water. Taking a handful of water, she cleared the sand from his airhole.

When they had circled about five times, the stunned dolphin suddenly gave a shake and lurched beneath the water, carrying Kathleen with him, shirt and all. The crowd waited anxiously, but Kathleen did not reappear. Nearing panic, Reed shook off his boots, preparing to dive in after her. Jean, who had also come to see what the disruption was about, put his hand on Reed's shoulder and pointed out to the center of the bay where the *Kat-Ann* lay anchored. With barely a ripple, the dolphin surfaced, Kathleen clinging to his fin. Throwing her had back, she laughed merrily; it was a light tinkling sound that danced eerily across the water to the shore. Her bright hair had come loose from its coiffure, and swirled around her like a golden cape in the morning sunlight.

As the throng looked on, Kathleen grabbed hold of the *Kat-Ann's* anchor chain and released the dolphin. It swam a distance from her, then turned and danced on its tail, chattering cheerfully at her. She laughed her tinkling laugh at him, and Reed shivered at the eerie sound it produced as it echoed over him. All watched spellbound as the dolphin returned and Kathleen reached for his fin. Together they frolicked in the bay, completely forgetting those on shore. Again they dived, and again everyone held his breath as they waited apprehensively. Moments later, just as Reed had again headed for the water, they surfaced so close to shore that Kathleen stood only chest deep. She murmured something to it, and the dolphin left her and headed out of the bay toward the entrance to the Gulf. About half way he turned once more and danced on the top of the water, chattering excitedly as though beckoning to her.

Kathleen smiled and shook her head. "No, my friend. Not this time," she said softly. "Perhaps another day."

She waded to the shore, her wet, nearly transparent dress clinging tightly, revealing every lovely curve. She stopped before Reed, gave him a cursory glance, and said calmly, "You'll get your feet wet standing around without your boots." She walked on up the hill to the house, leaving Reed gaping after her, stunned, and old Dan and Dominique smiling wondrously.

"Come, old friend," Jean said, laying his arm across Reed's bare shoulders. "Your sea goddess has given you quite a shock, and I believe you are in need of a long, stiff drink."

Handing his goblet to Jean to be refilled, Reed sighed, "Jean, I swear that little Irish minx is going to make a drunkard of me yet."

"She's a spectacular woman, Reed. Truly one of a kind," Jean said.

"I still can't believe I saw what I did today. Kathleen sporting in the bay with a dolphin as if it were a common everyday occurence! My God, I must have been hallucinating!"

"If you were, so was I," Eleanore spoke up, "and everyone else on the beach today."

From the doorway Kathleen called, "Hello! Has anyone seen Dominique? He is supposed to teach me how to play poker today." She had changed into a clean lilac gown and gathered her wet hair into a loose chignon at the back of her neck.

"He'll be along soon, I suppose. I saw him earlier on the beach," Jean told her.

"Oh. I guess I didn't notice him."

"Well everyone certainly noticed you," Reed grumbled. "All of you."

"Would you care to explain that comment?"

"My dear wife, you may as well have gone swimming nude for all your wet dress concealed!"

"And what was I supposed to do, let them harm that poor defenseless creature? Have you no heart, Reed?"

"I used to. It's that thing you are always walking on," he mumbled softly into his glass. Rising, he said aloud, "Thank you for the drink, Jean. I'd better go find a new shirt to put on. Mine rests on the bottom of the bay." He glowered at Kathleen and left.

Kathleen frowned at Reed's retreating back. "What is bothering him?" she asked no one in particular.

Jean waved her to a chair. "Sit down, Kathleen. Now might be a good chance for us to have a chat." He poured her a tall glass of lemonade, and when she had settled back, he continued. "You must try to be patient with him, *cherie*. This business of marriage is new to him."

Kathleen arched a delicate eyebrow at him. "Yes, but it is also new to me."

Jean chuckled softly. "That does compound the problem. It takes a while to get used to each other's ways. It will come in time." He went on to tell her of his earlier life and of his wife and family, which he sorely missed. He spoke of Reed and how they had first met in New Orleans.

Soon he had Kathleen laughing over their hair-raising escapades. Little by little he drew her out, getting her to speak of herself. As he got to know her, Kathleen also became better acquainted with Jean, and through him saw a new side of Reed. Though she already liked Jean, the more she listened, the more she appreciated him. There was a soft-spoken aristocratic gentility about him that appealed to her, and she admitted to herself that he was an extremely handsome man. Still, he was an admitted privateer, just a sneeze away from being a pirate, and Kathleen could sense the steel beneath the velvet surface. When it came to business and commanding his men and ships she knew he could be ruthless. Jean could be your friend or your enemy, and if it were the latter he would show no mercy, give no quarter; yet she found herself admiring him more and more.

Likewise, Jean found himself drawn to more than Kathleen's obvious beauty. Humor and intelligence sparkled from her large emerald eyes and added character to her faultless face. As much as he loved and admired Eleanore, he found himself envying Reed this fantastic woman. Intuitively he knew she was not telling him everything, that she was withholding vital information about herself, but it only served to make her more mysterious. Now he knew why Dominique was so enamored and Reed so befuddled. He found himself falling willingly under her magic spell, so softly spun he barely realized he was caught. The ruggedly handsome Jean Lafitte, terror of the Gulf of Mexico and the Caribbean Sea, was brought low by a woman he could only admire from afar, and he swore he would try to smooth the way for Kathleen and Reed and wish them all the happiness they could find.

A few days later, Kathleen needed some white thread and decided to walk down to the store. She was sorry she hadn't waited, for Rosita came in just behind her. Kathleen stood aside as Rosita flounced up to the counter.

"I need a box of cigars, Jake. Reed's special brand," she emphasized, shooting a superior look in Kathleen's direction. On her way out she said loudly, "It takes a real woman to know how to handle a man such as Reed. If you were nicer, skinny one, I might take pity on you and give you lessons."

"Should I ever decide to become a whore, I might take you up on that, Rosita," Kathleen shot back.

Reed was late again that evening. Kathleen awoke when he entered the bedroom. "Go back to sleep, Kat," he said softly.

"At least you're sober tonight," she decided as he undressed quietly in the dark and slipped into bed beside her.

"In spite of what you seem to believe, I do not make a habit of drinking myself into a stupor, my sweet. Besides, with Pierre laid up there is too much work to do and I will probably be working late quite often until we get caught up."

"Just don't wake me when you come in. I seem to require more sleep than you do, so I'd appreciate not being disturbed," she grouched. As an afterthought she added, "I'm sorry about your shirt, Reed."

"It's okay, kitten." He yawned sleepily. "Forget it."

Without Reed's company, each day seemed longer than the next. When Dominique could spare the time, he set about teaching her to play poker. Mornings usually found her in Jean's garden sketching the statues, until she discovered Pierre's bedroom overlooked the area. Twice she found him glaring from the window at her. The second time he shouted down, "I'll get even with you for this, you witch. You'll need eyes in the back of your head to escape me once I am well. Beware!" He drew a finger across his throat in a threatening gesture.

After that she avoided the garden and worked on her paintings at home.

Nearly every afternoon she and Eleanore strolled. The island was indescribably beautiful once one was used to the heat and the humidity. The swampy area to the north held the most beauty and the most peril, but if they avoided the marshy area frequented by alligators, snakes and insects, the two women had relatively little to fear. The large cat population Jean imported kept the rats and snakes to a minimum, and as long as the fresh sea breeze from the Gulf prevailed, the insects were no problem. It was during the occasional lull in the sea breeze, or when the northerly inland winds occurred, that the threat of disease and plague came to the fore. That, thank goodness, was rare.

The beauty of the island seemed to assault all the senses

at once. The vibrant colors of the flowers and birds were stunning, and Kathleen swore she had never seen such spectacular sunsets. No perfume on earth could compete with the fragrance of the island, and the small animals, insects, and birds created a symphony of sound for any and all to hear. The constant gulf breeze that wafted the sounds and fragrances about the island was as warm and gentle as a lover's caress, carrying with it the tang of the salty sea as well.

In the weeks Kathleen spent at Grande Terre, she came to love this tropical paradise. It seemed to her an island Eden, and to her mind should have been an island created especially for lovers instead of a pirate's haven. If not for the presence of the fort and busy docks, those unfortunate slaves, and the pirates, Kathleen would have been content to stay there with Reed forever, locked away in their own private paradise on earth.

One day Kathleen and Eleanore walked by the slave compound and found Reed busy trying to group the slaves by skills they had learned. It was then that Kathleen discovered that Rosita usually took charge of the females, thus working hand in hand with Reed every day. She watched as Rosita approached Reed, hips swaying. He leaned his dark head down to hear her words, and she rubbed her leg along his thigh intimately. Amused by whatever she had said, he threw back his head and laughed heartily, then sent her on her way with a smack on her ample rear. She turned and flashed him a brilliant smile, then noticing Kathleen, she tossed her long black hair over her shoulder and walked triumphantly away.

Reed followed her look, studying Kathleen with an unreadable expression. She darted him a grim look, spun on her heel, and strode off muttering oaths that brought a grin to Eleanore's startled face.

Another day the women came upon Reed and Rosita resting and lunching together under a shade tree. Rosita rose languidly, shook out her skirts, and flounced by Kathleen, saying, ''I'll take a strong man any day. You can keep your dolphins. Perhaps next time you can find a couple of hungry sharks to play with.''

"No, Rosita. I hate to disappoint you, but I have no fear of the sharks either. They would not harm me, I am sure.''

For an instant uncertainty reflected itself on Rosita's face, then haughtiness replaced it. "Ha! It is easy to be brave on dry land!'' she sneered.

Kathleen turned to find Rosita eyeing her strangely. "Aren't you supposed to be working?'' she snapped and walked away.

Arriving home earlier that evening, Reed found Kathleen, Dominique, Jean, and Eleanore playing poker in the library. Leaning against the doorframe, he commented acidly, "Well, what a cozy little scene we have here. Pierre must be recovering speedily to be left alone all evening.''

"Charles says his arm is mending nicely and he will keep it,'' Eleanore said in her quiet voice.

"Pierre is well enough to be spouting threats at Kathleen from his window,'' Dominique stated.

Reed swung toward Kathleen. "Is this true, Kat? When? Why haven't you told me?''

"Oh, Reed, settle down. The man is in no position at present to do me harm, so forget it. I have quit going to Jean's garden since then so as not to aggravate the situation.''

"What has he said?'' he asked heatedly.

"Other than warning me to beware of him and indicating he would love to slit my throat, he has said nothing. As for why I haven't told you, I haven't seen you long enough to do so, dearest,'' she said with venom in her voice.

Kathleen tossed down her cards. "Deal me out. I just can't seem to concentrate on the game any longer.''

After the guests departed, she walked with Reed to the bedroom. She lit a lamp near the chair, and taking up a shirt, sat down to sew.

"What is that you are doing, Kat? Aren't you coming to bed?''

"I want to finish this shirt for Dominique,'' she said calmly. "He said he is whittling a surprise for me and I thought it would be nice to give him something in return."

Reed threw up his arms in disgust. "This is just fine! My wife is sewing shirts for another man!" He strode across the room and stood over her. Gripping the arms of her chair, he leaned close to her face. "Just who are you married to wench, him or me?"

"I'm not so sure myself anymore," she replied evenly, looking up at him. "I am legally wed to one and get most of my companionship from the other."

"Damn!" he cursed. He said not a word more to her that night.

Kathleen and Eleanore stayed away from all the areas where they might encounter Reed or Rosita, trying to avoid another confrontation. They walked along the southern edge of the island along the Gulf beach, away from the docks, warehouses, and slave quarters. Still, as luck would have it, they chanced to meet them again. They were just coming up on a small grove of palms when Reed and Rosita emerged from the treeline. Rosita's hair was streaming down in tangles, her blouse was unlaced revealing most of her plentiful breasts, and her clothes had a very rumpled look about them. All in all, she looked as if she had just pulled herself hastily together after a healthy romp in the bushes.

Reed stopped short when he saw Kathleen. Rosita sauntered by her and said, "He does not love you, you know. He loves me."

"He is still married to me, Rosita, and there is nothing you can do about that," Kathleen countered softly with a glare.

"I won't have to. Pierre will see to you, and Reed will be mine alone when you are dead," the other girl whispered viciously.

Kathleen stared at her and then at Reed who was approaching her now. As Reed neared, Rosita gave him a parting smile and went her way.

"That's it, Reed! This is intolerable!" Kathleen exploded. "Is there no place on this island where I can walk without running into you and that dirty little slut?''

"We work together, Kathleen.''

"Yes, and it appears you play together, too. The least you could do is be discreet. I do not like having your cheap affairs flaunted in my face, especially before my friends.''

"Are you trying to dictate to me again, Kat? I told you before that it won't work.'' A nasty smile etched his lips.

"Obviously. Just leave me some pride, Reed. That's little enough to ask. After all, I am your wife!'' she answered tartly.

"It might serve to remind me if you started acting like one,'' he barked back, blue eyes snapping. Turning his back to her, he stalked away.

That same afternoon, Kathleen again walked down to the little store. As she was about to enter, she noticed that Reed was standing at the counter, Rosita at his side. Quietly backing away, she peeked through the window. From there she watched Reed purchase a bright red blouse and hand it to the smiling Rosita. Rosita threw her arms about Reed's neck and kissed him soundly on the lips.

Racing home, her skirts flying, Kathleen threw herself upon the bed and angrily cried herself senseless.

From then on Kathleen stayed home. She went for no more walks with Eleanore unless it was along the beach below the house. She spent her mornings painting or reading. She invited Eleanore in for lunch or tea, and sometimes they played cards or chess or just sat and sewed and chatted together. When Eleanore was busy, Kathleen would sometimes walk down to the beach and feed the fish and talk to the pelicans who came to beg tidbits. At these times she wished her dolphin friend would return, and often sat in the sand staring out to sea where she knew he had gone. She no longer went to Jean's for dinner, but had Mae serve her alone on the patio. Dominique would stop by often in the evenings and visit for a time. They would talk and he would whittle while she sang softly and played her guitar for him. She was almost always asleep by the time Reed came home.

Once in a while Jean would visit and try to draw her out

of her shell. They came to admire and respect one another, and in Jean Kathleen found another ally of sorts. Jean had known Reed for many years, and in some ways knew him much better than she did, and Kathleen tried desperately to unravel the mystery of her husband through Jean.

Jean's visits sometimes upset her in other ways as well. To her dismay, Kathleen felt drawn to the dynamic privateer as a man, and her intuition told her he felt it as well. It was only because of their mutual feelings for Reed and Eleanore, an iron will on Jean's part, and a wary reluctance on Kathleen's side that they remained friends and did not become lovers. A bond was formed between them, and things better left unspoken were never mentioned, though at times the air vibrated with tensions that both recognized and admitted to themselves in quiet reflection of things that could never be. If she had met Jean first—but she had not. Her heart had been stolen and was held captive by a tall black-haired pirate with sky-blue eyes, and she could not wrench it back from his grasp.

Dominique finally finished the piece he was whittling for her. She was delighted when he presented it to her, for it was a perfect replica of a dolphin.

"Oh, Dominique! It is beautiful!" she cried gleefully. "You have captured him perfectly!"

"I am glad you like it," he said quietly, dark eyes glowing. "It is good to see you smile again."

"I will smile even more if you like my present to you," she said, handing him the shirt she'd made. "I hope it fits."

He unfolded it, admiring her handiwork. He held it up to his broad chest, checking the sleeve length. "It looks fine, Kathleen. Just fine. I'm sure it will fit. I will wear it proudly."

"And I will treasure my gift from you as long as I live," she said softly, turning the dolphin slowly in her hands, studying the workmanship in detail.

Reed found her late that night asleep in a chair in the parlor, clutching the dolphin. He picked her up gently and carried her to bed. She stirred groggily, laying her head against his shoulder. Laying her on the bed, he stripped off her slippers and stockings and drew up the covers; kissing her tenderly on the forehead, stroking her silky hair back with his rough hand.

"Oh, kitten, why do we fight so all the time," he whispered softly. He gently pried her fingers loose from the figure. Studying it briefly, he placed it on the nightstand next to her head. "Dominique again," he sighed as he blew out the lamp. "Thank God this late work is almost at an end. This monk's life is not for me."

One evening when Dominique stopped by, Kathleen asked him to walk with her down by the warehouses. "Please, Dominique," she pleaded prettily.

"Why do you wish to go there?" he asked.

"I just want to prove something to myself, I suppose. Reed says he works late, but it is my guess he spends a lot of evenings at the tavern or with Rosita."

"Perhaps for a quick dinner he goes there, Kathleen, but I know he really does work late at the warehouses most evenings. I myself have helped him many nights. The work is nearly caught up. Just have patience."

"He wasn't working late the evening Pierre attacked me, and he stayed out all night. He was with her!" Kathleen exclaimed adamantly.

Dominique looked at her in surprise. "Didn't he tell you? He drank himself into a stupor and we had to bed him down at Jean's."

"By himself or with Rosita to keep him warm?" she asked snidely.

"Alone, of course," Dominique frowned. "I sent Rosita home myself after I said what I had on my mind."

"Oh, Dominique! He tried to tell me so, but I didn't believe him. What a fool I've been! Still, that was one night and he has been spending all his time with Rosita and none with me, and I must see for myself. Will you go to the warehouses with me or must I go alone?" She threw up her chin stubbornly.

"I'll take you, but it is against my better judgement," Dominique sighed loudly.

When they arrived at the warehouse, it was locked and dark. No one was there, or anywhere around. Kathleen looked up at Dominique and said in a tremulous voice, "See? I told you so. There is one other place I wish to check. The tavern." Whirling about, she marched briskly toward the docks.

"Kathleen, this is not a good idea! Please! Let me take you back to the house," Dominique pleaded, taking her arm. Kathleen responded by lengthening her stride, saying nothing.

The tavern door was open to the warm night air. Music and loud, raucous laughter, mingled with the clinking of mugs, could be heard long before they neared the place. Kathleen stood just outside the doorway, Dominique's large form behind her. She spotted Reed and several of his men seated around a table near the center of the smoke-filled room. Atop the table before them, Rosita was twirling about, her bare feet keeping time to the music, her skirts flying high about her brown thighs as she danced above them. As the music ended, she threw herself from the table into Reed's lap. The chair teetered for a second and then crashed to the floor as Rosita clung to Reed, her arms firmly about his neck. Laughter filled the room as she lay upon him and put her lips to his in a long, passionate kiss.

It was at this point that Kathleen stepped through the door. A deathly silence followed her as she made her way toward the couple on the floor. She stopped a few feet away, hands on hips, glaring down at them, emerald fire flashing from her eyes. At the sudden hush, Reed looked up in astonishment to see her standing there, resembling an avenging angel.

"I've heard it said that a picture is worth a thousand words, and now I think I finally know the true meaning of that statement," she proclaimed acidly.

She spun about and stormed out of the tavern, running blindly toward the docks, Dominique close behind. Behind her she heard Reed bellow her name. She ducked between two old tack sheds, flattening herself against the wall. Dominique followed suit. It wasn't long until Reed ran by. Still she didn't move, barely daring to breath. Soon she heard footsteps drawing near and knew Reed was returning, only this time he was moving more slowly, searching for her more methodically. She held her breath, every nerve aching with the strain, every muscle tense and ready to spring into action should he discover her. She heard him rattle the door of the shed next to hers. As he came to the opening between the two buildings, he stopped with his back to her.

"Kat!" he called. "Stop this foolishness and come out!" When he got no answer, he added, "I know what you are thinking, but you are wrong. Come out here and listen to reason." He still met only silence.

"All right, blast it! Have it your way! I'm tired of trying to understand you. At least Rosita is not so hard to figure out." He strode away into the night.

When she was sure he was gone, Kathleen relaxed. She gulped in air between threatening sobs. Dominique took her into his arms, holding her tenderly against his huge chest. "Kathleen, if you really love him you will not let Rosita do this to you. You should stand up and fight for him. You have so much more to offer than she does. You must prove to that blind fool that you are the better woman—and what a woman! I did not think you were the type to give up so easily."

"I'm not! Not normally," she choked out, "but I do not know how to compete with a woman like Rosita."

"I tell you truthfully, it is no contest. You are a very beautiful lady, Kathleen, and you could charm the birds from the trees. You are thinking Rosita is sexy, and so she is, in a very base way. But you are much more so. God created a masterpiece when he created you in such a tempting package. Can it be that you are not aware of the effect you have on a man? Just by being your natural self, you light fire in a man's veins. Other women work for years to perfect what comes to you without really trying. If you just employ your God-given attributes and follow your womanly instincts you cannot fail to win Reed, unless the man is utterly blind and stupid, which he is not. Rosita will not stand a chance if you choose to really expend yourself.''

''I cannot face him tonight, Dominique. Not yet. I need time to think, to straighten things out in my own mind, to solve this conflict raging inside me. I want to go aboard the *Kat-Ann* where I can be alone with my feelings. I'll be safe there and undisturbed.''

''I will row you out, but I want you to promise me something.''

''What is that?'' she asked hesitantly.

''There is to be a celebration, a fiesta of sorts, tomorrow evening at the auction arena. Everyone will be there, including Reed and Rosita, and possibly Pierre. Unless I miss my guess, Reed will hesitate to take you, not only because he is angry with you, but because Rosita will be there. I would like you to come as my guest. Since you would feel out of place in a fancy gown, I will bring you a skirt and blouse appropriate to the occasion. This will be a perfect opportunity for Reed to compare Rosita to you, and I can guarantee that you will emerge the victor. Will you do this for me; for yourself and for Reed? Promise me?''

''If you want so badly for me to be there, I will go with you,'' she conceded. ''I only hope you are right, Dominique.''

Once aboard the *Kat-Ann*, Kathleen headed straight for the captain's cabin. She lit a lamp and stood for a moment looking around at the room she and Reed had spent so much time in; the bed they had shared where he had first introduced her to love-making. Unable to stay with the poignant memories, she walked on into her cabin. There she undressed, donned her slacks and a loose top, and went topside. Going straight to the mizzen mast, she climbed the rigging until she reached the topsail. She settled herself in the shrouds, leaned back, and relaxed.

No moon shone down upon her. Only the stars shed their dim light. Alone with her thoughts, Kathleen struggled to sort out her jumbled emotions. ''As much as I hate myself for it, I cannot help loving him so. But do I want to fight for him or against him? Either way I know I want him to love me and only me. Dominique is right," she decided. "I must beat Rosita at her own game. Reed is not the only one who intends to keep what is his for himself. I do not know if he will ever love me, and I will die before I admit my love to him. I would be foolish to put that weapon into his hands. Whether he cares for me or not, I will at least make him desire me above all other women. If I use all my wiles and prove myself to be the type of woman he wants, can he go completely unscathed? And if I succeed, what then? Above all else in this world I desire this man, so if I gain his heart or only a part of it, I must give up my quest for revenge and try to believe in him as he asks me to. I must forget his treachery and deceit and trust that he will be honest with me and true to me. God, what an impossible task I have set for myself! To love him is as easy as taking breath; to conquer him is not entirely unthinkable; but to place my trust in him is not only the hardest thing I could do, but the most dangerous. It would place me in a very precarious position and make me so vulnerable that I'm not sure I can do it. But I must try. If he is mine, then I must be his as completely as possible without ever letting him know my heart is in his hands; for I fear what he would do if he knew he held such power over me."

Her decision made, she felt relieved. She stayed perched atop the mast until she felt relaxed and at peace with herself and the night. A soft breeze tugged at the reefed sails and cooled her flushed cheeks. She sat for a long while letting the night sounds serenade her. Finally she descended.

Upon reaching her cabin she sat down at her desk. After a while she unlocked the drawers and began to sort through the contents disinterestedly. At the back of the second drawer, her fingers closed upon a small box. Drawing it out, she set it gently on the desk and opened it slowly. In it were the few personal belongings of her father's that she could not bear to leave behind in Ireland.

His gold watch lay on top, the case closed to reveal the Haley family crest engraved upon it. This watch had been handed down through seven generations and would someday belong to her son if she bore one. Below it rested several sets of cuff links, which she sorted into pairs. Last of all, she brought forth two rings. One was a plain gold wedding band. The other had been a Christmas present to her father from her mother the year before her death. It was a beautiful square black onyx set in a silver mounting. From the center of it, like a star in a dark night sky, shone a single large diamond.

Turning the ring over in her hand, she regarded it thoughtfully. She glanced at the huge emerald Reed had given her on her wedding night. "If I should have this visible proof to the world that I belong to him, then he should have some symbol that he is mine," she thought. Replacing the other items in the box, she returned it to the drawer and locked it. The onyx ring she placed on the gold chain around her neck. She started for the captain's quarters, and then, changing her mind, she crossed the passageway to Nanna's old room and slept more soundly than she had in weeks.

Reed had returned to their house, where he paced each room in turn and finally settled into a swing on the front porch to await Kathleen's return. It was not long before he saw Dominique heading for Jean's. Reed leapt from the swing and charged across the yard to meet him in the road. "Dominique!" he roared. "Where is she?"

"She is safe, Reed," Dominique stated calmly.

"Where?"

"I cannot tell you. Let is suffice to say she is in no harm and wants only to be alone to think. You have hurt her."

"Kat believes only what she wants to. It is uncanny the way she always turns up at the wrong times, but I am not to blame if she deliberately interprets things for the worst."

"You have not helped matters, Reed. You should have protected her from Rosita's tongue and claws all along and set Rosita in her place. Instead, you stand back and seem to enjoy watching Kathleen suffer from Rosita's insults. In that you are to blame."

"Kathleen is my wife, and it is my affair, Dominique."

"I do not wish to interfere, my friend, or even to advise you, except to ask that you give her this time she needs."

"I have already given her more time than most men would," Reed said flatly as he stalked away.

He sat up most of the night waiting for Kathleen. He thought of her shining hair, her winning smile, her slanted, slightly wicked green eyes. "I love the little vixen, but I have no idea how to control her or understand her quicksilver moods. She is a beautiful mystery I have yet to solve. Blast her! Why does she affect me so, and why do I always show myself at my worst around her? Dominique says she needs more time. Maybe I have been unfair at times, but a few days more is about all I can stand. This situation must be resolved somehow."

He fell asleep in a chair, awaking with the daylight. He checked to see if Kathleen had returned. She had not. He washed and shaved and headed for the warehouse in clean clothes and a dark mood.

Kathleen returned to the house by way of the beach. As she entered the dining room through the rear patio doors, she heard Reed coming down the stairs. She ducked back out of sight until she heard the front door slam, and then scampered up to the bedroom. On hand and knee she dug around in the bottom of her wardrobe until she located a pair of black high-heeled shoes. They would go perfectly with the skirt and top Dominique would soon send. Rifling through her jewel case, she found several bangled bracelets of ornamented copper, silver and gold, a pair of large gold hoop earrings, and three long chain necklaces. "The perfect touch!" she thought excitedly. As an added accent she plucked out three large rings of handcrafted metal.

After a leisurely breakfast she called for a bath, and had Lally wash her hair with lemon-scented soap until it squeaked, and brush it dry until it cracked with life and shone like copper. Giving the girl a few coins, she sent her to the store, and she returned with four delicately carved hair combs.

Just before lunch, Dominique arrived, a package under his arm. Opening it, Kathleen squealed with delight as she unfolded a beautiful gold satin skirt. She held it up and watched as the sunlight reflected off of the lustrous fabric. It was very full, made for dancing, and had three rows of wide ruffles ending at her ankles. The blouse he had brought was a shimmering taffeta the exact shade of green as her eyes, cut very low across the bodice in a peasant style. The sleeves were short with very large ruffles which fell down to her elbows, and cut wide to trail below her arms as she lifted them. Last of all, he presented her with a long, lacy black mantilla for her head.

"Oh, Dominique! These are exquisite!" she exclaimed excitedly.

"I had the devil's own time trying to find a blouse that color of green. I had to go all the way to New Orleans and wake a few shopkeepers early this morning."

"You went to all that trouble for me? Dominique, you are such a darling, but you shouldn't have."

"I want you to be the most desirable woman at the fiesta tonight. Reed's eyes will fall from their sockets when he sees you in these."

"The most homely woman on earth would seem beautiful dressed in these clothes, my friend. You are truly a prince!" she said with glowing eyes.

All through lunch Kathleen had to make a supreme effort to appear normal and concentrate on Eleanore's conversation. Afterward she had Lally press the skirt and blouse, and hung them carefully away in the recesses of her armoire. She lay down and tried to rest for the evening ahead, but found herself too tense. Unable to rest, she settled herself in a chair on the patio and worked on a black silk shirt she was sewing for Reed.

She was sitting there in the late afternoon sun when Reed came home. He stood for a minute in the doorway looking out at her. She glanced up and immediately resumed her sewing. He slipped quietly behind her.

"Where were you last night, Kat?" he asked quietly.

Laying aside her fabric, she gazed down at the bay. "I

was on the *Kat-Ann*," she replied evenly.

"Alone?"

"Yes."

"What is this you are sewing?" He reached down to finger the soft material. "Another shirt for Dominique?"

"No, it is a shirt for you if it turns out well." Rising, she smoothed her skirts and retrieved her sewing. "I'm going in to dress for dinner. Will you be eating here this evening?"

"Yes. I have some work to do in the library, but I'll be finished in time to dine with you."

"Fine. I'll tell Tess to set a place for you then."

Kathleen returned to her room where she had Lally help her arrange her hair into a shining coil atop her head. She dressed carefully in a melon-colored silk gown, and was just finishing when Reed came up to wash for dinner. "I'll meet you in the dining room," she commented, and hastily exited the room.

Dinner was a very strained affair, the two of them only speaking when necessary. Several times Kathleen caught Reed staring at her as if about to tell her something, but each time he merely resumed his meal. As the meal ended, he rose. "Dinner was delicious. Thank you, Kat."

"Thank Mae. She cooked it."

"Regardless, it was good." He cleared his throat and continued hastily, "I am going back out as soon as I change clothes. I'll probably be late, so don't wait up."

"That's fine," she said, trying to maintain a calm she didn't feel. She dared not look at him for fear her eyes would give her away.

As soon as Reed had changed into a crisp black shirt and pants and left, she hurriedly dressed in the new skirt and blouse, adding her jewelry. She slipped into the black heels, inserted a tall tortoise-shell comb in her hair, and draped the delicate mantilla over it. Viewing herself in the mirror, she looked like a Spanish dancer. A satisfied smile crossed her face.

"Good luck!" she told her image.

Chapter 9

THE auction arena where the slaves were viewed and sold was a large round platform outdoors, with a planked floor a few inches above the ground. Potential buyers usually stood or were seated outside the huge area while the slaves were paraded around the inside of the circle. Later, each slave was brought individually to the center of the arena, stripped, examined, and sold to the highest bidder.

Tonight the platform would be used for dancing. Lanterns were strung on ropes over it, and torches were lit all around the perimeter. At one end, a small band of musicians played softly, and on another side were tables laden with drink. Men and women sat on chairs arranged at smaller tables scattered around the edges of the circle, or sat on the planks themselves. The women resembled brightly colored peacocks in their gay skirts and blouses and the men were not far behind. Dominique had worn breeches of royal blue and the white shirt Kathleen had sewn for him, and had tied a bandana rakishly about his head, making him indeed look a pirate.

As they approached the fiesta, Kathleen noticed Jean, dressed completely in white, with Eleanore, who was resplendent in a turquoise dress with white eyelet lace. She stopped abruptly as she caught sight of Pierre sitting at a nearby table, his right arm in a sling. "There's Pierre," she said.

"No one will let him bother you, *cherie*," Dominique

calmed her. "Come. Pull the mantilla close about your face so that everyone will wonder who is with me. Let us keep them in suspense for a little while."

Pulling her close to him, he led her to a table. When he had seated her, he sat across from her. As she looked toward him, she realized he had chosen the table next to Reed, Jean, and Eleanore. Rosita was standing at Reed's elbow.

Jean had noticed Dominique's arrival. Nudging Eleanore, he nodded toward Kathleen, who looked quickly away. She heard him say, "Who is the intriguing wench with Dom?"

All eyes turned toward her as she let the mantilla fall forward to shadow her face.

"Dominique, where have you been hiding your shy lady? Who is she?" Eleanore asked curiously.

Kathleen leaned forward and whispered, "Dominique, perhaps this was a mistake."

He shook his head, frowning. "No. Why should you sit at home while the rest of the island celebrates? Fight fire with fire, Kathleen." He reached over and brushed the mantilla from her head.

"It's Kathleen!" Eleanore exclaimed.

Kathleen looked up into Reed's diamond blue eyes.

Reed turned to Dominique and said churlishly, "While I did not object to your escorting my wife about the island, I hardly expected you to carry it this far."

"Kathleen, you simply look ravishing!" Eleanore babbled, ignoring Reed's dark mood. "You look so Spanish and festive!"

Tearing her eyes from Reed's, Kathleen answered quietly, "Thank you, Eleanore. You look quite lovely yourself. Turquoise sets off your hair so wonderfully."

"Won't you two join us?" Jean offered.

"No, thank you. We are fine right here, Jean," Kathleen replied, eliciting a broad wink from Dominique behind Reed's back.

"It is plain to see that they do not wish to be disturbed," Rosita said snidely. "Come *querido*, dance with me." She pulled Reed to his feet and led him out onto the dance floor.

"This may not work, Dom," Kathleen said moodily. "We are not off to a very good start."

"Have more confidence, little sister. At least he did not drag you back home."

"No, he would rather I see him dance with Rosita," she said as she watched them whirl by.

"While he is occupied, I will take this chance to get you some wine." Dominique slipped off.

Eleanore took his chair almost immediately, her brown eyes sparkling with glee. "Kathleen, what are you up to this time?"

"I think I'm up to my neck in hot water," Kathleen muttered.

"Oh, really? Why do I get the impression Rosita is about to get some competition and a well-deserved lesson in the bargain?"

"Am I so obvious?"

"In all but your attitude, my dear. It seems your courage needs a little bolstering, but the outfit speaks loud and clear. You are without a doubt the most beautiful woman here tonight. Every man's eyes are drawn to you."

"All but Reed's it seems," Kathleen retorted.

"So do something about it," Eleanore quipped.

"Such as?"

"Start by dancing with Jean, and then Dominique and any man who asks you. Then let intuition be your guide." With that Eleanore dragged Jean from his seat and shoved the two of them onto the dance floor.

"I think we are meant to dance," Jean said, shaking his head at Eleanore and smiling down at Kathleen.

"I got that impression also." Kathleen smiled back.

"Shall we, madam?" Jean invited. Bowing gallantly, he proceeded to sweep her into the tempo of the dance.

Kathleen was hard put to keep track of the number of dance partners she had after that, or the glasses of wine she had pressed into her hand. Soon she took to handing the wine to anyone nearby before she overstretched her limit.

Of one thing she was achingly aware. Reed had not once asked her to dance and was pointedly ignoring her, although she would turn and catch him staring at her with a quizzical expression when he thought she was not watching.

They were well into the evening, and Kathleen and Dominique were sitting at their table catching their breath, when she suddenly noticed the couples deserting the dance floor. Only Rosita was left standing a few feet away from Reed in her red skirt and bright yellow blouse. The musicians started playing a lively Latin number, and Rosita began to dance. She lifted her skirts, switching them about her brown legs, and whirled off on fleet bare feet. She danced once around the circle and came back to face Reed where she stomped her feet and leaned forward, teasing him with her heaving breasts. Hands on hips, she undulated before him, swaying her hips temptingly, and then whirled around, causing her skirts to flare out around her bare brown thighs.

Kathleen could not fail to notice the interest showing plainly on Reed's face. Excusing herself, she quietly made her way around to where the musicians were. Tapping the guitar player on the shoulder, she whispered something to him and he nodded affirmatively with a broad grin. Kathleen reached her seat just as Rosita finished her dance and sat down across from Reed, a pleased look on her flushed face.

Winking at Dominique, Kathleen walked to where Rosita had started her dance as the musicians played a fanfare. She struck a traditional pose from the flamenco; back arched, head back, one arm curved over her head, the other extended from her side. A hush felll over the crowd as they waited expectantly. Kathleen ignored her when Rosita commented acidly, ''What a farce this will be.''

With the first fast beats of the music Kathleen's feet started tapping out sharply the staccato beats of the flamenco, keeping time perfectly with the guitarists. As her heels rapped swiftly on the wooden floor, she began snapping her fingers in an alternate rhythm, bringing her arms out in front of her. Green eyes blazed into those of blue as she danced toward Reed, hands extended in a beckoning gesture, wrists twisting in entreaty. When she stood just inches from him she spun away, skirts swirling high to reveal her bare thighs for a brief instant. Her eyes boring into his, her feet constantly tapping, she pulled the comb from her hair, flinging it at him. One by one she removed the pins from her hair until it fell in a shining cascade down her back. Lifting it high off her neck, she arched her breasts toward him, swaying her hips provocatively as she moved toward him, her gold earrings swinging with each step.

Again she let her hair fall as she spun away, and it spread out like copper flames as she danced. She stopped, arms extended upward, hands weaving their own intriguing language as her hips enticed him. She was caught up in the dance, and the rhythm of her body spoke to him in a language as old as time itself. Open lust showed on Reed's face, and he found it impossible to tear his eyes from hers. He felt as one bewitched.

Turning sideways of him, careful not to break eye contact, she swished her skirts about to the rhythm of the music, teasing him with flashes of her slim thighs, kicking her heels up one at a time; never missing a beat. Bringing her arms again over her head, she started clapping out a new cadence, twisting her body from side to side, bracelets jangling as she inched backward. The message of her movements and the spark in her eyes were clearly a dare to him. Suddenly, as the music increased in tempo to a building crescendo, she whirled toward him, falling on her knees before him, head to the floor, hair spread out at his feet in a momentary submissive gesture. With the last tremendous beat of the music, she straightened her back, thrusting her breasts forward, one arm reached over her head, the other extended toward him, her emerald eyes flashing their challenge.

Silence reigned as she sat posed at his feet. Then, slowly, he reached out, taking her outstretched hand in his, blue eyes smoldering with desire. Standing, he pulled her to her feet, drawing her close to him, enfolding her in his arms. Bending his head to hers, he took her lips in a searing kiss of passion as he swept her into his arms. The wild applause and calls of her audience fell on deaf ears as he carried her off without a word.

Kathleen wrapped her arms about his neck and snuggled against his warm, broad chest as he carried her silently home. He took the stairs as lightly as a feather, kicked open the door, and walking to the bed, dropped her onto it. In the soft lamplight she watched as he shrugged out of his clothes. Quietly she slipped off her necklaces and blouse, dropping them to the floor along with her skirt.

Reed came to the side of the bed, gazing deeply into her dark green eyes, and she held out her arms to him in welcome. He sank on the bed next to her, clasping her to him in a fervent embrace. He kissed her deeply, thoroughly, leaving her gasping for breath. He nibbled at her ear and whispered, "Kat, my beautiful, delectable kitten. How I have missed you and your sweet body beneath mine."

The week that followed was a delightful interlude filled with love and laughter. Reed had finished his work at the warehouse and told Jean flatly to find someone else to take over any last-minute details. He would be there to help with the slave auction in two weeks, but until then he wanted time alone with Kathleen.

The two lovers could not seem to get enough of each other. Dawn tinged the eastern sky that first morning after the fiesta before Kathleen curled up exhausted but content beneath the sheets and closed her heavy eyelids. Reed's deep voice jolted her awake a few hours later as he complained, "What a sleepyhead I've married. Come, wench! It's eleven o'clock and I'm dying of hunger."

Kathleen nestled deeper under the covers, burying her head in the pillow. Suddenly the cover and pillow both were gone and Reed was planting kisses up and down her body. "Just what kind of heartless monster are you to interrupt a girl's beauty sleep?" she giggled groggily.

"First of all, you are far too beautiful already, and secondly, you are a woman. No mere girl could arouse a man the way you do, my sweet, or satisfy him so well."

"You've not only kissed the blarney stone, Reed Taylor, but you must have swallowed the blamed thing whole!" she laughed.

He smiled down at her, blue eyes twinkling. "I love to hear you laugh, Kat. You have a merry, tinkling laugh that rings of mischief." He kissed her lightly on the nose and rose from the bed. "Stop lying there looking so tempting and get dressed before I forget my empty stomach and join you there. It wouldn't surprise me if one day they find us together in bed expired from starvation."

Kathleen was dressed and brushing her hair when her eyes lit on the black onyx ring lying on her dresser top. She picked it up and turned to him. "Reed?"

"Hmm?" He looked up from his task of pulling on his boots.

Taking his left hand, she slipped the ring on his finger, delighted that it fit. He looked from the ring to her face in confusion.

"I want you to have it," she said softly. "Do you like it?"

"It is splendid! Yes, I like it very much." He examined the ring closely. "Where did you come by it?"

"It was a gift from my mother to my father, and now I want you to wear it if you will."

"I will wear it with pride and pleasure because you gave it to me, Kat. Thank you." He pulled her close and gazed into her large emerald eyes. "Someday perhaps it will be passed down to our son."

"Our son?" she echoed. "I hadn't thought that far, I guess."

"Well, at any rate we are going, I'm surprised we haven't one on the way already." He eyed her questioningly.

"Are you so anxious to be a father? I had no idea that the thought might appeal to you. We've never spoken of it before. Do you like children?''

"Yes, I like them. I get along with them well, and I want children of my own.''

"You mentioned a son. What about daughters?''

"Daughters with slanted green eyes and long copper tresses will be welcome, too,'' he smiled down at her.

"And if they should have black hair and eyes like silver-blue stars?'' she asked softly.

"As long as they have a face as lovely as their mother's and the figure of a sea nymph like hers.''

"I've heard they arrive smaller and with pudgy cheeks and bellies and lots of wrinkles,'' she giggled impishly. "Come on. Let's get you some food before you lose all the strength you are going to need to start this family you speak of.''

After breakfast they walked to Jean's where Reed told him to have someone else take over his remaining duties. As they were leaving, Eleanore asked, "Will I be seeing you for lunch, Kathleen?''

Reed turned to the petite woman and said jauntily, "My dear Eleanore, Kathleen will not be available for lunch or for tea, nor much else for at least a few days. She will be much too occupied in entertaining me.'' He laughed at Eleanore's amused look and Kathleen's red face. ·

"Reed!'' Kathleen exclaimed, trying desperately to hide a smile. Taking him by the arm, she steered him toward the door. "Please excuse us,'' she said with a hint of laughter. "I must take this brute home and teach him better manners. He is not fit for polite company just now. He takes these fits every now and then, I've heard.''

"And the only cure is a long bedrest with a ravishing cat-eyed goddess!'' he shot over his shoulder as she dragged him down the steps.

The time they spent together was idylic, almost a honeymoon. They had long talks, long walks on moonlit beaches, ate dinners alone by dim candlelight, and made love so often that they were rarely fully dressed. One afternoon, after a particularly satisfying session of lovemaking, Reed lay next to her, toying with the long strands of her hair.

"There is something special about you, Kat, beyond your lovely face and exquisite body," he said quietly. "No other woman has ever affected me the way you do, or satisfied me more. What strange power do you hold over me? Could it be I have married a witch who has cast her magic spell on me?"

"Only if it is possible that I have wed a tall, dark god who moves with the grace of a panther and can see into my very soul with frosty blue eyes, and set my entire being aflame with a mere touch." She looked at him lovingly, running her fingers lightly through the curly black mass on his broad chest.

"Are you content with me then?" he asked lightly.

"More than content, my darling," she sighed, drawing him to her once more.

For five days they shut themselves away from the world, seeing no one but the servants. They talked and laughed and learned about each other. Reed told her more of Savannah, promising to take her there soon. He told her about his family and the plantation called Chimera, which meant fantasy or strange dream. He spoke of how after his father had died last year, his mother had begged him to give up the sea and settle down. Though he went home periodically to see that everything was running smoothly under the overseer's hand, he would not stay, not even when his younger sister added her pleas. He described his mother and sister in loving terms, and Kathleen knew he felt very close to them even though they did not understand his love for the sea.

When Kathleen asked him about her aunt's family, he depicted her aunt as a lovely woman, her uncle as very congenial, her cousin Amy as very petite and pretty, if somewhat spoiled, and cousin Ted a likeable young whelp of nineteen inclined toward practical jokes.

"I think I'll probably like Ted the best," she said with a twinkle in her eye.

"I'm sure of it, and just as sure that Amy will be pea-green with envy when she sees you."

"Why?"

"Because you are more beautiful, more interesting and

less affected than most young woman, and married to me."

"Did the two of you have an understanding, Reed? Were you engaged to be married?"

"Heavens, no! Amy is cute, but always primping and fishing for compliments. Always playing games and trying to trap me. We grew up together, or to be more exact, she and my sister did. We attended balls and barbeques together, but I was never serious about her."

"Perhaps she is serious about you, though," Kathleen stated.

"To be honest, I believe she has set her cap for me, but she is not my type at all," Reed declared.

"And what is your type, Captain Taylor?" she teased.

"I like my women tall and lithe with long legs, nicely rounded breasts, and slim hips," he enumerated as he stalked her about the room. "And with long luxuriant hair to their hips, slanting green eyes, and naked in my bed," he finished with a laugh as he pounced on her, knocking her backward onto the bed.

In turn, Kathleen told him of a peaceful childhood on the Emerald Isle, and how close she was to her mother and father. She related her mother's death and how indulgent Papa had been after her return from school in England. She spoke of how her father and mother had met and loved the land and each other, and described the estate for him. Not once did she mention the shipping firm with the other seven ships or her sailing and fencing instructions.

"I will tell him later when I feel more sure of him," she mused.

She described herself as a willful tomboy in her younger years, and spoke tenderly of Nanna and all her helpful ways, and she told him of her love for the sea and how thrilling it was to sail on the *Kat-Ann* to England with her papa so often.

"It is unusual to find a woman who possesses such a love of the sea and ships," Reed told her. "I have noticed it in you, though I've kept silent. What do you think of so deeply when you stand so long staring into the waves?"

"It's hard to explain. I'm not sure you would understand," she said hesitantly, eyeing him speculatively.

"Try me."

"Well, I think of the strength of the sea, of how awesome it is, of its changing moods and currents. After a while it becomes an entity as real as you or me, only more powerful. It speaks to your soul, whispers in your ear, envelopes your being and hears your heart and its troubles. At those times I feel very much a part of the sea, at home and at peace, as though I belong to it. This huge commanding force, be it gentle or angry, has a hold on me and won't let go. When I used to think of Savannah I wondered if I would shrivel up and die after a while, not being free to sail so often." Looking intently at him, she questioned, "Does that sound strange to you?"

"No, kitten. To someone who has not felt it or witnessed it, it would seem odd indeed, but knowing you as I do, it is rather mysterious and intriguing."

"I'm glad you understand," she told him, snuggling close.

"What I don't understand about you would fill a book," he added. "Where did an Irish lass learn to dance such a sensuous flamenco?"

Kathleen felt her face flush. "I had a Spanish roommate at school who taught me to dance and play the guitar."

"And did this roommate tell you how alluring it would be to a man?" Reed quirked a dark eyebrow at her.

"Was it really?" she countered flirtatiously.

"You little vixen! You knew exactly what you were about every second! Tell me, Kat. What would you have done if I had gone off with Rosita instead?"

"I really hadn't thought about that. Perhaps I would have done nothing, and then maybe I would have tried my charms on Dominique," she taunted.

His handsome face darkened. "You are mine, Kat. Mine alone. Remember that. What is mine I keep," he growled.

Kathleen tilted her chin stubbornly at him. "So do I," she challenged, green eyes flashing.

Late afternoon of the fifth day of their self-imposed exile, Jean broke in on them. They were sitting on the rear patio talking when he strode out of the house past a flustered Joe. "Reed, Kathleen, my apologies for the intrusion." Then to Reed he said, "We have a big black buck who tried to escape this afternoon. He is a valuable slave and will be worth a lot of money on the auction block. I cannot afford to have him marked up, but he must be whipped, Reed. I want you to administer the lashes; forty of them. You are good enough with the cat not to break his black hide," Jean stated authoritatively. "I trust no one else with the task."

"How soon?" Reed requested simply.

"As soon as you can get down there," Jean replied.

Kathleen considered this new aspect of her husband. All ship's captains and quartermasters knew the use of the cat, but few men were proficient enough to administer the lashes and not break the skin. Usually a man's back ended in shreds and tatters, sometimes to a point that muscles were severely damaged. She had seen many men whipped, but never by a master with the cat. Her curiosity was boundless.

"Reed! May I go along, please?"

He looked at her astonished. "It is not a pretty sight, Kat. Not exactly suited to a lady."

"I know that."

"Then why would you want to see it?"

"Just humor me, please, Reed. Lecture me later with the 'I told you so's.' "

"All right, you red-haired wildcat, come along." He grabbed her upper arm, pulling her forward. "Maybe you should see this, for it will be an example of what is in store for you if I ever catch you bedding another man."

Kathleen caught Jean's pondering look. "Possessive lout, isn't he?" she quipped lightly, arching her eyebrows.

The huge black man had been stripped and roped between two stakes set into the ground. Sweat glistened on his bare back. A crowd had gathered, and Kathleen noticed that the other slaves had been herded out of their huts and were being forced to witness the punishment.

Kathleen stopped at the inside edge of the circle as Jean and Reed walked toward the would-be escapee. Dominique spied her and came to stand at her side. "What are you doing here, *cherie?*"

"As crass as it sounds, I am curious, Dominique."

He shrugged and said, "If it gets to be too much for you, I will walk you home."

Jean announced to all the penalty and the reasons for it. Reed walked to the edge of the circle and accepted the whip from a man there. The whip was a cat-o'-nine tails. It was made of nine leather straps braided a third of the way down with knots tied at intervals the rest of the length of the straps. Reed flexed the whip, testing the weight and swing of it, making the leather sing. Taking a stance several feet behind the prisoner, he swung the whip from his side instead of extending his arm back from his shoulder as many men would have. The leather thongs whistled through the air, making a solid thwack as they made contact with bare flesh. The black man gasped in pain as Jean counted, "One."

Reed swung the cat again and again in steady measured strokes as Jean continued to count. On the sixth stroke, the slave screamed aloud in his agony, his body convulsing with each contact of the whip. Angry red welts appeared on his back, but no blood flowed. Kathleen noted that with each swing Reed caused the tails to connect in another area of the man's back. He would be welts from neck to legs by the time forty lashes had been applied. Unable to stand under the searing pain, the man's legs collapsed beneath him, leaving him dangling by his bound wrists. By the twenty-fifth stroke his shrill screams had ceased. The slave had fallen unconscious and gave only grunts of pain when the whip stung his skin. Reed glanced questioningly at Jean who shook his head, motioning for Reed to continue. Reed reapplied the cat for the remaining lashes. Kathleen realized that this was done as an example to the watching slaves.

The fortieth lash was meted out and not a drop of blood ran from the man's back. Still, he would be in much pain when he awoke, and probably unable to move without agony for quite a few days. The welts would disappear by auction time, but the memory of them would remain for months.

Reed tossed down the whip and strode to where Kathleen waited. "Still here?" he asked archly. "Well, my bloodthirsty Irish wench, what ran through your head as you viewed the flogging? Do you enjoy hearing a man scream with pain until his voice gives out and his mind refuses to function?"

She jutted her chin at him. "I did not relish hearing the man's shrieks, nor observing his suffering. I did, however, admire your expertise with the cat. He will have no scars to show because of your skill. I had heard of men who could apply the lash so expertly, but doubted the reality of such talk until now."

"Sometimes you astound me, Kat," Reed said quietly. "At times you seem so tender and vulnerable, and at others you accept life's harshness in stride, hardly batting an eyelash."

"There are times when it does not pay to reveal your emotions, Reed. It is a hard fact to learn, and even harder to enforce sometimes when feelings run deeply, but you lay yourself open to much heartache if you do not steel yourself against your own weaknesses at times."

"From one so young that strikes me as a very strange attitude. Where is your trust in the world's goodness, my sweet?" he asked, gazing tenderly into her serious face.

She shrugged and said, "I have my periods of trust too, when the time feels right to me, and when it doesn't I draw on my own resources, trusting only my own abilities."

Reed walked with his arm about her waist and thought, "You shield yourself from me, too, Kat. One day I hope you will not feel the need and will yield to me totally, without reserve." Aloud he said, "I am here to protect you, kitten."

"I know," she replied, but thought silently, "And who will protect me from you, my love?"

Reed walked her home and left to talk matters over with Jean. "I won't be long," he promised. "We'll eat as soon as I return."

Kathleen waited for him on the patio. She renewed sewing his shirt while the light held, and as the sun set behind the trees beyond the bay, she sat quietly strumming her guitar. All at once the fine hairs on the back of her neck seemed to stand up, and a shiver ran through her. Some sixth sense told her danger was near. Although she could see nothing, she sensed something was about to happen. A sudden insight led her to lift her guitar from her lap and hold it before her. Just as she shielded herself with it, something hard struck it, splintering the wood and snapping the strings. She looked down to see a long dagger protruding from the shining body of the instrument. Her eyes scanned the darkening area beyond the garden and her ears picked up the muffled sound of running footsteps just as she perceived the shadowy form of a man disappearing into the darkness.

Kathleen leaped from her chair, overturning it in her haste. "Damn you, Pierre! I know it was you! That's just your style, you coward!" She stood shaking with anger and relief.

"Kat! What's wrong!" Reed came running out onto the patio. "I heard you screaming as soon as I hit the front gate!" His gaze took in the overturned chair. She turned and he saw the knife blade stuck firmly in the guitar. "My God!" His face went pale. "What went on?"

"A man—I can only guess who—just tried to kill me," she shouted angrily. "The yellow-livered vermin stood out there in the shadows and pitched a knife at my heart. God knows only why I felt his presence in time to shield myself!"

"Pierre!" Reed snarled. "I'll kill the skulking scum!" He wheeled around and started for the house.

Kathleen ran after him, grabbing his arm. "Reed! We can't prove it was him!"

"Who else could it have been?" he stormed.

"Reed, please! Use your head! Without some proof it would be murder! You could hang! Please listen to me! Let it pass until we can prove it was him!" she begged, eyes brimming with tears.

Suddenly she was shaking uncontrollably, sobbing brokenly. He gathered her tenderly against his broad chest, stroking her back comfortingly. At length her sobs lessened and he led her to a chair. "Kat? Are you all right, sweetheart?"

"I will be." She hiccuped, brushing at her wet cheeks. "Just don't leave me, not yet."

He kissed her sweetly. Her lips tasted of salt. "Oh, sugar. When I think how close I came to losing you!"

She clung to him, wrapping unsteady arms about his neck. "Hold me, darling. Let me feel your strength surround me. I need you so just now."

He held her tightly and whispered, "I'm here, my sweet. I'm here."

Chapter 10

AFTER a restless night and precious little sleep, Reed had reached a decision. He would take Kathleen to Savannah where she would be safe. Propped up on his elbow, he studied her sleeping form. "How I will miss her! This past week has been marvelous! I have learned so much more about her and yet, in a way, I find her even more of a mystery." Reed thought of what her reaction would be when he told her he must take her to Savannah and return to Grande Terre alone, and he winced inwardly. "Damn Pierre's dirty eyes! Kat and I were just beginning to pull together in this marriage instead of apart. Now this happens! Last night she needed me, reached out to me, and it felt so warm and right. I think she is finally starting to trust me. Will it destroy that awakening trust when I tell her we must be parted for a while? I must find the right way to tell her today."

Kathleen stirred and slowly opened sleepy green eyes. A questioning look crossed her face when she saw him looking down at her, his tan face lined with concern. She reached up a slim hand and gently caressed his cheek. "Why the worried look?" she inquired sweetly.

"Not now!" His brain flashed a warning. "Don't tell her yet. You already know what her reaction will be. Savor her sweetness while she is a warm and compliant kitten purring in your arms. Relish the taste of her offering herself openly and willingly to you one last time before you shatter this fragile illusion of love."

He leaned over her, kissing her tenderly. "How lovely you look this morning, with your hair all tossled about your face, your cheeks still flushed." One copper curl wound about his hand as if to capture it.

"Are you trying to seduce me with honeyed words, my prince?"

"Ha! You've found me out!" he joked, flashing her a brilliant smile. He nibbled her ear and she shivered deliciously. His lips found their way along her throat to one rose-tipped breast. Tangling her fingers in his hair, she pulled his head close to her as he tugged gently at her nipples. White-hot flames licked through her as his hand snaked across her stomach and between her legs. The rhythm of his fingers started a throbbing deep within her. His mouth found hers in a fiery kiss of desire as his tongue mated with her own. Her passions built as their lips blended and his fingers worked their magic. Her body writhed in agonized ecstasy as she moaned and cried out for release, a release which came in a mind-shattering explosion that left her shaky and trembling.

He caressed her tenderly, her head on his chest, until her breathing gradually returned to normal, and then he pulled her up and to astraddle him. His blue eyes twinkled merrily as he commented, "Since you are sitting here so pertly, perhaps you would like to ride on top this time." He helped her position herself. His hands firmly about her waist, he slowly lowered her onto himself. She gasped as she felt herself filled with him. Hands on his chest, she steadied herself as he continued to assist her movements. Lifting his head, he located her breast and teased the hardened tip with his tongue. Kathleen let out a strangled cry at the combination of emotions coursing through her like heat waves. Acceding to the wild, building need in both of them, he drove faster, and deeper within her as her eyes widened in wonder and passion. His thrust speared to the core of her being as he crashed fiercely into her. Her body and mind cried out. He heard her small sounds emitting unbidden from her throat as she whipped her head back and forth in her torment. Their torture ended as wave after wave of delerious rapture throbbed through them, flooding them with wondrous ecstasy that knew no bounds and seemed to want never to subside. She cried out with the force of it and he clutched her tightly to him as their passion slowly diminished. Their bodies slippery with sweat, they lay sated in each other's arms.

"Will I never fail to marvel at the heights you take me to? The enormity of the feelings you incite in my body?" she said weakly. "Just when I think you have taught me all there is to know, you find some new way to arouse me to a fever pitch."

He laughed softly, and turning her face to his, he assured her, "There is more yet I can teach you, kitten. In time we will explore all the possibilities together."

After a leisurely breakfast Reed suggested a walk on the beach, intending to tell her of his decision. They crossed the road and walked the beach along the outer edge of the fort, heading toward the southernmost tip of the island.

"Dominique was telling me of a ship that sank just off the point. He said that it was one of their most richly laden captures. Do you know where it lies?" Kathleen asked him.

"Just out there." Reed pointed to the location. "On a clear day with calm seas you can see the shadow of it in the water." Reed shaded his eyes from the glare and studied the surface. Then he shrugged. He'd thought for a moment he'd seen movement in the water, but realized he must have been mistaken.

He looked back at Kathleen as she questioned, "Why haven't they recovered the treasure if she lies so shallow?"

"There is no man willing to try. These waters are heavily infested with sharks. Why do you think I was so upset the day you saved your dolphin? They frequently enter the bay."

Kathleen peered at the area he had pointed out. It was calm enough to see the wreckage. She too thought she saw movement, and frowned as she searched the water. Tugging at Reed's sleeve, she pointed to a dark shadow, just beneath the surface of the water. They both watched as a swimmer's head popped up on the surface.

"Who could be stupid enough to be swimming out there?" Reed wondered.

"It's Rosita," Kathleen informed him, a chilly calm entering her voice, "and from the looks of it, she has company arriving." Four dark forms were silently skimming through the water in Rosita's direction.

Reed cupped his hands around his mouth and yelled a warning. "Rosita!"

The girl looked toward the shore and waved, still not aware of the approaching danger. Again Reed shouted. "Sharks! Swim for shore!"

The frightened girl looked around her, trying to determine the direction from which they came. Because of the glare of the water, she saw nothing until two of them sliced the water's surface to the north of her with their fins. Rosita's heart froze in her chest, and she let out a blood-chilling scream. Vaguely she heard Reed calling to her.

"Swim, Rosita! Hurry, *chica!* Don't panic now!"

By sheer will she managed to get hold of herself and start swimming desperately for shore. The sharks seemed to sense her urgency, and one by one they swam between her and the shore and silently started a circling pattern. As Rosita churned the water, the sharks circled her slowly, ever tightening their circle.

"My God! She'll never make it!" Reed exclaimed, pulling off his shirt. Before he could get his boots off, Kathleen had skinned out of her clothes, grabbed his shirt, and was running toward the sea. He ran after her. Catching her arm, he spun her around.

"Kat! What do you think you are doing!"

"I am about to save Rosita's worthless skin!" she retorted.

"No! I cannot let you! I will see what I can do."

"No, Reed," she said firmly, shaking off his grip. Gazing deeply into his eyes, she said simply, "Trust me. I know what I am doing. You would not make it either. Those sharks would tear you apart, but they will not harm me. Believe what I tell you!"

Reed felt mesmerized by the strangeness of her tumul-
tuous emerald eyes, the soft assurance of her voice. Before
he realized what had happened, she pushed him down on
the shore, kicked sand into his eyes, and dashed headlong
into the water. Partway out, she turned to look behind
her. All their shouting at Rosita had brought the islanders
rushing to the beach. She heard Reed bellowing at her and
saw he was still struggling with the sand in his eyes.
Dominique and Jean were standing at his side watching
her.

"Hold him, Dom!" she yelled back. "Don't let him
go!" Dominique waved to show he had heard her.
Kathleen swam on toward Rosita.

By this time the sharks had closed their circle considerably about the hysterical Rosita. Occasionally one of
them would bump her with his snout or body, signaling
the last stage before they would attack in earnest, tearing
her body apart.

With the aid of a wet handkerchief, Reed had now
cleared his eyes of the sand and would have gone after
Kathleen had it not been for Dominique, who held him
firmly back. As he fought desperately to free himself, Jean
grabbed him also, and together they held him in an iron
grip, ignoring his pleas. All eyes were trained on the
drama taking place before them.

As Kathleen neared the circling sharks, she called to
Rosita, "Rosita, are you harmed?" She got no answer.
Again she cried, "Answer me! Can you swim?"

"*Diós!* I am so scared! What can I do? I cannot swim if
they circle and strike me!"

"I am going to lure them away from you, and when I do
you must swim for shore as fast as you can. Do you understand?"

"*Sí!*"

"You must keep swimming. Don't stop! Your life
depends on that! Are you ready?"

"*Sí!*"

Kathleen dived beneath the surface, and as she swam
around the circle, each shark in turn broke off to follow

her. She surfaced and started heading away from shore with strong, sure strokes, the sharks spread out behind her like an escort. Reed was frantic as he watched helplessly.

As soon as Kathleen saw Rosita struggle up and collapse on the beach, she turned for shore once more. When she reached the spot where the sunken ship lay, she drew in a deep breath and dived deep. All four sharks followed her pattern.

She stayed down a long time—an impossibly long while. On shore, Reed thrashed about furiously trying to throw off his captors. An anguished cry broke through his clenched teeth. "Kat! Oh, my God! I've lost her! My sweet love!" he cried hoarsely as he strained against the hands that bound him.

Suddenly she sprang up in the water before them. The sharks were gone. She waded to shore. In each hand she held a large oilskin bag. She plunked them down on the sand as Reed swept her off her feet, kissing her madly. When finally he set her down, she asked calmly, "How is Rosita?"

"She is fine. Shaking with fright, but otherwise unharmed," Jean told her.

Kathleen looked around her. Everyone was staring at her in mute amazement, Reed included. Only Jean seemed unconcerned, as if for her he considered it all quite natural.

Having recovered from his fright, Reed stormed at her. "That was a fool thing to do! You could have been killed, and there for a while, we all thought you had been!" Glancing at the bags at her feet, he eyed her suspiciously. "Is that what I think it is?" he demanded.

Kathleen gulped and nodded. "I'm sorry, Reed. The temptation was so strong and I just couldn't resist it."

Kneeling, he untied the drawstrings on the bags. One was filled with gold, the other glittered with jewels of every description. "So!" Reed exploded. "While we were all holding our breaths in suspense, you are on a blamed treasure hunt! I ought to wring your stupid neck!"

Glaring back at him, her eyes brighter than the jewels,

she snorted, "After saving Rosita's life and risking my own, I guess I'm due any reward I choose, Reed Taylor! Especially since these bags would have rested on the ocean floor forever and done no one any good if I had not recovered them!" Almost at once she softened. "I'm sorry I caused you alarm, and I'm sorry about kicking sand in your face. I'll put these things back if you like," she teased blithely as she bent to pick up the bags.

Abruptly he pulled her to him, holding her tightly, showering her face with kisses. "You little minx! One thing is for certain. You are one of a kind. The world could stand no more than that!" Standing away from her, he realized for the first time that she wore only his black shirt. It clung to her like a second skin, hanging just halfway to her knees, giving everyone a spectacular view of her long limbs.

"Good grief, Kat! The entire island is agape and it's little wonder! Look at you!" At her surprised look, he added glibly, "I do wish you would quit borrowing my shirts, my dear, before my closet is completely spent." He bent, placed the oilskin bags in her hands, and lifting her into his strong arms, carried her home.

If loving could be called wild and worshipful all at once, that is how Kathleen would have described what occurred after they arrived home. Reed acted as though he expected her to disappear before his very eyes at any moment. When Jean stopped by just before lunch, he and Reed closeted themselves in the study for a private conference, and when Jean left without his customary farewell, Kathleen got her first indication that all was not as it should be. She was not in suspense long, however, for Reed called her into the room almost at once.

When she had seated herself, he turned from the window and stared at her gravely with the look of someone who is about to impart bad news. Finally he sighed and sat across from her. "There is no easy way to say this, Kat, so I'll just come right out with it. I have instructed Lally to begin packing your things. We are sailing on the evening tide. I am taking you to Savannah."

Cold fingers of dread iced their way up her spine. She tried to keep her voice even. "To visit?" she asked falteringly.

"Not to visit. I am taking you to live with my mother and sister at Chimera."

She could not meet his eyes, fearing what she would read in them. "And you? Will you be staying there, too?" She could not keep the trembling from her voice.

"No. I must return in time for the auction."

"Then why take me all the way now? Why not wait until after the auction?"

"Because I intend to go back out to sea. I am outfitting the *Kat-Ann* as a privateer, and I will be operating with Grande Terre as my base."

"Then take me to Savannah for a visit. I want to meet your mother and sister and my aunt and her family. But I want to stay here if this is where you will be returning. I would never try to tear you away from your life at sea, Reed. I know how closely it can hold you. Let me stay here," she pleaded prettily. "I am comfortable with the people here. Jean, Eleanore, Charles, and Dominique are all my friends."

"And Pierre is your enemy," Reed stressed, his voice starting to rise. "I want you in Savannah, safely away from him. Surely you can see that after what happened last night."

"Jean and Dominique will protect me while you are gone," she argued.

"Jean has an island to run and his business in New Orleans. He will be much too busy to babysit my wife. Dominique is another matter altogether. He will also be sailing often, and when he is not, I am not sure I want him around you that much. While I am here he behaves the gentleman, but left alone who knows what might happen between the two of you. You were awfully chummy the night of the fiesta," Reed said sarcastically, raising a black eyebrow at her.

"It is not that way between Dom and I, Reed, and you know it!" she retorted hotly.

"Isn't it? You sew his shirts. He brings you gifts. I can see how convenient it would be for the two of you if I were out of the way," he said churlishly.

"Blast you, Reed!" Kathleen shouted. "Blast you to hell!" She leaped from her chair. "I will not stay here and listen to this idiocy one minute longer!" She stormed toward the door.

"As long as you are leaving, you may as well begin your packing."

"No! I am staying, Reed."

He, too, jumped from his chair, reaching her in two long strides. "You will sail tonight, Kat, if I have to truss you up like a Christmas turkey to accomplish it," he assured her in a low growl.

She tried a different tack. "Then take me to New Orleans. I will stay with Eleanore and Charles."

"And what good would that do with Pierre in the city so often? He does visit his wife and children and look after his interests there at intervals. It would be child's play for him to arrange some accident without ever implicating himself. In fact, it would be easier there than here. No, Kat. Savannah it will be."

"Why can't I sail with you? I don't get seasick and I do love the sea, you know. I promise I would keep to the cabin and quarterdeck and not be a nuisance."

"That would be the day!" he snarled. "Besides, men do not like sailing with women on board. Many feel it is bad luck. Above all else, this is not a pleasure cruise, madam. We will often be fighting for our lives, and you would be in the way and a constant worry to me." Reed's voice softened as he continued, "Kat, I do not mean to hurt you, but you must go. You cannot sail with me or go to New Orleans, and you cannot stay here for several reasons. First there is the danger from Pierre. And, again, I do not relish the idea of you and Dominique together while I am gone." His voice hardened again and his eyes became as piercing as arrows. "Savannah!" he announced with finality.

"If I go to Savannah, I will be dumped off like so much

discarded baggage. I refuse to be shut away on some plantation with your mother and sister. If you expect me to sit
patiently by like some well-trained dog and wait for your
occasional appearances and a brief pat on the behind, you
are sadly mistaken, sir!'' she declared vehemently, eyes
glittering like bits of green glass. ''And furthermore,
unless you want all of Savannah to know of your unsavory
conduct and the fact that you are in league with pirates, in
reality a pirate yourself, you will escort me to my aunt's
and say nothing of our marriage!'' She glared at him
defiantly as she flung down her challenge.

''And what do you suppose will become of your
precious reputation then, my dear,'' he sneered, glaring
back at her.

''You are so adept at lies that sound convincing, I'm
positive you can assure them all that I came through the
voyage unscathed—quite virginal in fact!'' she goaded
him. ''If I am to be discarded in Savannah, then make it a
permanent condition, Reed. Let us make a clean break of
it and forget we've ever laid eyes on one another. It is
better we stop the pretense now and let this farce of a
marriage die a quick death.''

''I'll not give you a divorce, Kat. Never! You will die a
lonely old woman before I release you to marry someone
else,'' he ground out between clenched teeth. ''You are
mine for as long as I wish it!''

''I do not recall asking you for a divorce, Reed. Just let
me live my life and you live yours, preferably as far away
from me as you can manage it! In the bargain, you still
have the *Kat-Ann*, which was your objective in the first
place.''

''And I suppose you intend to pass yourself off as a
delightful little ingénue fresh off the ship from Ireland,
pure and untouched?''

''I intend, husband of mine, to enjoy myself to the
fullest. I shall go to balls and barbecues and parties. I will
be sought after and squired about and courted to my
heart's content by decent, upstanding gentlemen.''

''To no end, Kat. You will either ruin your good name

or force me to kill some unsuspecting fellow if you carry through this charade. I'll not be cuckholded, wench!" he bellowed.

"I will be careful of my reputation, to be sure, Reed. At least for a while. Someday, when I tire of the game, perhaps I will meet someone who will not care if I am married or not. Then I might become his mistress and escape your clutches under his protection," she goaded further.

"Over my dead body, you green-eyed witch!" he roared.

"Perhaps that can be arranged, too. Or by some chance you may meet your demise on one of your daring pirate raids and solve all my problems!" she taunted. She walked toward the door. "I'll pack now."

"And if I decide to ignore your insufferable demands and take you to Chimera after all?" he demanded.

"Then I shall expose you for the pirate you are and watch you dangle from a gibbet with a noose about your neck," she answered with a cold, wicked smile.

Kathleen bid a tearful goodbye to Eleanore later that afternoon. "Please write to me, Eleanore. I'll miss you dreadfully."

"As I will you, Kathleen," Eleanore sniffled. "I promise to write."

"I can't tell you how much I appreciate your friendship these past weeks," Kathleen added.

"Perhaps we shall see one another again when you and Reed have straightened out this misunderstanding."

"No, Eleanore. It is much more than that. It is over. Reed wanted to deposit me at his plantation and leave me to pass my days as a docile, obedient wife awaiting her wayward husband's return. He is not really even fond of me, but wishes to hide me away and forget me. All he was ever interested in was the *Kat-Ann.*"

"Kathleen, I've seen you two together and although you have your fights, I'm sure he cares a great deal for you. Jean told me how distraught he was this morning when he thought he had lost you."

"I am his property, Eleanore. Nothing more. A convenient plaything whenever he chooses to take me off the shelf. Well, I am tired of his games and his deceit. I can stand no more. If nothing else, I'll salvage my pride.''

''Pride won't keep you warm at night when your bed feels too large and lonely, my dear,'' Eleanore counciled.

''It will have to do, I'm afraid. I have made up my mind to stay at Aunt Barbara's and try to forget this entire disastrous episode ever occurred. Most of all, I shall try to forget that lousy blackguard and our very regrettable marriage,'' Kathleen declared.

''And Reed agrees to this?'' Eleanore queried. ''That does not sound like him.''

''He is not happy about his property defecting, especially since I plan to enjoy myself, but there is little he can do. I have convinced him he has no other choice.'' Kathleen did not elaborate further.

Saying goodbye to Dominique was harder still. This time Kathleen did not reveal her plans to be introduced to Savannah society as a single maiden. That would involve telling him of Reed's original confiscation of the *Kat-Ann* and her own threats of blackmail. She simply told him that Reed was taking her to her aunt's in Savannah to insure her safety.

''I hate to see you go, *cherie,*'' he told her, his dark eyes full of sadness. ''I have grown very fond of you. You have made this island sing the short time you have been here.''

''I do not want to leave, Dominique. You are very dear to my heart and I shall miss you. If you are ever in Savannah, look me up.''

This brought a grin to his face. ''I am sure your aunt would love that idea, little one!'' he laughed.

''Then write to me, big brother. Let me know how things are going here.''

''Reed will keep you informed. He will surely be in port there often if you are there.''

Kathleen winced inwardly. She had not thought of this. Her only thoughts were of avoiding Reed entirely. ''I would still adore hearing from you,'' she said sincerely.

Before they sailed, Rosita came by the house to see

Kathleen. "I want to thank you for saving my life this morning," she said humbly. "I regret the trouble I caused you."

"It is best forgotten, Rosita."

"Someday I will repay you for rescuing me. I am in your debt."

"If ever that time comes, Rosita, I will gladly accept your aid." Going to one of the oilskin bags, Kathleen shook out a handful of jewels and gave them to the Latin girl. They lay sparkling in her open palm. "I want you to have these since you were risking your life for them earlier. In return I want only your word that you will never attempt anything that foolish again."

"That goes without saying," Rosita replied breathlessly, her brown eyes dancing. "I also promise never to bother your husband again. He is yours."

"If only you knew!" Kathleen thought grimly to herself.

Her last encounter before boarding the ship was with Jean. He took her face in both hands and gazed silently into her face for a very long time, his eyes roving over her features as if to memorize them for all time. In his hazel eyes was all the love and all the questions he dared not voice aloud.

"It is best you go, Kathleen, best for all of us, for reasons Reed has not even thought of." His voice was huskier than usual. "We will meet again, little sea witch. Reed was so right to compare you to Venus."

She tried to turn her face from his view. "Oh, Jean," her words caught on a sob as her eyes filled with unshed tears.

He jerked her head back around with more force than necessary, and his tone was now harsh. "Listen to me! Reed has told me all of it. You are hurt. I understand that, but what you are doing is wrong. You are fighting dirty, like a little alley cat! You are using the lowest form of revenge on the man who is your husband and my best friend." His look was now condemning.

Kathleen could not face that look, and lowered her eyes

to the floor. "I'm sorry," she whispered. "I'm sorry to hurt you and Dominique in what I am doing. I don't really think of you as pirates. Surely you know that."

"I am glad to hear it, but it does not whitewash what you are planning. What you are doing to Reed is hateful and spiteful and will come back to haunt you. Believe me when I say that you misjudge him. I know him very well, little love." The endearment slipped out.

"Do not sing his praises to me, Jean. I do not want to hear them." Her chin jutted out angrily. "I thank you for what you are trying to do, but just let it be. He and I are like oil and water. It's regrettable, but we are just not compatible."

"There you are wrong again. You and Reed are like fire and ice. Whenever you come together it is with a hiss and a cloud of steam, never quietly. One day the fire will melt the ice and only the love will blaze on."

Kathleen shook her head. "Perhaps the ice will kill the fire."

Now it was Jean's turn to shake his head. "No, Kathleen. As much as I should wish it for my own sake, I think not. The fire is too great. To deny it is to fool only yourself." His face was sad and gentle now, and he released her face from between his hands. "Go now. Do what you must as we all must do. The Spanish have a saying for it. *Que será, será,* what will be will be."

Kathleen nodded past the lump in her throat. "Fare thee well, Jean. I will think of you often."

"I will hope so," Jean answered softly with a smile.

When they boarded the *Kat-Ann*, Reed headed straight for the bridge. Kathleen secretly instructed Bobby to take her trunks to Nanna's old room. Hurriedly she cleared most of her remaining things from her cabin and the captain's quarters, and instructed Dan to install a lock on her door immediately. She shut herself up there and breathed a sigh of relief. At least on this voyage she would not have to bear Reed's odious presence!

Kathleen had just finished her dinner when there came a furious pounding on her door.

"Kat! Open this door at once!" Reed ordered her.

"Go hang, Reed!" she screamed back at him. "As of now I am Kathleen Haley once more, at least as far as I am concerned."

"I care little for you silly games, wench. I haven't the time to argue the point with you now, but rest assured I shall deal with you yet—and soon!"

She released her breath in a rush as she heard his footsteps receding. "Surely after all the harsh words we have exchanged, he does not mean to carry this ridiculous mockery clear to the shores of Savannah before he relents?" she speculated tremulously.

Much later she discovered how wrong she was as her door burst open to admit Reed's tall form. He strode to where she lay staring at him wide eyed. Silently he gathered her into his arms, ignoring her protests, and carried her struggling into his cabin. He kicked the door shut and threw her onto the bed, pinning her down.

"You are still my wife and always will be," he stated tersely. "You have backed me into a corner for the moment, but not until I deliver you to your aunt does your little deception begin. Until then I intend to enjoy you in my bed at my will."

Before she could reply he brought his mouth down savagely on hers, bruising her lips beneath his. She struggled ineffectually beneath his weight. He caught her hair in his hand and jerked her head back painfully. Thus caught, she could do nothing as he raised himself from her, and with his free hand removed his clothing. With one mighty wrench he ripped her gown completely to the hem and threw himself on her.

Kathleen screamed up at him, "I hate you, Reed Taylor! Do you hear me? I loathe you!" She tried to claw his face, but he caught her wrists and jerked them above her head.

He grinned down at her devilishly. "It matters not to me whether you hate me or not as long as you respond to me in bed."

"You insidious cur! You rutting beast!" she spat out.

161

Already he was kissing her body with his warm, insistant lips; already her treacherous body was responding to his mouth. She hated herself for having so little control, but when he caressed her as he was doing now, she was lost. Soon she was moaning in her desire, her body craving his.

He raised his frosty blue gaze to meet her eyes of luminous green. "Admit you want me, Kat. Let me hear you say it!" he demanded triumphantly.

She hesitated only a moment and then hissed at him, "All right! I want you, you damned devil!"

He laughed wickedly and shook his head. "No, Kat. That won't do. Beg me prettily for what you want."

Her body aching for relief, she whispered defeatedly, "Please, Reed. Make love to me."

Chapter 11

SAVANNAH was everything Reed had told her it would be, and more. They sailed sixteen miles up the Savannah River, passing plantations along the way, and suddenly the port of Savannah lay before them. Ships lined the docks for miles along the shore, and even though trade was limited to coastal trade by the Embargo Act, the port bustled with activity. At this point in the river, a high bluff rose above the water. The shoreline was packed with numerous warehouses built against the white bluff and rising above it. Savannah itself lay just beyond, extending southward from the river.

After the relative quiet of Barataria Bay, the bustle of this teeming port seemed almost confusing to Kathleen. She stood on deck observing the activity while Reed docked the *Kat-Ann*. As the gangplank was put into place, she suddenly felt unsure of herself. After weeks of coming to rely on Reed and drawing from his strength, she felt alone and weak, and more than a little frightened. She clasped the rail so tightly that her knuckles turned white. With all her being she regretted having to leave the *Kat-Ann* and Reed.

"I would give ten years of my life to turn around right now and go back to Grande Terre. Why does Reed have to be so stubborn? Everything was working out so well. I was beginning to think Reed might even care for me. Even now, if he were to tell me he loved me, I would back down and go live with his mother," she admitted silently.

Kathleen started as a warm brown hand covered hers on the rail. She looked up into icy blue eyes and blinked back her tears. Reaching up, she automatically brushed back the stray black lock from his forehead.

"Your hands are cold, kitten," Reed stated gently, searching her face.

"I know," she replied lamely. "It's foolish, I know, but I'm scared."

"Of what?"

"Scared of leaving what is familiar and venturing into the unknown; new people, new faces, a new town, actually a new country," she admitted.

"What? My brave Irish vixen with the sea-green eyes afraid of anything? I can scarcely believe it! You brave storms and calms at sea with a daring few men possess, rush headlong into a duel, and charm the very sharks of the deep, yet you balk at this! Kat, you truly amaze me! Could it be you are regretting your threats?" Reed asked curiously.

"The only thing I regret is leaving Grande Terre," Kathleen pouted. "Oh, Reed! Take me back to that beautiful island and let's forget any of this has happened!" she pleaded.

"I can't do that," he said firmly.

"Why?"

"You know why. Let's not go through it all again." Turning her to him he said, "Let me take you to Chimera, Kat."

"I would, Reed. I would if only—" her voice trailed off as she turned her head to hide the tears gathering at the corners of her eyes.

"If only what, Kat?" he asked tenderly.

She smoothed her mint-green dress and straightened her shoulders. Throwing out her chin, she answered, "Nothing, Reed. Forget it. Some things are just meant to be in a million years, I suppose. Will you take me to Aunt Barbara's now?"

The light went out of his blue eyes and they became shielded once more, his face stony. "Have Bobby bring your trunks up. I'll go hail a carriage," he said gruffly and strode away.

Dan ambled up and stood beside her. "Cap'n." He nodded.

"Hello, Dan," she mumbled. "What is it?"

"The men and I have told Captain Taylor that we'll be leaving him here and takin' the first ship home to Ireland. In the meanwhile we'll be stayin' in a hotel on the wharf. We'll be near. If ye need us just whistle, Cap'n."

"Thank you, Dan. It's comforting to know I still have someone I can rely on. I'll pay your lodgings until you can find a ship to sign on, and thank you for all you've done for me." Turning her attention to the other ships, she added, "Mr. Kirby said the *Starbright* would be sailing to Savannah soon. I wonder if she's here yet? We spent over six weeks on Grande Terre. She could be in any day now."

"I'll scout around fer ye and let ye know the minute she docks."

"I'd appreciate it Dan, and don't let Reed know, or he'll have two ships."

"Aye, Cap'n. I'll start checkin' right now."

The day was bright and hot, but the breeze from the river kept the residents of Savannah from suffering too greatly this first week of August. Most of the more affluent families had gone to country homes or the plantations of friends or relatives for the summer to better escape the heat. Reed had told Kathleen that there was a chance her aunt's family would not be in Savannah for a few weeks yet.

They rode together in the open carriage as Reed pointed out landmarks to her. The streets were wide and shaded, made of cobblestone brought to Savannah as ballast on ships and put to good use in this and other ways. Many of the beautiful homes they passed had fancy iron gates and fences. Some had iron grillwork on windows and doors and around balconies, giving a delicate, intricate, and novel look to the city. The iron had also been brought as ballast and put to use decorating this fabulous southern port.

Away from the docks, the streets were quiet, the people they passed pleasant and unhurried. The entire atmosphere was one of ease and dignity.

Reed explained as they passed square after square that James Oglethorpe, who founded Savannah in 1733, had planned the layout of the city. As it grew, parks and squares were to be constructed at spaced intervals, enhancing the charms of Savannah and making her the gem of the south. The residents of Savannah were proud of their unique city, and justifiably so. These quiet green oases were a balm to the spirit, with trees and tropical plants and flowers to delight the senses. Birds and squirrels flitted about totally unconcerned with the humans about them. Spanish moss hung from tall oak trees, creating the impression of an emerald fantasy world veiled in mist.

Kathleen fell in love with Savannah on that short ride to her aunt's home. "I've never seen anything like it, Reed." She sighed in awe, her eyes as wide as green crystal saucers. "It rivals even Ireland!"

"You would never guess to look at it that twelve short years ago, in 1796, almost two-thirds of Savannah was leveled by fire," he commented. "I was just a boy then, but I remember it well. Even for Chimera you could see the entire sky red with flame. The devastation was unbelievable! But the people rebuilt quickly, and more and more homes were built of brick and stone and stucco instead of wood, especially in the town itself."

"It is very beautiful. No wonder you are proud of it."

"I can take no credit for it, of course, but thank you. I've traveled a great deal, but I consider Savannah one of the most beautiful places I've seen."

The carriage stopped before a wide iron gateway on Oglethorpe Square. A negro servant looked out, and upon seeing Reed, opened the gate with a wide grin. "Aftanoon, Cap'n Taylor, suh!" he cried.

"Good afternoon, Willy. Is the family at home or have they gone to the country for the summer?"

"They's heah, suh. Been waitin' on some relation from 'cross the sea, so's I heah." He glared quickly at Kathleen.

"Then they'll be glad to know the wait is over. This is," Reed hesitated slightly and continued in a tight voice,

"Miss Kathleen Haley, Mrs. Baker's niece."

Willy bobbed his head toward Kathleen. "Ma'am. I'll run on ahead and tell 'em yoah heah." With that he loped down the drive and out of view around a curve.

Kathleen looked at Reed's set jaw and said, "You'd better get used to it, Reed. From now on I am Miss Haley to you."

"For now, Kat," he said bitterly, "but mark my words, someday you'll have to pay the piper."

The carriage rounded a curve and came upon an impressive red brick home trimmed in white and shaded by huge oak trees. It sat on a well-trimmed lawn as smooth as green velvet and dotted with small flower gardens. A sparkling fountain lay directly opposite the house on the outer side of the drive. Six huge brick pillars supported the house's upper veranda and were covered with twining ivy. Curtains fluttered at the open windows. As they approached, the wide white double doors flew open and a petite blonde woman of about forty walked briskly across the porch and down the steps toward the carriage.

Reed handed Kathleen down, and the lovely lady threw her arms about her and kissed her lightly on the cheek. "Kathleen, it is so good to see you at last. We were beginning to think something had happened to you." She held Kathleen at arm's length and studied her with a smile. "You definitely take your coloring from your mother, but the set of your jaw and those high cheekbones are your father's." Turning to Reed, she extended her hand. He bowed slightly, much to Kathleen's amusement, and kissed her fingertips.

"Barbara, it is my pleasure to see you once more. You are as radiant as ever."

"And you are just as flattering, Reed," she said with a smile. "If I had known it was your ship Kathleen had sailed on, I needn't have worried so."

"Thank you, madam. We had a little trouble at sea, which is why we arrived in Savannah so late, but nothing too serious. As you can see, your niece has arrived safe and sound into your care."

167

Barbara glanced back to the carriage and turned to Kathleen curiously.

"Where is your Nanna, Kathleen? I was sure Kirby said she was accompanying you."

Kathleen bit her lip and answered shakily, "That is one of the problems Captain Taylor was referring to; and to me it was very serious. We lost Nanna part way over, Aunt Barbara. She was drowned in a storm when she washed overboard."

"Oh, my poor dear! How terrible for you, especially so soon after losing your dear father." Barbara took Kathleen's hand in hers and patted it sympathetically. Suddenly a frown crossed her brow. "That must mean you crossed the ocean alone, with no one to look after you, or did some kindly matron take you under her wing?"

"Ahem!" Reed cleared his throat, looking abashed. "If I might explain the situation to you," he addressed Barbara. Out of the corner of his eye he saw Kathleen stiffen and pale visibly. "Lady Haley was the only woman on board after Mrs. Dunley's death. I assure you that I gave her my protection throughout the voyage. None of my men attempted to bother her. I even gave her the added comforts of the captain's quarters for her own," he added innocently.

Kathleen choked at his half-truths and quickly covered her open mouth with her hand and coughed delicately. From behind her aunt's back, she glared at him as her aunt said, "Reed, dear, that was so gallant of you. I really must commend you for being more of a gentleman than I have given you credit for. You seemed to have a wild streak in you, which is why I have not especially approved of you as a beau for my Amy."

"I still have that wild streak, dear lady. Sometimes, though, I manage to restrain it for a short while," he admitted, grinning rakishly down at her.

"Oh, you bad boy!" Barbara laughed, shaking her head at him. Quickly changing the subject, she said, "Come, let's all go into the house where it is cooler. Reed, you will stay for dinner, of course."

"I really have some other matters to take care of at the docks. I will send Lady Haley's trunks along directly."

"Fine. Take care of your business and return for dinner at eight. William and Theodore will be home by then, and I'm sure they will want to see you. I won't accept no for an answer," Barbara insisted firmly.

"Yes, ma'am. Eight o'clock," Reed answered. "Till then, ladies." He bowed again, climbed into the carriage, and drove off without a backward glance.

All at once Kathleen felt bereft. The urge to call him back was nearly choking her. Her aunt was leading her into the house. "My dear, I suppose you will think me very rude, but there is something I simply must discuss with you," she was saying as they seated themselves in the parlor.

"Yes, Aunt Barbara?"

"Reed Taylor is as charming as any of our boys, perhaps more so, but he is much bolder, more of a rogue. That is why I have tried to discourage Amy in her interest of him. However, if the two of you should become enamoured of one another, I would understand. I shall treat you like my own and love you, Kathleen, but I shall try not to dictate to you over much. After all, any young woman who has had to endure the hardships you have suffered these last few months becomes mature much faster than a sheltered, pampered young girl like Amy. She has her mother and father to guide her, but then you have your station and title to consider, and I'm sure you don't take the responsibility lightly."

"What is it you are trying to say, aunt?" Kathleen interrupted.

"Just that if you become interested in Reed, I will try not to interfere merely because my daughter fancies herself in love with him. I have other plans for her, but you may be mature enough to handle him where Amy is not. She is headstrong, and I cannot make her see that she could never hold him. He would make her very unhappy."

"And he would not make me unhappy?" Kathleen asked quietly.

"Perhaps, but as lovely as you are, he would be a fool to do so. Besides, my dear, I suspect you have him half hooked already from the way he was looking at you. He has rarely displayed much interest in our local girls. It would take quite a woman to tame that rascal where others have failed," Barbara suggested with a sly look at Kathleen.

"I don't think I'm interested," Kathleen sighed. "I just want to take one day at a time, make a few new friends, relax and enjoy myself. It seems so long ago that Papa was alive and boys were squiring me about and vying for my attentions. I want to be courted by many before I make my choice. I want to dance and flirt and forget the serious side of life for a while."

"I can understand that, Kathleen, and I applaud your wisdom. There is one thing more I must ask you, and I pray you don't take offense."

"Yes?"

"Did anything happen between you and Reed on the trip over?" Barbara looked Kathleen straight in the face.

"To what are you referring?" Somehow, Kathleen met her aunt's look squarely, without her usual telltale blush.

"If you have to ask, then I feel assured that nothing took place." Barbara smiled, satisfied. "In case you have wondered where the others are, Amy has gone to Chimera to stay with Reed's sister, Susan; Theodore is out riding with some friends; and William is still at the office. He is a very respected lawyer, the best in Savannah," she announced proudly.

"Amy is at Chimera?" Kathleen inquired hesitantly. "Will she be coming home soon?"

"Not until the weather breaks. In fact, if we had not waited for you, we would be there already ourselves, but William had work to catch up on and Theodore refused to go until you arrived. Mary Taylor is one of my dearest friends, and we usually spend our summers with her. She in turn spends part of the winter season with me. We will join them in a few days. You will love Chimera! It is so beautiful! And there will be parties and barbecues and picnics; all sorts of delightful activities. You will meet all the young people your age and have such fun, I am sure!''

"I'm sure," Kathleen echoed, silently fuming. *So I end up at Chimera regardless! What a joke on me! Reed knew this all along and said nothing! I'm not the only one who has kept secrets, it seems.*

"Now let me show you to your room where you can rest and freshen up," Barbara continued. "I'll send you your trunks when they arrive."

"One thing more, Aunt Barbara," Kathleen requested. "Mr. Kirby told me that Grandmother O'Reilly lives near here. You said I can make most of my own decisions, and I would like to spend time with her and get to know her."

"Of course. She lives on the plantation next to Chimera. Emerald Hill, I believe it is called. We see her at nearly all the social functions. She and the Taylors are close friends, and she will undoubtedly be at Chimera often while we are there."

"Oh? Mr. Kirby was under the impression she led a fairly solitary existence in her advanced years."

"Oh, my, no, Kathleen!" Barbara said in amusement. "Your grandmother is quite active in Savannah society. Very high on the social roster, in fact."

"I was also led to believe that you did not approve of or acknowledge any relationship with her," Kathleen ventured.

"True, we do not admit to any relationship or family ties, but neither does she. We do not avoid one another, nor seek each other out, but we do respect one another."

"Then I shall see my grandmother when I wish, but if it would make it easier for you, I will not reveal my relationship with her right away, but let it come out in due course."

"I appreciate that, Kathleen," Barbara commented as she led the way up the stairs.

The room that was to be Kathleen's was decorated in gold and blue. The walls were a light gold with dark blue draperies and matching bedspread and canopy, and a light blue carpet that was so thick she felt she was wading through it. The bed, dresser, desk, and dressing table were a light cream color etched in blue, and the bed hangings were a sheer pale blue. The entire mood of the room was peaceful and dreamy. Even the fireplace was of light yellow stone. Barbara opened a door to one side to reveal a walk-in closet, and another to reveal a small water closet complete with brass tub and commode.

"Do you like it, Kathleen?" she asked.

"It is lovely! Elegant enough for a princess! I shall be very happy here, Aunt Barbara."

"Please, Kathleen. Call me Barbara. Somehow Aunt makes me feel so matronly. Even the children calling me Mother doesn't make me feel so old. Besides, we are going to be friends, aren't we?"

"I'm sure of it," Kathleen avowed.

Kathleen freshened up, and then, opening the doors, walked out onto the veranda. The view from her room was as tranquil as the room itself. A huge moss-draped oak stood off to her right, and below lay a beautiful, well-tended rose garden. Roses of every hue lent their fragrance to the light breeze blowing through her windows. A rolling green lawn spread itself into a row of trees shielding from view the neighboring property.

Kathleen was standing at the iron railing enjoying the sunshine when her door opened to admit an enormous black woman. She wore a starched white apron over her black dress, and a spotless white bandana over her frizzly grey hair.

"Honey chile, you git yourself out o' dat sun dis minute! You gonna freckle lak a hen!"

Kathleen grinned and stepped obediently into the room.

When the black woman stepped closer, she shrieked, "Land o' Goshen! Look at you! You'se nearly as dark as me! Chile, I can see I gots my work cut out for me!"

Kathleen stared at her and started to giggle. The woman put her hands on her abundant hips, and ordered, "Turn aroun', gal. Hold your ahms out from your body."

Kathleen obeyed, giggling even more. The black

woman frowned and shook her head. "Yo is brown all ober! Surely dat ain't your natch'ral color! Hike up your skirt a bit."

Kathleen at last found her tongue. "No, it is not my natural coloring. My legs are white. See?" She pulled up the corner of her skirt.

"Tha's a relief!"

"Why?" Kathleen ventured.

"'Cause I'm gonna hab 'nuff trouble tryin' to bleach da rest o' you white agin."

"But I like my skin tan like this."

"Lordy me! You cain't mean dat! Ebery young lady wants skin dat's lily white. Da men folk laks dere ladies lak dat. What dey want wif a woman what looks half fried?"

Kathleen laughed until her sides ached while the black woman continued scolding her. She was sitting on the edge of the bed with tears streaming down her face when Barbara entered the room.

"I see you have met Mammy," she said with a knowing smile.

"Yes," Kathleen managed between laughs. Drying her eyes, she continued, "Mammy does not seem to like my tan. Will you please help me convince her that I prefer my skin this way? She seems determined to soak my body in milk and lemon and heaven knows what else to correct the problem."

"Mammy has definite ideas about how a lady of quality should appear," Barbara explained as Mammy scowled at them. "She has been with us for over twenty years and has virtually raised Amy. She is the one who runs this household and everyone in it."

"I'm glad to meet you, Mammy. I am Kathleen." Kathleen nodded to her. "I am sorry I laughed, but you remind me so of my old Nanna. She was always scolding me, too. But I warn you, she had little success for all her efforts, so don't expect much. I am very stubborn when I set my mind to it, and I intend to keep my tan." Kathleen grinned.

Mammy studied her with a sharp eye and finally said,

"All right, but why a beautiful young lady wants to go aroun' lak dat is beyon' me. When you gits tired o' bein' a wallflower let me know, an' I'll bleach you white agin, Miss Kathleen."

"Fair enough, and thank you for the offer."

Mammy smiled and shuffled from the room.

"Oh, dear!" Barbara exclaimed. "I nearly forgot! I came to tell you that a strange old man has arrived from the ship with your luggage, and insists on talking to you. He says his name is Dan. He is waiting in the kitchen. Do you wish to speak with him?"

"Yes. I'll be right down. Thank you, Barbara."

Dan was seated at the kitchen table drinking coffee when Kathleen entered the spacious kitchen. He rose instantly. "Thought ye might want these," he stated flatly, handing her the oilskin bag with the *Kat-Ann's* log books in it.

"Thank you, Dan."

"That other thing ye wanted me to scout around for is in port, too," he advised her warily, looking about him suspiciously at the cook and her kitchen help. "Just in yesterday. Got a letter fer ye." He dug into his shirt pocket and handed her the letter.

It was from Mr. Kirby. Kathleen tucked it into the pocket of her dress. "How long will it be here?" she asked ambiguously, her brain working fast.

" 'Bout three days, I reckon."

"Make sure of it. Use my name and authority, but I want it for my own use. Ready and outfitted soon. Do you take my meaning?"

"Yes, ma'am."

"As soon as a certain gentleman leaves port, we'll figure out the rest. Stay close."

"Count on it," Dan assured her. He tipped his hat and left.

Back in her room, Kathleen's mind was racing. The *Starbright* was in Savannah! Her own ship, waiting for her next move. What sport it would be to sail her out to sea. Here lay the key to her revenge on Reed! Opportunity was knocking if she could just iron out the details.

First she needed a place to hide the ship, not too far away, but safe from view, especially Reed's. Dan would already be looking for a place, she knew. Next she needed an excuse to be away from the house for a few hours alone. She would have to figure that out soon, before they left for Chimera. Later she would need more time away, perhaps a week or two. It seemed impossible, but where there was a will there had to be a way. Something would occur to her.

Finally she needed money. This reminded her of the letter from Mr. Kirby. She pulled it from her pocket and tore it open. Included with a short personal note was a check for a staggering sum. According to Mr. Kirby, this was her allowance for the next two months. Kathleen stared at it in disbelief. It was enough to keep her in new dresses for a year! Now she had the money she needed and plenty to spare! If only her other problems could be solved so easily. Dan would be rounding up a loyal, close-mouthed crew, probably a combination of Irishmen from the *Kat-Ann* and *Starbright* who had sailed under her command before. Her main problem was going to be how to disappear for a week or more at various times. She needed a reliable alibi, but how, when she was new in town and knew no one yet? If Kathleen could solve this dilemma, she would know exactly how she was going to extract her revenge on Reed. She would pirate him! Each time he sailed out, she would follow in disguise, and lie in wait. When she was sure he had a hold full of booty, she would relieve him of his spoils until she drove him to ruin! She would show him that it didn't pay to steal ships or anything else from Kathleen Haley! If only he had loved her, she could have forgiven him, but now he would pay royally!

Barbara gave Kathleen a tour of the house before they dressed for dinner and introduced her to Cook, who evidently sampled most of her delicious dishes; Rose, a small delicate girl who would be Kathleen's maid when Mammy was busy elsewhere; and most of the other house servants. As they completed their tour and were heading back up the stairs to their rooms, the front door flew open. A slim young man of about twenty, with curly light brown hair and soft brown eyes, stood staring up at them.

"Close your mouth, Theodore," Barbara commented dryly. "It is not polite to stare, son."

He blinked, and a slow smile spread across his handsome young face. "You are even more beautiful than I'd imagined!" he said softly. Then he added with a chuckle, "Amy is going to hate you! At last she has some worthy competition around here. The belle of the county is about to lose her throne to a more deserving queen!" He gave an exaggerated bow, grinning from ear to ear.

"Theodore!" Barbara warned lightly, not raising her voice, but letting him know by the tone of her voice she was displeased. "Both Kathleen and Amy are beautiful in different ways, and it is unfair to predict how Amy will feel about her. They may become very close friends."

All this time Kathleen stood silently by, a delighted smile on her face. Yes, she was going to like Ted. Who knew what mischief they could dream up together! The prospect was exciting after her gloomy moods.

Ted shrugged at his mother's comments. "Anything is possible, I suppose." He had not once taken his eyes from Kathleen, and now he asked, "Do you ride, Green Eyes?"

"Theodore!" Barbara gasped again, a look of desperation on her face.

Kathleen just laughed and answered with a jut of her chin, "Better than most, cousin dear. I wouldn't be Irish if I didn't."

Even Barbara laughed and said, "Well, Theodore, you asked for that! I definitely think you have met your match, you young scamp, and may finally get your come-uppance!" Taking Kathleen's arm, she led her up the stairs. "Come, Kathleen. Theodore will have to wait until dinner for your company. Perhaps by then his brain will catch up with his tongue and he can be civil."

Kathleen washed her hair, then allowed herself the luxury of a long hot bath full of scented oils that made her skin feel smooth and silky. She dressed carefully in a lemon colored satin dress, knowing full well that it emphasized her tan and would send Mammy into spasms. The dress was etched in rows of fine white lace and clung low to her full breasts. The sleeves were short, fluffy layers of lace. She chose a delicate white fan to complement the dress, and with Rose's help managed a coiffure of thick braids intricately woven between waves and curls. About her neck was stretched a band of white lace from which hung a single emerald teardrop that matched her earrings and her sparkling green eyes.

Rose was delighted with the final results, exclaiming over Kathleen's figure and small waist. "And those legs! Lawsy me! You got the longest, prettiest legs I ever done seen on a gal! It's almost a shame to hide them under a skirt. If the gents ever guessed whut they cain't see, it would drive 'em crazy!" Rose exclaimed.

Kathleen smiled, thinking that Reed already knew, and she hoped it did indeed drive him to distraction. It would serve him right, the conniving snake!

Downstairs Kathleen found everyone gathered in the parlor, including Reed. Barbara came forward and, taking her by the hand, presented her to a portly, balding gentleman in a grey waistcoat and trousers. "Come meet your Uncle William." And to William she said, "This at long last is our niece, Kathleen."

Kathleen curtsied politely as he kissed her hand, his brown and grey mustache tickling. "My pleasure entirely, Kathleen," he said, his quick brown eyes flicking over her.

"Thank you, sir," she replied, feeling she had just been thoroughly assessed by this sharp old lawyer.

They stood for a moment silently measuring one another, Kathleen never wavering, and then a satisfied smile broke over her uncle's face, altering his stern demeanor to one of congeniality. "I see you are not afraid to look a person straight in the eye," he observed with a chuckle. "You are no shy little miss, are you?"

"No, sir, that I'm certainly not!" Kathleen smiled in return.

"Good. Then we'll get along fine. If there is anything more annoying than a flighty, fainting young female, I have yet to encounter it."

As Kathleen sat down, Ted handed her a frosty mint julep and seated himself next to her. "Mother is the only one who has had any time to talk to you as yet, and I for one am dying of curiosity about you. What are you really like behind that lovely face and fabulous green eyes?"

Kathleen noticed Reed's dark frown and laughed lightly. "I'm sure that only time will tell you, Ted, for I certainly shall not. A lady never willingly reveals all her secrets." For the first time she addressed Reed, "Captain Taylor, I want to thank you for sending my trunks around so promptly."

Before Reed could respond, dinner was announced and Ted leaped up at Kathleen's side. "May I escort you to the table?" he requested, already propelling her forward.

"The boy has misplaced all his tediously learned manners," Barbara apologized, taking Reed's arm. "She has him spellbound." Reed's barely concealed scowl was his only answer.

The dinner of roast duckling and dumplings was delicious, and for dessert, a mouth-watering peach cobbler appeared. At first the conversation centered around Kathleen and her life in Ireland, and finally on the ocean voyage. Reed seemed to enjoy listening to her explanations of the trip, knowing her discomfort. Several times she felt his eyes on her and his unvoiced laughter, and she feared he would contradict her story at any moment. Finally, almost out of pity, he did step in and finish the tale himself, adding a few embellishments of his own.

"Following Mrs. Dunley's demise, Miss Haley was fairly well in a state of shock, even into the next morning, but thankfully by that evening she had recovered remarkably and seemed quite a different person," he related, eyes twinkling.

Kathleen shot him a withering look over the rim of her goblet. He merely nodded, flashing his strong white teeth in a daring smile. His gaze fell to her hands, where she was clutching her wine glass, as if she were visualizing his neck between her hands. His smile widened even more.

"Have you been introduced to everyone in our household?" William was asking.

"Oh, yes," Barbara cut in. "I gave her a tour of the house earlier. Mammy was extremely upset over Kathleen's tanned skin. She is just itching to get out her milk and lemon recipe!"

"Don't let her, Kathleen," Ted broke in. "I think you look marvelously healthy that way. Personally I prefer your tan over the pale look of all of our Southern ladies, don't you, Reed?"

"I'm not sure," Reed said thoughtfully. "I rather like creamy white complexions in a way."

"Oh, so it's that kind of game we're playing now," Kathleen thought furiously. She faced him squarely and commented sweetly, "Really, Captain Taylor? I'd pictured you as the type who would favor the dark Latin look."

"I enjoy many different types of women, Lady Haley," he countered lazily.

"Beware of spreading yourself too thinly then, Captain," Kathleen replied smugly. "It is said that he who burns his candle at both ends stands in danger of being burned."

"I'll try to remember that," Reed assured her, obviously amused.

"And what type of gentleman strikes your fancy, my dear?" Barbara interjected, taking advantage of the underlying current between Kathleen and Reed. "I'm sure Theodore has many friends from which you could choose a suitor."

"I find I have a decided preference for the tall, dark sort, Barbara," Kathleen answered, once more looking at Reed. "One such gentleman I met had the most interesting dark eyes, and though most people would find him fearsome, I found him quite gentle, with a warm, sincere heart. We became quite good friends and may have become more than that but there were circumstances against it, and he was too much the gentleman to cross the bounds of propriety," Kathleen continued, deliberately reminding Reed of Dominique.

"Such a pity!" Barbara mused. "But perhaps you will meet again and the fates will be kinder."

"I doubt it, but regardless, I know that we shall always

remain friends. He is a very loyal person, to trustworthy, and he has the most remarkable brother!''

"What does he do for a living?" William asked, totally unaware of the play going on.

"His family is in the business of transporting trade goods," Kathleen improvised, chancing a look at Reed who seemed as if he were about to strangle.

"It is rare to find such a fine friend," Barbara noted.

"Yes," Reed agreed. "Usually when you do you will find they also have a few unsavory relatives skulking about to balance the scales," he pointed out, his jaw muscle twitching in a manner Kathleen had come to associate with his dark anger.

After the men had finished their brandy and cigars, they joined the ladies in the parlor.

"Will you be seeing your mother soon?" William was inquiring of Reed as they entered the room.

"I'll be riding out this evening."

"Please tell her we will be following in a few days, as soon as I can close up the house," Barbara requested. "Your mother has invited us for the summer as usual."

Reed's dark eyebrows jerked upward on hearing this, and he switched his gaze to give Kathleen a quick look of barely hidden humor with a definite touch of gloating thrown in. Kathleen stared back at him defiantly.

"It is indeed a pity that I shall miss you, but I'll be gone before you arrive," Reed told Barbara. "I've pressing business to attend to, and shall only be staying a day at Chimera. I would dearly love to be there when you arrive, but I shall endeavor to return soon, believe me." The double message of his words was not lost on Kathleen, nor was the swift look of triumph he shot her as he left.

Chapter 12

Two days later Dan came by and told Kathleen that Reed had sailed that morning. "I've found a cove at the mouth of the river b'tween a couple o' islands thet are well hidden. The passage at first looked ta be too shallow, but on closer inspection I'm sure a ship could clear it if she has a sharp captain," he reported.

"Is the *Starbright* in readiness, then?" Kathleen questioned.

"Crew an' all," he assured her.

"Fine. I'll sneak out tonight. Meet me at the end of the drive."

Late that night Kathleen sailed the *Starbright* down the Savannah River. Careful of the tides and currents, she skillfully maneuvered the ship into the narrow channel Dan indicated and around the backside of a small island. The crew held its breath the while, praying they would not run aground. There they dropped anchor.

"Congratulations, Cap'n. Not many could have navigated thet channel durin' the day, let alone at night," Dan said proudly.

"I hope you remembered to have someone meet us with horses," Kathleen said as Dan rowed them toward the mainland in the dinghy.

"He'll be there," Dan alleged.

"The crew looks to be all men who have sailed with me before," Kathleen commented. "However, I do not want too many from the *Kat-Ann*. They will be too easily recognized. Bobby especially must go. Also, I should like it if a few of the men are willing to be assigned to Reed's ship in case we should need their aid there. You and I must devise costumes to disguise our identities.''

''Jest what is this plan ye have in mind, Cap'n?''

''You will have to sound out the men and make absolutely sure they are all loyal to me,'' Kathleen went on, ignoring Dan's questions. ''Explain about the *Kat-Ann* and find out how each man feels about practicing a bit of piracy.''

''Piracy!'' Dan exploded.

''Yes, piracy,'' Kathleen reiterated with a secretive smile. ''It is the perfect way to pay back that pirate devil I'm married to! What better way can you think of, Dan?''

''Then ye mean to steal the *Kat-Ann* back from him?'' he asked.

''In a way, old friend, but I shall do it slowly. I have given it a lot of thought. I shall relieve him of all his stolen booty every chance I get until I finally drive him to ruin. That ship means a lot to Reed, and it will be very painful to him to be forced to sell it.''

''You'll be drivin' yerself to the poor house too, then, won't ye lass?''

''Not as long as I keep Papa's shipping line a secret. I'll have the company buy back the *Kat-Ann* through an agent.''

''It's a dangerous game ye'll be playin'. What if ye're caught?''

''That is why I must be sure all our men are loyal. Any who do not wish to join us must return to Ireland immediately and reveal nothing if they value their necks! I've too much at stake to risk harboring an informer. Also, I must find a way to disappear from here when Reed sails without arousing suspicion. As soon as I have a plan worked out I'll let you know. In the meantime, prepare the ship and crew.''

They reached the shore and were climbing out of the dinghy when a man approached leading two horses. Kathleen and Dan mounted up, leaving the boat for the young sailor.

"Have ye realized the *Starbright* is registered in several countries includin' this one?" Dan questioned.

"Aye. That is why we must disguise her too. The figurehead must be removed and a new one affixed."

Dan looked startled. "Beggin' yer pardon, Cap'n, but ye know that's bad luck."

"What else do you suggest?" The figurehead they were discussing was that of a buxom young maiden with long flowing locks curling about her bare breasts. Her arms were thrust above her head, and in each hand she held a shining star.

"Let me think of it a bit," Dan suggested past a wad of tobacco. "I'll figure somethin' out."

"You do that," Kathleen agreed, "and while you're at it, set the sailmaster to making up a new flag to fly. I should like a pot of gold with crossed swords through the handle on a backdrop of emerald green. I want you to have the entire ship painted green, sails and all. The green will blend with the sea by day, and the sails will not show up in the moonlight. Also, her name must be painted over so she can't be traced to me. We shall give her a new name," Kathleen announced with pride. "We shall call her the *Emerald Enchantress.*"

Dawn was nearing as Kathleen climbed the outside veranda stairs to her room. She felt she had just closed her eyes when Mammy shuffled in with her breakfast tray. "What time is it?" Kathleen mumbled sleepily as Mammy drew aside the mosquito netting.

" 'Most nine o'clock," Mammy answered with a smile. "Since you're usually up by now I got your breakfast fo' you. 'Sides, we gots to leave 'bout 'leven fo' Chimera."

"Thank you, Mammy. Just set the tray on the table." The aroma of fresh coffee was already working its magic, and after all her activities the night before, Kathleen's stomach felt as if it were touching her backbone. "Is Ted up yet?" she inquired.

"Master Theodore jes' got up, too."

"Will you ask him to stop by my room after he's dressed? I need to consult with him about a horse."

It was a short while later that Ted appeared at her door. He eyed Kathleen appreciatively, noting how the forest green of her riding habit accented her eyes. "You wanted to see me?"

"Yes, Ted. Come in." She led him to a chair. "I wish to buy a horse and I need you to tell me the best places to start looking."

"What kind of horse do you have in mind?"

"A young thoroughbred or an Arabian, I suppose. Fast, surefooted, good lines and breeding," Kathleen enumerated. "I'm not sure if I want a mare or a stallion. I'll know when I see the horses."

"Well, Kathy, horse racing is a popular sport here in Savannah. We have races nearly every week. I'll take you with me Saturday afternoon if you want. There will be horse traders and breeders there showing some of their available stock."

"That is only three days from now, Ted. Will it be possible to leave Chimera so soon after arriving and not seem rude?"

"Oh, of course," Ted said confidently. "It isn't that far to ride, and there will be others who will want to go too."

"Then I'd be delighted. Meanwhile, do you have a horse you can loan me for today? I hate to ride in a carriage on a beautiful day. Barbara won't mind, will she?"

"Father will ride in the carriage with her. He always does. Come along and we'll fix you right up," he assured her as he ushered her out the door.

As soon as they'd saddled the bay gelding for her, Kathleen asked Ted to accompany her to the bank where she cashed the check from Mr. Kirby and hurried home again.

It was a leisurely hour and a half ride to Chimera over pleasantly shaded country roads. Kathleen enjoyed the ride, but regretted not being able to ride astride. Despite all of Mammy's houndings, however, she refused to wear a bonnet. Instead, she tied her hair back loosely with a ribbon and revelled in the feel of the sun on her face.

As Barbara had predicted, Kathleen was enchanted by her first glimpse of Chimera. It sat on a knoll a good way back from the road, surrounded by shade trees and the greenest grass Kathleen had seen since leaving Ireland. The air was filled with the scent from flowering trees and bushes. The house itself was a huge white wooden mansion with two wide wings angled back on either side of the center section. Eight immense pillars dominated the front of the structure, running the entire height of the house and supporting the second story veranda, and, farther up, the widow's walk. Nearly all the windows and doorways sported fancy black iron grillwork. Beyond the main house, the kitchen, stables, barns, and other out-buildings could be seen, as well as neat rows of white-washed slave cabins. Hidden from view, Ted told Kathleen, was an inner courtyard formed by the angle of both wings of the house, with flowers planted all around the edges, a beautiful fountain in the center, a lovely gazebo at the far end, and a high wall closing off the rear where the wings ended.

They dismounted in front of the house where a servant took charge of the horses and baggage, and were ushered into the cool, darkened entryway by a young slave girl. When her eyes adjusted to the light, Kathleen was surprised to find that what she had assumed to be an entryway was actually a very large reception room from which two broad, open marble staircases curved majestically from either side of the room to a wide upper hallway. Intricately designed wrought-iron railings enhanced the beauty of the staircases and lined the outer edge of the hallway. Two magnificent crystal chandeliers hung from the high ceiling of the room itself, with another to match centered in the upper hallway. This main hallway, with doors and lesser halls leading to other areas, extended completely around the upper edge of the reception room, while the entire front of the upper level opened onto an outer veranda. On ground level, more doorways opened along the left side of the room and beyond the staircases. On the right side, wide, lacy ironwork doors could be folded open to join this room with an even larger elegant ballroom. Divans, chairs, and shining cherry tables were arranged on the highly polished marble floor for the comfort of guests, giving an aura of welcome in the midst of pure elegance.

As Kathleen stood stunned at the splendor of the scene before her, doors to the left of her slid open and a graceful dark-haired woman glided toward them. She was of medium height, but her gleaming black hair was arranged atop her head, making her seem taller. Her grey eyes glowed in her delicate pale face as she approached them.

"Barbara! I was so happy when Reed told me you would be coming soon. I've missed your company this summer."

The two women embraced each other lightly. Standing back, Mary Taylor accepted a light kiss on the cheek from William, and acknowledged Ted's presence with a smile and a nod.

Taking Kathleen by the arm, Barbara introduced the two ladies. "Mary, dear, I would like you to meet Kathleen, my dear brother's daughter. She came over on Reed's ship. I suppose he told you."

Mary Taylor faced Kathleen with a curious smile. Quickly her grey eyes assessed the young woman before her. Kathleen wondered how much Reed had told his mother, and it took all her willpower not to sqirm under the woman's evaluation.

"Reed did mention something about having you aboard," Mary said in her soft southern voice. "I'm surprised he didn't mention how beautiful you are. Welcome to Chimera, Kathleen."

"Thank you, Mrs. Taylor. I appreciate your including me in your invitation. I hope it is not inconvenient for you," Kathleen said politely.

"Not at all," Mrs. Taylor assured her. Turning to Barbara, she continued, "Now, I must get back to my guests. I have a few ladies in for luncheon. One of the servants will show you to your rooms and help you get settled. As soon as you have freshened yourselves, please join us in the parlor, ladies. Amy and Susan are there as

186

well as Kate O'Reilly and a few of the other neighbor women. Kathleen will have the opportunity to meet these select few in a quiet gathering.''

As they ascended the curved staircase, Barbara explained, ''The rooms at the head of the stairs are ladies' resting rooms and a dressing room for the men. To the right are six suites, including Mary's rooms to the front of the house, with Susan's next to hers. Amy usually rooms near Susan or in with her. The other wing also has six suites, one of which is the master suite, belonging to Reed. That entire wing is off limits to single ladies, as of course, the other is to unwed gentlemen. That way reputations are not in danger of becoming tarnished or questioned.''

Kathleen nodded. ''Rather a good means of keeping temptation out of anyone's way.''

''Exactly,'' Barbara stated firmly. Then, turning to eye Kathleen shrewdly, she added, ''That is a very astute observation, Kathleen. Are you sure you are only seventeen?''

They followed the servant to their rooms. Kathleen was delighted that her suite faced the open courtyard to the rear of the house. Barbara and William took the corner rooms farther down the hall. The apartment might have been designed with Kathleen in mind, it was that plush and pleasing to her. The bedroom, which opened onto the outer veranda, was done in shades of green with glossy cherry furnishings, and had a small bath with a tub. The sitting room was done in beige and apricot with delicately designed walnut furniture, and was very cheery.

Quickly, Kathleen changed into a peach linen gown with deeper peach trim. The neckline was low, and a wide ribbon accented the tiny waistline and was tied in a huge bow in the back. Mammy arrived in time to help her sweep her shining hair up into a loosely curled mass atop her head. Devoid of any jewelry but her wedding ring and a pair of diamond earrings, Kathleen waited nervously in the hall for Barbara, unwilling to face the ladies in the parlor alone.

Upon their entrance into the parlor, Mary Taylor rose

and came to greet them. Heads that swiveled in idle curiosity remained turned to view the new arrivals. Kathleen stood quietly, hoping she appeared more calm than she was. Somewhere in this room, she realized, was her grandmother, Kate O'Reilly, whom she had not seen in fourteen years. Desperately she hoped that Mr. Kirby or someone had informed her grandmother of her impending arrival in Savannah so that the older woman would not be shocked into a heart attack.

"I wanted to go to her house quietly first. This is no way to reacquaint oneself with one's grandmother," Kathleen thought anxiously. "How will she react? Will she understand that I am living with Aunt Barbara's family? Will she despise me for it? Will she immediately declare that I am her granddaughter before all these women and embarrass Barbara in the process? What will she think of me?"

Further worries on that subject were pushed aside as Kathleen realized she was being introduced to the group by Mary Taylor's softly accented voice. Immediately two young girls detached themselves from a group of young women standing around the piano, and started toward them.

One was a younger replica of Barbara, blonde, petite, and blue eyed. Kathleen knew this must be Amy. The other girl had dark hair and resembled Mary Taylor. Kathleen could also see something of Reed in her features.

Barbara stepped forward and embraced her daughter. Then, turning back to Kathleen, she said, "This is my daughter Amy, Kathleen, and her friend Susan Taylor, daughter of our hostess."

A friendly smile lit Susan's pretty face as she murmured a quiet welcome. Kathleen experienced a strange, hurtful tug at her heart as she realized that the smile was exactly like Reed's. Her attention was drawn to Amy as the petite blonde spoke to her in an odd accent combining England and southern Georgia.

"Well, cousin Kathleen, finally we meet. I'm sorry I wasn't in town to greet you upon your arrival, but my delicate skin just can't take the heat of summer as well as yours obviously can,'' she said in a loud, overly sweet voice. ''I do hope you had a quiet voyage.''

Some little imp inside Kathleen prompted her to reply, ''I quite understand, Amy, dear. Your family greeted me nicely, and although the ocean crossing was not without its perils, Captain Taylor was extremely attentive and competent, and brought us to port safe and sound.'' Kathleen smiled sweetly at Amy.

Barbara grimaced inwardly, knowing full well that war had been declared by both girls in those first few seconds. Now Kathleen was being led around and introduced to each of the ladies individually. Within a few minutes, she was standing before a stately woman with grey-streaked auburn hair and bright green eyes. As their eyes locked, a strong current of love and understanding flowed between them, and Kathleen's smile became warm and genuine for the first time that afternoon. The older woman was introduced as Kate O'Reilly. She smiled and nodded at the introduction, and before Kathleen could wonder how to greet her, she said quietly in a commanding voice, ''I would appreciate having ye walk with me in the courtyard after lunch, Miss Haley.''

''Yes, ma'am,'' was all Kathleen had time to reply as she was led on to the next lady.

Kate O'Reilly was seated on a small divan with another of Savannah's leading dowagers when Kathleen and Barbara entered the parlor. She had thought she was prepared for Kathleen's arrival in Savannah, whenever it would occur. Mr. Kirby had written to her, explaining that he was sending Kathleen to live with the Bakers so as not to burden her with the lively young girl. Also, through her social contacts, she had heard that Kathleen had indeed arrived a few days prior, and would be coming to Chimera with the Bakers soon for the remainder of the summer season.

What Kate was not prepared for was to see her replica standing in the doorway. It was as if time had turned back the clock and she were gazing at her reflection in a mirror forty years prior. If there were any difference at all between the way she had looked and the young girl standing before her, Kathleen was even more beautiful. It was all Kate could do to remain calmly in her seat. She knew from Mr. Kirby's letter that Kathleen knew her grandmother resided near Savannah. What Kate was unsure of was Kathleen's feelings about her Irish heritage. The girl's father had been an English lord and Kathleen would have undoubtedly been educated in England. Perhaps Kathleen would hesitate to reveal her Irish ties; perhaps her loyalties lay totally with the English. Maybe her old Irish grandmother would be an embarrassment to her.

Kate O'Reilly decided to reveal nothing to this group of gossipping women as she and Kathleen were introduced. She would talk privately with Kathleen later. Then as her own bright green eyes gazed up into Kathleen's, she felt that silent pull, that quiet yet turbulent and undeniable alliance of Irish blood meeting Irish blood, and she knew that all would be fine. As Kathleen's delightful smile grew warmer and her eyes took on added sparkle, Kate felt the love flow between them and knew that before her stood a proud young Irish woman, and that the Irish far outweighed the English bloodline. Kathleen was truly the granddaughter her heart had longed for all these years.

After luncheon, Kate motioned to Kathleen and the two of them escaped into the courtyard. Kate guided her toward a far corner where a lovely little white gazebo stood beneath an old weeping willow. Once they had settled themselves on the cushioned benches, Kate turned to Kathleen and asked, "Ye know who I am, lass?"

"Yes, ma'am," Kathleen answered awkwardly.

"Good. How do ye feel about running into yer old Irish grandmother here?"

"I was hoping to call on you privately as soon as possible and save you any shock, instead of popping us so suddenly. Did Mr. Kirby write to you?"

"Aye, lass, he did. I knew ye were coming and would be living with the Bakers." A touch of bitterness crept into the old woman's voice. "Do ye think ye'll like living there?"

"I suppose so. Ted and Barbara and Uncle William seem nice enough. I wish I could live with you, but no one seemed to think that was a very good idea," Kathleen added.

"Because I'm so much older and ye need to be around young people yer own age. Am I right?" At Kathleen's reluctant nod, she continued, "Ye failed to say what ye think of Amy Baker. How do ye think the two of ye will get along together?"

"Amy, I fear, is definitely going to be a problem!" Kathleen declared adamantly.

Kate laughed heartily. "I'm sure ye are right, dear. Now, tell me all about my lovely Ireland. I want to hear about ye and yer mother and father. I want to know about yer childhood, yer likes and dislikes, yer skills and education, and yer heart's desires. I want to get to know me only grandchild."

Kate listened with interest to all Kathleen had to say. They shared the sorrow of Ann O'Reilly Haley's death. They shared the joy of Kathleen's pranks, which the young girl told with relish. Kate readily accepted the fact that Kathleen preferred to ride astride a horse rather than side-saddle. "I always did too, love," she commented.

When Kathleen related tales of her fencing prowess, the old woman's green eyes glittered. " 'Tis proud of ye, I am, Kathleen. There's more of yer old grandmother in ye than I thought. I always had a yen to learn to fence, but the opportunity never presented itself. I'm a crack shot with a pistol, though," Kate added proudly. "What other talents are ye hiding behind those pretty silk skirts?"

"Papa taught me to sail a ship, and I'm truly good at it, Gram. I love it!"

"But ye didn't sail yer own ship over. Why?"

"Mr. Kirby warned me not to start off on the wrong foot in Savannah. I could see how most people would be shocked, so I agreed to hiring a captain." Kathleen's voice hardened perceptively at the thought of Reed.

"I heard ye say earlier that Reed Taylor captained the ship," Kate said, observing Kathleen's face closely. "What do ye think of him?"

"As a captain he is superb," Kathleen stated firmly.

"And as a man?" Kate pressed.

"As a man I thoroughly despise him!" Kathleen almost shouted. At her grandmother's questioning look, she blurted, "He stole my ship! He stole the *Kat-Ann* from me, Gram! He's a blasted pirate, and I mean to pay him back for it someday!"

Kate moved closer to Kathleen and pulled the sobbing young woman into her arms. "Tell me about it, darlin' How did he steal yer ship?"

Wiping at her teary eyes, Kathleen almost whispered her answer. "He tricked me into marrying him." Kate gasped, but did not interrupt as Kathleen went on. "All that I owned is now his, the estate in Ireland, the shipping firm with all eight ships, everything."

Throwing her chin out, she gazed into her grandmother's face and added, "So far he only knows about the estate and the *Kat-Ann*. I have not told him of the other ships. It is bad enough giving up one unwillingly. You won't tell him, will you?"

"I'll keep yer secrets, Kathleen. Now tell me how this marriage came about so I can better judge the situation. Living next door to the Taylors all these years, I have come to know Reed quite well. We like each other and respect each other a great deal. I have listened to him and understood his love for the sea when his own family stood against him. He confides in me like kin. I can not imagine him doing anything so hateful, yet if ye say it is so, it must be."

Fighting for her composure, Kathleen launched into her story, telling of Nanna's drowning and Reed's deception. "He let me think he had defiled me while I was in a state of shock, and I couldn't remember any of it, but what else was I to think after waking up completely naked?" Kathleen explained. She went on to describe their weeks aboard ship, and then asked her grandmother, "You said Reed confides in you. How much has he told you of his sailing activities, Gram?"

"If ye are referring to his association with certain

persons on Grande Terre, he has told me of his activities. I can not say I approve, but I share his excitement in that life."

"I hesitated to say anything of it to you because of the pact Reed and I made," Kathleen explained. "You see, when Reed tired of me, he wanted to bring me to Savannah and dump me off here at Chimera to rot. Little did I know I'd end up here anyway!" Kathleen grimaced slightly. "I told him I would not be shoved off in a corner and forgotten like some hound waiting patiently for a kind word and a pat. In fact, I blackmailed him, Gram. I told him I would say nothing of his pirate activities if he said nothing of our marriage."

"Kathleen!" Kate was stunned. "And Reed agreed to this?" she asked incredulously.

"Not willingly, I assure you, but in the end he considered it only slightly better than hanging." From there Kathleen went on to explain about the chain of events that took place on Grande Terre and led up to the situation she now found herself in.

By this time Kate was beginning to better realize what had occurred and why Kathleen felt as she did. "Then no one else in Savannah knows of this, lass?"

"Only a few of the ship's crew. I needed to tell someone who would understand. I felt sure you would, Gram."

"Aye, love, but I think ye are too set on revenge. Oft' times revenge backfires on ye and yer hasty words and actions come home to roost. I speak from experience, believe me."

"It can not be helped, Gram. I must avenge myself and I need your help," Kathleen beseeched. "Knowing and liking Reed as you do, can you side with me against him?" she asked hesitantly.

"Ye are my granddaughter and I will always help ye in any way I can against anyone else, Kathleen," Kate promised her with a warm hug. "Now, what kind of help can an old Irish woman be to ye?" she said, her green eyes dancing in merriment.

"First of all, I must ask you not to reveal the fact that I

am your granddaughter.'' At the hurt look on Kate's face, Kathleen hurried to explain. "At least not until I can have a chance to strike back at Reed. The Bakers will not say anything, I am sure. It would require much explaining on their part. Besides, Reed would no longer confide all his secrets and plans to you if he knew, and I need all the information you can squeeze from him. Also, since I must hide my plans from Aunt Barbara's family, I need an alibi to cover my absences for a week or two at a time while I pirate Captain Taylor,'' Kathleen revealed with a triumphant grin.

"Pirate him!" Kate exclaimed with a start. "Saints alive, Kathleen! I never dreamed ye'd go so far! Surely ye can not mean to kill yer own husband?''

"No, indeed! That would be far too merciful and quick to suite me! I intend to ruin him enough financially to make him sell the *Kat-Ann*, and I will have an agent from Papa's firm buy her back from him.''

"Ye could be caught, lass, or even killed. Can ye be sure it is worth it? Ye might regret yer actions in time. Besides, where will ye get a ship and crew?''

"The ship is one of Papa's that docked here about a week ago. I have commandeered her for my use and had her repainted, and have a loyal crew standing by. I will need your help in devising a disguise for myself, though. I cannot let Reed know it is I who is ruining him, much as I would like to. He has no idea I can captain a frigate, though I almost let it slip that I am a master at fencing when I bested Pierre Lafitte.''

"Ye saved Reed's life that day, Kathleen. Surely ye must love the man if ye would risk yer life for him,'' Kate commented softly.

"Oh, yes. I love him more than I have ever loved anyone else or ever will again, I am sure,'' Kathleen declared sadly, her voice catching in her throat. "But he does not love me in return. I am a possession, just like his precious ship, only not as important to him. He has already told me he will never let me go, never release me from this loveless marriage. He swears I'll die a lonely old

194

woman before he'll grant me a divorce, and threatens violence on any man who would bed his wife."

"And did ye expect differently, girl?" Kate defended Reed huffily. "I am Catholic even if you are not, and I can not condone divorce. I have to agree with Reed at least on this, granddaughter."

"I did not ask him for a divorce and never will," Kathleen placated her grandmother. "Oh, Gram, if he would only love me even a little I would gladly give up all thoughts of vengeance and be a dutiful wife. I would shut myself away out here and raise his children and do all else he asked of me. I would go anywhere, wait for him a lifetime if need be, if I was assured of a place in his heart."

"And are ye so sure he does not love ye, my dear? Some of what ye have told me indicates to me that the possibility exists. Ye are both strong-willed and temperamental. It could be that all that fighting and spatting disguises a lot of love. Are ye still willing to risk destroying that chance?"

"On this you are wrong. Reed enjoys dominating me, but he does not love me. He delights in my body and my looks. He takes pleasure in bedding me and he is very jealous at times, but I do not honestly believe he loves me at all."

"Have ye ever really tried to make him love ye?" Kate inquired with a wry smile and a speculative look at Kathleen. "There is more than one way to skin a cat, my girl." At Kathleen's brooding look, she went on. "Why not put that infamous O'Reilly beauty to work for ye? Entice him, bait him and lure him. Ply your feminine wiles and wits against him. Flutter yer lashes, sway her hips, throw out yer chest, and bat those big green eyes at him. What flesh-and-blood man could resist!"

"True enough, he succumbs to my charms at times, but only for a while. Then it seems he becomes immune to them except when it suits him," Kathleen mused, chin in hands. "Still, the thought is interesting. After all, what do I have to lose?"

"Aye, lamb, and look at all ye stand to gain," Kate emphasized. "Will ye not try charm instead of force? 'Tis the safer route.''

"No! I shall extract my ounce of flesh with one hand while I attempt seduction with the other. That way I stand a better chance all the way around," Kathleen reasoned.

"Beware ye don't get burned in the process. The stakes are high and the risks even greater in this game ye plan to play," Kate warned, shaking her head sadly. "I'll help ye all I can, but pray be very careful, lassie." Kate's face split in a mischieveous grin. "Ye remind me so much of meself when I was young! How anyone can look so much like an angel and scheme like the devil is truly amazing! I have a feeling ye are really going to stand Savannah on its ear! Nothing this interesting has happened me way in years. Aah, it's good to have ye here, Kathleen. Ye are making me feel young again. This could turn out to be fun after all, if we don't get caught in our own trap.''

After a moment of silent comradeship, Kathleen said quietly, "Gram?"

"Aye?"

"I love you."

"I love ye too, lamb."

"I'm so glad! I need someone to love and understand me, especially right now."

Taking Kathleen into her arms once more, Kate answered softly, "I know, dear. So do I. I thank the dear Lord for sending ye to me. Now, regardless of all else, we'll have each other.''

Chapter 13

TED took Kathleen riding across the plantation, and she was amazed at the amount of land it encompassed. Not only were there acres of croplands, but many of forests as well. They stopped a couple of times to water their horses in delightful sun-dappled glades where the streams ran merrily across rocks and reeds. At one point the property was bordered by a river so wide and fast that they dared not cross on horseback there. Ted showed her where Kate O'Reilly's property started so she could travel to her grandmother's home the back way if she preferred.

By the time they returned to the house, Kathleen barely had time to dress for dinner. She was so tired and sleepy that the meal passed almost in oblivion, and shortly afterward she asked to be excused and retired to her room, where she slept like the dead until morning.

After breakfasting at a small table near the open windows in her bedroom, she dressed and went in search of anyone else who might be up so early. She encountered the slim servant girl who had answered the door the day before, and after some gentle persuasion, the girl agreed to give Kathleen a tour of the house. The left wing consisted of four large lower rooms. The parlor Kathleen had seen the day before, as well as the elaborate dining room with tall windows and French doors leading into the courtyard. Down a wide hall and beyond the parlor was a well-stocked library with shelves reaching from floor to ceiling. It was a comfortable room; one Kathleen felt she could spend a lot of time in.

Across the hall was Reed's study. The slave girl hesitated to let Kathleen into this private male sanctuary, but Kathleen persisted. The room was large and airy, also opening onto the courtyard. A huge stone fireplace dominated the room. Anyone entering here could catch the nautical aura. Ship's models, trade maps, mementos from all of Reed's travels were scattered hodge-podge about the room in neat but illogical order. The large leather-covered furniture was slightly worn and comfortable looking. The entire room emanated masculinity. The room smelled of tobacco, whiskey, and Reed. A deep feeling of longing assaulted Kathleen's senses.

Next the maid escorted Kathleen through the forbidden upper left wing, except for Ted's suite, of course. At the rear corner of the wing were Reed's rooms. Kathleen could have found them blindfolded, for here too they smelled of cigars and the lime cologne Reed usually wore. Once more she fought down an intense wave of nostalgia. The furnishings were masculine, and in the bedroom doorway, Kathleen stopped short. She had never before seen a room decorated in this manner. The entire room was done in red and black in an Oriental fashion. The dresser and highboy and wardrobe were all painted in a glossy black lacquer with Eastern scenes on the doors and drawers. Against the white walls, deep red velvet curtains hung at the windows. The carpet was an Oriental delight. The immense four-poster bed of dark wood was overlaid with a deep red bedspread and hung with sheer black draperies with fierce red dragons interwoven in the fabric. Altogether, the effect of the master's room was absolutely stunning, almost overpowering.

Kathleen walked back to her rooms in a daze. Her husband was indeed a man of distinct and powerful tastes, and endless surprises.

A short time later Kathleen entered the stables looking for someone to saddle the bay for her. In the far stall, she noticed a large dark horse standing quietly. Approaching closer, she saw a magnificent black stallion. By his fine head she knew he was an Arabian. Kathleen reached out her hand and stroked his soft nose. "Hello, Beauty," she said softly. "If I were a betting person, I'd say you have to be Reed's horse. That's too bad, because you are just the type of horse I am looking for. Why aren't you out in the sunshine this fine morning frolicking with the mares?" She continued to talk softly and stroke the big animal's neck. He nuzzled her hand, blowing his warm breath in her palm.

Suddenly he pulled sharply black, baring his huge teeth and laying his ears back. Startled, Kathleen stepped aside as he lunged forward, neighing and snorting angrily.

"Miss! Miss! Git back away from dere! Massa Reed don't cotton to folks bein' 'round Titan. He's dangerous!"

"Who's dangerous, the master or the horse?" Kathleen challenged as she turned to see the tall negro in charge of the horses.

"De hoss, ma'am, ob course."

"Nonsense! He was as gentle as a lamb before he smelled your fear. Who cares for him while your master is away?"

"I does, ma'am."

"Who exercises him?"

"A couple ob de field hands helps me wif him ebery week. He lets him out to pasture fo' a day, but it ain't easy to catch him again. Massa Reed would hab my skin if'n dat hoss eber got away."

"Saddle him for me," Kathleen ordered quietly.

"Ma'am, please! I cain't do that. Massa Reed will whup me fo' shoah! Yo' might could git kilt!" the stable man pleaded. " 'Sides, ain't nobody kin git a saddle on dat black debil but Massa hisself!"

"Then I'll do it myself, but first you must leave. This horse dislikes you for some reason." At his wide-eyed, fearful look, Kathleen added, "I will be fully responsible if either the horse or I come to any harm. Now go."

As soon as the man had gone, Kathleen calmed the huge horse. "Titan. That's an apt name for you, boy. How would you like some fresh air and sunshine?" Talking softly and moving slowly, she found a saddle and eased the door of the stall open. ''Easy, Titan.'' The big stallion balked slightly at the bit, but allowed her to saddle him. He seemed anxious to be off and running as she led him outside the stable to the mounting block.

Taking a deep breath, Kathleen drew herself into the saddle. Barely had she seated herself on the brute's back then they were off like a shot. Seeing a fence up ahead, and sensing that Titan was going to jump it, Kathleen somehow managed to hike up her skirts and swing one leg over the horse to sit astride. Giving him his head, horse and girl sailed into the air and over the fence, racing across the grassy field. Finally Titan slowed, and Kathleen gained control and turned him toward Emerald Hill and her grandmother's house.

Kate O'Reilly was sitting on her side veranda when Kathleen burst around the corner of the house on Titan's back, skirts flying. After an initially startled look, Kate gave a hearty laugh and motioned for Kathleen to join her on the porch swing. ''If me eyes don't deceive me, that looks like Reed's horse, me dear,'' she said with a chuckle.

''Titan, I believe his name is,'' Kathleen supplied. ''He's just the type of horse I am looking for. Such a beauty! I doubt I'll find his equal anywhere close. You don't suppose Reed will consider selling him to me, do you?''

''Ye're a good judge of horseflesh, Kathleen. 'Tis the Irish in ye. Reed is sure to have a royal fit when he finds out ye've been riding Titan. And be sure he'll know about it five minutes after he's home. Which reminds me, ye'd better call me Kate or Mrs. O'Reilly around here if ye'll keep yer secret.''

''I'll remember that, Kate.'' Looking about her, she asked, ''Is it safe to talk to you now?''

''If ye keep yer voice low, love.''

''I've thought a lot about my disguise, and I am going to sew some shirts and breeches up, but I need some good heavy material. I can't fight in anything flimsy. I particularly want green cloth. I'll need to darken my hair, too, or Reed will recognize me immediately. Also, it will have to be something that will not streak in salt spray and water, but will still wash out readily when I return to port."

"Come with me into the house, Kathleen. We'll look in the attic and see what we can find." Kate led the way into the house. It was cool and dark in the wide hallway as Kathleen followed Kate up the stairs. Three and a half floors up, they came to another hallway with servant's quarters on one side and a huge attic on the other. Kate went directly to three huge trunks against one wall. Piling boxes aside, she opened the first, rummaged through, and closed it again. Out of the second trunk filled with old dresses and bolts of cloth, she selected a length of forest-green broadcloth, and at the bottom of the chest found a bolt of leaf-green canvassing. Finally, in the third trunk, she pulled out a length of leather dyed emerald. Spreading it out, she eyed Kathleen critically and said, "There is enough here, I think, if it suits yer purpose."

"Oh, Kate! It is perfect!" Kathleen said, eyes glittering like jewels. "I'll be the best-dressed pirate on the high seas!" she whispered with a giggle. "I'll even make a mask of it to be sure Reed does not recognize me."

Gathering up the fabric, Kate took Kathleen's arm. "Come. We'll see about a room for ye for when ye come to visit and when ye are just supposed to be visiting," Kate grinned. "It would be safer to leave yer horse here and ride one from my stable when ye go, and perhaps color yer hair and change yer clothes here, too."

On the second level she led Kathleen to a corner room with an outer entrance from the veranda. "Ye can come and go without notice easily from here as long as ye avoid the servants. A few I will trust with limited knowledge for yer safety's sake and yer aid."

"Thank you, Kate. You are a love!" Kathleen exclaimed, hugging the older woman.

"I know!" Kate laughed. "Now! To yer hair! I will look up in my books and prepare a rinse to darken it. Black would be best. A totally opposite look from yer own. What do ye think?"

"I will rely on your judgement there. Black sounds just fine."

"Can ye stay for luncheon, lass?"

"I really should get back. When the stable keeper tells them I rode off on Titan, they will probably think I'm lying dead or at least terribly wounded. I've been gone quite a while."

"I can send a boy over to tell them ye are safe—unless ye are anxious for another encounter with the darling Amy," Kate added slyly.

"If you think Mrs. Taylor would not think me too rude, which she probably does anyway after I absconded with Titan, I would love to stay."

"Fine! I'll send a note explaining that I have invited ye to spend the afternoon. Since I tend to throw me weight around a bit anyway, no one will think it strange. I am usually very outspoken and apt to have me own way in the community. At my age I can get away with such things. Ye would laugh to see how I have them cowed most of the time, especially the young folk. Ye would think I was quite a fearsome old dragon who breathed fire and ate children for my dinner! Ha!"

During lunch Kathleen again expressed her desire for a horse of distinction. "I don't consider myself an average person, and I'll not settle for the average horse, Kate," she declared adamantly. "Where will I find such a fine animal as I desire? Ted says he'll take me to the races Saturday to talk to the horse traders, and there are always the horse fairs, but I could look for months before I find the right one!" she pouted.

"Well, we'll have to see what we can come up with, won't we?" Kate said mysteriously. "Did ye know that Reed bought Titan from us?" At Kathleen's startled look, she laughed and said, "To be sure, lass. Besides the cotton and rice, yer grandfather used to raise the finest horses in Georgia. What else is an Irishman best suited for, I ask ye? Of course, since that dear man passed away I've had a foreman who sees to that end of things. He doesn't run things as well as Sean used to, but we still sell fine horses. Perhaps not the best in the state any longer, but superior stock."

"Have you any that are outstanding?"

"Only one ye might be truly interested in. He's a special breed that Sean worked for years to develop. He never lived to see his dream become a reality, but he felt so strongly about it, and we had discussed this idea so thoroughly, that after his death I made a special point of seeing that his final steps of breeding were continued. The result is a very rare breed called a palomino. The Spaniards developed it first, but have never done much with it. It is an unusual breed, one ye don't see often."

"What is he like?"

"He's a beauty, Kathleen. He is a pale gold Arabian with a white tail and mane, four white socks, and a white blaze down his face. He's a powerfully built three-year-old stallion about fifteen hands high, smart as a whip, fast as the wind, but wild as a banshee," Kate explained with pride. "Yer grandfather, if he could see the result of his years of loving labor from heaven, would be so proud!"

"He's not been broken?" Kathleen asked, her curiosity fully aroused.

"Nay. Not that many have not tried, Reed included. He's thrown them one and all. Reed was sure he could break him if he had time. He planned to try again this summer, but has been too busy. Now I know what kept him so occupied!" Kate commented with a wink.

"May I see him, Gram?"

"Kate," her grandmother reminded her.

"May I, Kate?" Kathleen repeated excitedly.

"I'll have him brought up to the paddock for ye." Kate smiled.

The palomino was indeed a majestic sight to behold, his golden coat shining in the sunlight like a newly minted coin. He trotted around the paddock, his head held proudly as if he were aware of how special he was.

"He's a truly glorious sight!" Kathleen sighed ecstatically. "I've never seen anything like him!" Turning to Kate, she asked, "If I break him, will you sell him to me?"

"Ye wound me, Kathleen. I will not sell the horse to ye." At Kathleen's crestfallen look, she continued. "I can think of no one I would rather see riding Sean's pride than ye. I will not sell him, rather I'll give him to ye in exchange for a promise.''

"What would you have me promise?'' Kathleen inquired warily.

"First ye must care for the horse as yer grandfather would have wished. Secondly, ye must never sell him or give him to anyone outside the O'Reilly family. Third, I must have him available to service the specially selected mares here so that we can produce more of his line. He must not be gelded.''

"You have my word, Kate. I agree to all you ask.''

"Now, lass, I have one more request that does not hinge on yer ownership of the horse.'' Gazing lovingly into Kathleen's shining emerald eyes, she said, "Since ye are me only grandchild, I will naturally bequeth all this to ye alone.'' She waved her hand wide, indicating her estate. "I would like for ye to come often and begin to learn horse breeding. 'Tis an art, and one that involves many years of learning. Ye can not learn it overnight, and I cannot bear to see all Sean's efforts go for naught.''

"I'll come as often as I can sneak away or visit and not arouse suspicion.'' Returning her attention to the horse, she announced. "He looks like a great golden god! I'll call him Zeus!''

"An apt name for him, to be sure. Now all ye have to do is break him. Ye'll need a bit of luck there, darlin'.''

"Nay, Kate. Skill and the right approach is all I need. This is Thursday. I'll wager I'll be on his back by Sunday, and not for just a second or two before my rear hits the ground. He'll not throw me.''

For the rest of the afternoon Kathleen tried her strategy. Slowly, gradually, she moved from her place atop the railing into the paddock. She walked slowly about talking or singing softly to Zeus, letting him get used to her presence. After a time, she sat on the ground and waited patiently, a lump of sugar held in her open palm near her knee. It seemed an enternity before the palomino's curiosity overcame his wariness. He approached Kathleen from the rear. When she did not move, he nudged her coppery head with his nose. Not turning, she slowly raised her open palm, offering him the sugar. She smiled slightly as the soft velvet of his nose touched her palm. As he took the sugar lump, she talked soothingly to him. As she turned toward him, he bolted away to the far end of the paddock.

Kathleen rose and walked leisurely back to the fence. Leaning against it, she drew a carrot from her pocket and broke it in half. Holding it out to him, she waited. Again the golden giant approached cautiously as Kathleen urged him on in a low voice. As he accepted her offering, she stroked his neck and forehead gently. He quivered slightly, but allowed her touch for a short minute. When she stepped toward him, he fled. Once again she produced a piece of carrot. This time she let out a loud whistle for him. Zeus jerked up his head and listened. Again she whistled. He eyed her apprehensively, but sighting the carrot, decided to chance it, and trotted up to her.

"Good boy, Zeus," she said as she patted him lovingly. "Pretty boy. You are such a smart fellow." Unlike before, Kathleen started walking away from him, calling him as she did. "Come, Zeus. Come boy." The big animal trotted after her like a tame kitten. Periodically she stopped to stroke him, running her small hands along his neck and shoulders and finally his back and flanks, getting him used to her touch and her voice.

Kathleen climbed from the enclosure and walked around the corner of the stable out of sight. Entering the stable, she found a rope halter and lead. When she was sure that Zeus had wandered away from the fence, she reappeared. She hung the halter on a post and whistled for the horse. His head flew up, as did his ears. He hesitated only a moment, then trotted up to Kathleen, confidently searching for his reward. She gave him a palm full of oats. "Don't get too used to expecting sweets, love," she told him. "You'll get spoiled and absolutely useless." Zeus whinnied, shook his proud head, and ran off.

At her next whistle he came readily. Talking calmly to

him, she carefully fitted the halter over his head. He pawed the ground and tried to pull away, but she held him firmly, constantly speaking softly to him. When she had the halter adjusted, she rewarded his trust with another bit of sugar, then spent three-quarters of an hour leading him in all directions inside the paddock. At last she led him into the stable stall, rubbed him down, fed and watered him, and left.

When Kathleen rode into the stable yard at Chimera on Titan, Ted ran out to greet her. He followed her into the stable where she unsaddled the black Arabian and quartered him in his stall, standing a safe distance away as she curried the horse. "Kathleen! You had us all scared to death when Gus came to the house and told us you had taken Reed's stallion. We could imagine you broken and bleeding out there someplace!"

"Calm yourself, Ted. This horse is a lamb!"

"Well, I'll tell you, Reed will not be happy at all to find out you have been riding his prize stallion! No one else has ever dared mount him but Reed."

"And the horse has been made to suffer because of it! Ted, I know it was presumptuous of me, and Mrs. Taylor must think me horrid, but this animal needs to be properly exercised regularly during his master's long absences. He can not be shut up in a stall, let out to pasture occasionally, and never ridden. No doubt Reed must break him all over again each time he returns!"

"You are right in that assumption," came Mary Taylor's voice from the stable door.

Startled, and feeling suddenly very guilty, Kathleen said apologetically, "Mrs. Taylor! Please forgive me for my actions today. It was impulsive of me and terribly rude. No one was about this morning or I would have asked your permission and told you where I was going. And when I saw this gorgeous creature shut away in a stall because everyone was in fear of him, the temptation overcame me. I'm truly sorry for worrying everyone or taking advantage of your hospitality, but I can not honestly say I am sorry Titan and I got our exercise, and I would like to continue to do so in your son's absence for the horse's benefit."

Kathleen looked at Mary Taylor hopefully.

"I will not forbid it. However, the horse belongs to Reed, so the permission to ride him is not mine to give, Kathleen. If you decide to continue doing so, you must answer to Reed when he returns," Mary said firmly, but kindly.

"Thank you, Mrs. Taylor. I appreciate your understanding."

Mary smiled and said, "You are not to feel restricted here, Kathleen. The house and grounds are open to you as a guest. Please feel free to come and go as you wish, but leave a note or word with a servant. You will not be required to dine with us or attend the social functions if you do not wish to, but we are anxious to get to know you and will welcome your company at any time."

Kathleen returned the smile and replied, "You are very kind, Mrs. Taylor. I did not really know what to make of everything here or what would be expected of me altogether. It is all so new and different to me —the homes, the people. I know so few people as yet, and though I'm not normally shy, I feel out of place. At least the horses seemed familiar ground."

"Yes, and unlike people, they never comment on your dress, or hair, or manners, or your tan," Mary answered with a wise look. "Don't worry, Kathleen. Amy's opinion of you will not color anyone else's judgement. From your lunch invitation today with Kate O'Reilly, I would say you are well on your way to charming all of Savannah. With Kate's open endorsement, which is a miraculous feat in itself, invitations will be forthcoming from the finest homes in the county. You need not worry about being accepted into Savannah's elite society. You already have your pretty little slipper in the door, and Barbara and Susan and I will be there to help you along."

"But Susan is Amy's closest friend," Kathleen interjected before she had time to think.

"I like to think my daughter is open-minded and fair. It would not surprise me if the two of you get along famously. We must just wait and see."

"It *would* be nice to have a close friend to exchange

confidences with.''

"My door is always open, Kathleen, but I agree that you need someone closer to your own age.''

"Won't I do, Kathy?'' Ted offered.

Both women laughed at him. "Now, Ted, I just can't see myself asking your opinion of hairstyles and dresses and petticoats,'' Kathleen retorted with a giggle.

Dinner that evening was much more pleasant for Kathleen. Susan was seated across from her and chatted in a friendly manner with her. Ted was as devoted as ever, but Kathleen thought she noticed his attention wandering quite often toward Susan, who blushed prettily each time she caught him watching her. Mary and Barbara kept up lively conversation and alternated at artfully steering William's comments away from business. Amy retrained her haughty attitude, but everyone ignored her jibes and enjoyed the meal.

Afterward, they all gathered in the parlor. Susan commented thoughtfully, "If Reed were here we could have a double game of whist.''

"We can all play euchre instead,'' William suggested.

"Kathleen probably doesn't know how to play it, Father,'' Amy put in.

"I learn quickly, Amy dear. It can't be any more difficult than learning poker. That, too, is an American card game, but I enjoy it.''

"Ladies don't play poker in Savannah, cousin,'' Amy retorted snidely.

"Such a pity! Men do know how to enjoy life, don't they?'' Kathleen said with an affected sigh. Then brightening, she said, "Oh! By the way, did I mention that Kate O'Reilly is going to sell me a horse?''

"She raises mostly Arabians, doesn't she?'' Barbara inquired.

"Yes, but this one is special!'' Kathleen told them, her face alight with joy. "She is going to sell me her palomino as soon as I break him.''

"Kathleen!'' they all chorused, and then everyone was trying to talk at once.

"He's dangerous, dear!'' Barbara said fearfully.

"He's definitely not a lady's horse," William was heard to say.

"Who will you find to break him?" Ted asked.

"Does that mean you won't be exercising Titan anymore?" Susan questioned.

"He's a beautiful horse, Kathleen," Mary supplied.

"Reed wanted him! He was going to break him and buy him. He's Reed's horse! He won't like it one bit if he comes home to find you own him. But I am forgetting that you have to break him first, and everyone around has already tried and failed. You won't find anyone who will be able to do it. You are lucky, really, for I doubt you've ever seen Reed in a temper. It's bad enough you are riding Titan without his permission, which he would never give if he were here," Amy predicted.

Kathleen smiled wryly at Amy's remarks. "Please! Please!" She held up her hands for silence. "One question at a time! First, he's a fabulous horse and I want him. I'm sure I can handle him. Reed already has Titan. He has not been able to break the palomino yet, and cannot begrudge me if I manage where he has failed. I shall break him myself. I have already begun this very afternoon. All that aside, Captain Taylor's temper will not keep me from having that horse!"

"You'll be hurt!" Susan burst out, her grey eyes wide with worry.

"Kathleen, please. Be satisfied with a gentler animal," William counciled. Even Barbara wore a concerned frown.

Mary rose and stood beside Kathleen, her arm about her shoulders. "Let the girl be. She has ridden Titan today, a feat none of us would dare to try. If she says she can gentle the palomino, I believe her. As far as Reed is concerned, Kathleen is right. He already has a fine Arabian and should not be greedy." Changing the subject, she said, "Let's play cards. Ted, Barbara, and I will play with Kathleen and teach her the game. Amy, Susan, and William can either join in a seven handed game or start one of their own. The winners get a double serving of that delicious chocolate fudge cake the cook has baked this afternoon."

Chapter 14

THE next morning early found Kathleen again at Emerald Hill. Once more she practiced whistling for Zeus, and found him responding favorably to her. He pranced and whinnied and begged for treats while she talked and sang, cajoled and laughed at his antics. He stood quietly allowing her to pet and stroke him at will. She led him around the paddock for a while and finally tied him to a post to see what he would do. At first he was annoyed and tried to pull loose, but as he heard Kathleen's gentle voice, he calmed down and accepted the situation and his sugary reward.

All morning Kathleen worked with him, leading him, stroking him, making him respond to her commands, tying and untying him. Near lunchtime, Kate came out of the house to watch. "He is already starting to trust ye. What is next, Kathleen?" she wanted to know.

"Next we remove the halter and put on a bridle. Then comes the saddle blanket. I want to accomplish both yet this afternoon."

"Ye'd better give him a rest and come in for lunch. Ye'll both be better off for the break."

The afternoon grew hot and humid as Kathleen worked diligently with Zeus. He balked at the bridle with its bit, trying unsuccessfully to spit the hateful metal from his mouth. He stomped and jerked and protested loudly, but made no move to harm Kathleen. Kathleen was firm but gentle, refusing to give an inch. She let him stew for a while tied to a post and sat on the nearest rail, alternately ignoring and comforting him. At last he gave in and stood quietly. Hoping she wasn't rushing things too much, Kathleen got the saddle blanket. Approaching him slowly, she eased the blanket onto his back. Zeus rolled his eyes, flattened his ears, and as Kathleen stepped hastily aside, he bucked the blanket off.

"No, Zeus! Behave!" she commanded him in an unyielding tone. Again and again the scene repeated itself until Kathleen thought she would scream. At long last Zeus decided that the blanket would reappear each time he threw it off, that it was a useless battle. The nasty thing was here to stay. Tired and dusty, Kathleen nevertheless untied Zeus from his post and spent the next hour leading him, blanket and all, about the paddock. Finally, at four o'clock, she stabled him, brushed and fed him, and called it a day.

"Tomorrow the saddle," Kathleen told Kate tiredly.

"But tonight the dinner party in yer honor," Kate reminded her.

"Oh, no! I almost forgot! I'm so tired I could sleep for a month!" Kathleen grimaced.

"Well ye can't, so ye'd best hurry back there and get cleaned up. My, ye're a sight, ye are! If ye hurry, ye might find time for a quick nap to take the edge off. It wouldn't do for the guest of honor to fall asleep over her soup! Ha!" Kate started into the house, then turned and said, "Oh, Kathleen. I suppose yer Aunt Barbara would appreciate if ye took special care with yer appearance tonight as ye'll be introduced to several eligible young bachelors this evening." She grinned.

"But, Kate! You'd be surprised to learn that Barbara would much prefer that I charm Reed and save her dear Amy from a tearful marriage!"

"Is that a fact?" Kate was amused. "How droll!"

"Aye, but I'll charm the others while I'm about it just for the practice and to bolster my own ego," Kathleen said with a sly wink.

"That's me girl! I'll enjoy watching ye take Amy down a peg or two!" Kate claimed proudly.

After a leisurely bath and a short nap, Kathleen felt restored once more. She dressed carefully in a copper satin gown with loosely ruffled sleeves that came to her elbows and trimmed with cream colored lace. It was in the latest style, with an empire waist and smooth folds to the floor. The neckline was daringly low and her firm, full breasts rose high in the bodice, a tempting sight to any man. As she went downstairs to join the others in the reception room, she looked beautiful and felt confident.

As Kate had predicted, there were several good-looking young swains to whom she was introduced. They nearly fell over themselves dancing attendance on her all evening. None could seem to keep his eyes off her bosom or face for long, undecided as to which was more enticing. From time to time, Kathleen would catch Kate's eye and both were hard put not to burst into gales of laughter. Many times Kathleen had to hide her face behind her delicate lace fan.

All through dinner, seated between two likeable but boyish young men, Kathleen silently thanked Mrs. Bosley from her old English boarding school for teaching her to appear to be listening when she wasn't the least bit interested. After dinner the men repaired to the library for cigars and brandy, and the ladies took coffee in the parlor.

"Which of them do you like best, Kathleen? They are all enchanted by you. You are so beautiful!" Susan observed generously.

"I've only just met them, so I really can't say as yet, Susan," Kathleen replied. "Besides, I think you are too modest. Many times I noticed them looking your direction with wistful expressions. You look lovely in that yellow dress. It contrasts so nicely with your dark hair, and sets off your grey eyes. You are a very attractive girl, Susan. Don't sell yourself short."

Susan's eyes took on a gleam of gratitude. "I hope we can be very good friends, Kathleen. I like you."

"I'd like that more than I can tell you," Kathleen responded warmly.

The men joined the ladies a short time later, and the older persons gathered at one end of the parlor while the younger set grouped around the piano for a songfest. By the end of the evening, Kathleen was beseiged with pleas to escort her to the races the next afternoon. To each she replied that she had already promised to go with Ted, but perhaps another time.

It was apparent to all that Kathleen was well on her way to becoming the new and undisputed belle of Savannah, and Amy was livid. After everyone had departed, Amy approached Kathleen and said spitefully, "You realize, of course, that you've nearly disgraced us all with that disgusting display of bosom tonight. That is the only reason the men were attracted to you. They are not used to ladies who reveal themselves so boldly. We are more discreet in Georgia, Kathleen."

Kathleen turned on her cousin with wickedly glittering green eyes. "It pays to be discreet when you have so little to display," she baited, looking pointedly at Amy's shapely, but less mature bust.

Ted smothered a burst of laughter as his sister stormed up the stairs.

Early the next morning Kathleen rode Titan to Emerald Hill. By nine o'clock she had rehearsed Zeus in all he had previously learned. She returned from the stable with a saddle. Ever so gently she eased it onto his back and stepped away to view his reaction. Surprisingly, all he did was turn to look at her. She stroked his fine neck and spoke reassuringly to him. Then she proceeded to tighten the cinch and adjust the stirrups. Taking the reins, she then led him around the paddock. Gaining courage, she unlatched the gate and led him out. They walked together down the lane and part of the way across the field and back again. Putting him into the paddock once more, Kathleen removed the saddle. She let Zeus run a bit, then whistled for him. When he came to her she saddled him again, this time with swift sure movements, and led him about for a few minutes. She repeated the procedure twice again to assure herself of his acceptance, then, pleased with her morning's work, she exercised Titan for a while and returned to Chimera to prepare for the races.

As much as Kathleen had looked forward to watching the races, she saw very little of them. There were at least a score of young men grouped around her wanting attention, hanging on her every word. All the men she had met the previous evening were there as well as more of Ted's friends and acquaintances. They clustered about like bees to honey, and as the afternoon wore on, Kathleen was fast becoming irritated. Out of all of them, only one really caught her eye. He was introduced as Gerard Ainsley. He was tall, slim, blond, a little older than the others, and more mature. Kathleen guessed him to be around twenty-two. Ted told her his father owned the bank in Savannah. When they were introduced, he kissed her hand, holding it slightly longer than was necessary, and gazed deeply at her through liquid brown eyes. Then he stepped to the edge of her throng of admirers and only now and then directed some comment to her in his deep voice. Every so often, Kathleen would glance up and catch him eyeing her as a cat does a mouse hole. Of all her would-be suitors, he alone ingrigued her.

Sunday morning found Kathleen seated next to Amy in the First Presbyterian Church of Savannah on Church Street in the Baker family pew. The pastor delivered a rather dull sermon, droning on in a monotone which nearly lulled Kathleen to sleep. Following an unhurried carriage ride back to Chimera, they dined sumptuously on fried chicken and dumplings, which left them all comfortably full and sleepy.

Rather than succumb to the urge for a nap as the others did, Kathleen donned her breeches, saddled Titan, and rode to Kate's. She alerted her grandmother that she was going to attempt to ride Zeus, and headed for the paddock, Kate hurrying along after her. From the stable she brought the saddle, bridle, and blanket. Zeus cantered up at her whistle. Tying him to the post, she saddled him. She walked him about the paddock for a few minutes, then led him back to the fence where she climbed the first two rails and eased herself into the saddle. There she sat holding her breath, waiting for Zeus to throw her from his back. The golden horse skittered sideways a few steps and shook his head about, but as Kathleen's voice came down to him, he calmed considerably. She dismounted, waited a few seconds and mounted again, this time from the ground. Her landing in the saddle was not nearly as gentle as her mount from the rail. She stroked his neck as he pranced about. This procedure was repeated until Zeus stood perfectly still with her on his back.

Now she took up the reins. Nudging him with her toe of her boot, she urged him into a walk. He responded immediately, walking slowly, as if he realized he carried a very precious burden on his back. Patiently she taught him to answer the pressure from her knees and the guide of the reins. He learned quickly. They stopped for a while at tea time, then went right back to work. When Kathleen felt confident that he understood her directives, she let him trot, then canter, and finally the two of them were galloping around the paddock. She drew him up to the fence where Kate was watching. "I did it!" she cried out with passionate pride.

"Was there ever any doubt in yer mind?" Kate queried in a proud voice.

"Today," Kathleen said truthfully. "Today I thought he would reject my weight on his back and try to throw me."

"Nay! The horse trusts ye. He knows ye are his friend. He trusts ye'll do him no harm. Ye have done today what no one else has been able to do, and ye did it better. I've watched each day from the window and saw ye winning him over with love. He's yers now, body and soul, and his proud spirit's not destroyed as it may have been if he'd been broken with force."

"He's a very intelligent horse, Kate. He learned so quickly!" she praised him.

"Aye. That he did. 'Tis in the blood and the breedin'. Still, I have no doubt that Reed could have broken him in

215

time, though not as lovingly as ye did. 'Twas a beautiful thing to watch ye! Yer grandfather would have been proud o' ye, lass,'' Kate alleged with a catch in her voice.

By the time Kathleen stabled Titan and entered Chimera, she could see everyone seated at the dining room table. She crept quietly up the outer stairway to the veranda and slipped into her room. Washing and dressing as quickly as she could, she joined the others.

"You almost missed dinner entirely,'' Amy pointed out as a serving girl placed a steaming bowl of soup before Kathleen.

"I'm sorry for being so late,'' Kathleen apologized to the others present, "but I had a project to finish. I rode the palomino today,'' she added with pride radiating from her flushed face.

"Honestly?'' Ted exclaimed in amazement. "Did he throw you?''

"Are you all right?'' Barbara inquired anxiously. "You're not hurt?''

Laughing outright in her joy, Kathleen explained, "No, he did not throw me, and I'll be perfectly fine as soon as I come down from this cloud I'm on.''

"Won't Reed have a fit now!'' Ted whistled softly.

The next week flew by for Kathleen. Each day she worked out with Zeus until she was satisfied enough with his performance by Thursday to stable him at Chimera. Monday was Kate's birthday, and Mary Taylor threw a party that evening in her honor. Kathleen presented her grandmother with one of her finished paintings of the statue of Venus. Kate loved it, exclaiming over Kathleen's talent. There were scores of people in attendance, and Kathleen met many young ladies her age and accepted several invitations to teas and galas. Kathleen was introduced to numerous people and her dance card was filled within minutes, but she chose Gerard Ainsley to escort her to dinner. The rest of the week was filled with afternoon teas and garden parties and evenings of quiet dinners at home and carriage rides about the countryside, with Gerard unobtrusive, but seemingly ever-present.

Finally Friday arrived, and all day was spent primping and preparing for the ball Barbara and Mary had planned to officially introduce Kathleen to Savannah. Everyone of importance had been invited to Chimera, and the ballroom had been opened and decorated with flowers and greenery. Servants buzzed around all day cleaning and polishing, dusting and shining until everything sparkled. The game room behind the right staircase was aired and prepared for the gentlemen's use, and the greenhouse beyond the ballroom was the source of countless bouquets and potted ferns. The cook evicted anyone who dared to trespass into her domain as she created delicious delights for the evening to come. The entire household was in a flurry of activity.

Kathleen, in an effort to escape the confusion, took herself up to the third level of the house into the long abandoned schoolroom. There Susan and Reed had learned their lessons from various governesses and tutors in years past. With her she took her small sewing basket and the green material Kate had given her. Carefully she cut the cloth, using her old riding breeches for a pattern. Next she patterned a short vest to match the trousers, and a small mask just large enough to cover her eyes and hide the curve of her cheekbones. The problem arose when she found her needles and thread too flimsy for the canvas and soft leather.

"Drat!" she swore to herself. "Don't tell me I've done all this cutting and measuring for nothing!" Crossing the hall into the sewing room, she searched everywhere for sturdier utensils, but discovered nothing to suit her need. Finally, out of desperation, she hid her material in the rear of her armoire and hurried downstairs.

There she found Ted wandering about harassing the housemaids and generally getting in everyone's way. "Ted! Leave those poor girls alone. They've work to do!" she chided him with a smile. "Have you nothing else to better occupy your time?"

"Not until you arrived. Everyone seems to be busy but me."

"Well then," Kathleen said, taking him by the arm. "I think I've solved your problem. How would you like to ride into Savannah with me?"

"I'd love to, Cinderella, but shouldn't you be getting ready for your ball? Susan and Amy have been cloistered away in their room for the past hour with all sorts of mysterious creams and lotions designed to make them soft and lovely." He grimaced playfully.

"Oh, there is plenty of time, Ted. It's only eleven, and if we go on horseback instead of taking the carriage, we can be back with time to spare. I absolutely must get some things in town, and I assure you that you will be pleasantly missed around here," she added jokingly. "Maybe a little work will be accomplished in your absence."

"I see. You are doing your good deed of the day by removing me from the premises. Very well, cousin, I'll see to the horses while you dress, but hurry. I don't want Mother blaming me if you are late for your own welcoming party."

Ten minutes later they were on the road to Savannah, Kathleen proudly atop the golden Zeus. On reaching the outskirts of the city, Kathleen headed directly for the docks. "Where are we going?" Ted queried confusedly. "I assumed you intended to shop."

"I do, Ted, after I attend to another matter first," she answered. She pulled up before an old hotel and started to dismount.

"Just a minute, Kathy! You can't go in there alone. Besides, Mother would have my hide if she even suspected I let you come to this end of town. This is no place for a lady!"

"Oh, brother!" Kathleen retorted huffily, then retreated partially. "All right. I'll stay here. You go in and ask after a Mr. Dan Shanahan. If you find him, tell him I wish to speak to him immediately."

"What about? Who is this man?"

"It is personal, Ted, and if you refuse to do it I'll have to see to the matter myself," she declared stubbornly, jutting out her chin in a determined manner.

"Oh, all right," he grumbled. "I just hope you know what you are doing."

The next few minutes Kathleen spent uncomfortably perched upon Zeus deliberately ignoring sly looks and catcalls from several grubby men and a couple of rather blowsy women from one of the nearby brothels. She breathed a sigh of relief when Ted appeared at last, followed by a grumpy-looking Dan. Dismounting, she tossed her reins to Ted and pulled Dan aside. Quickly she gave him her instructions, and before Ted had time to collect his thoughts they were mounted once more and riding into a more respectable section of town.

For the next three hours they scurried from one elite shop to the next as Kathleen chose dress patterns and materials, was measured and remeasured for outfits. Deftly, she explained styles she wanted created especially to suit her, and the colors and fabric she desired. By offering bonuses to already staggering prices, Kathleen obtained promises of swift completion and delivery of most of her purchases. Not only did she order several new gowns and the accessories to match, but also a beautiful moss green velvet cloak and bonnet for cooler weather. She left with hasty promises to return in two weeks for fittings.

As she and Ted headed toward her aunt's home on Oglethorpe Square, Ted exclaimed, "I don't believe I've ever seen anything like it before! In just three hours you have purchased what would have taken Mother and Amy a whole week to select, and most of the designs were unique, so especially you. I'm very impressed, Kathy! I didn't think women could make up their minds so quickly or be so—so organized."

"Thank you. I'll take that as a compliment. You know, Ted, contrary to popular opinion, women do have quick minds and the ability to use them," Kathleen answered indulgently.

"I didn't mean to imply that they don't," he said hastily, "Only that they don't often show it. At least not in front of me."

"Yes, I suppose you are right," she conceded. "Many

219

men would be scared off by any hint of brains in a woman. Silly, isn't it? They expect women to be beautiful but dumb. One would think they would value intelligence in a mate."

"Some do. Especially the men who are intelligent enough themselves not to worry that their wives might be smarter than they are." Ted snickered.

"I like the way you think, cousin!" Kathleen said with a laugh.

Dan was waiting for them outside the gate when they arrived at Aunt Barbara's. He approached Kathleen with a bundle in his arms. "Got the stuff ye asked fer," he grunted.

"All that?" Kathleen exclaimed, eyeing the bundle.

"Found some other things, too. There's a note inside thet explains," he said shortly, eyeing Ted suspiciously.

Kathleen nodded. "Thank you, Dan. I appreciate the trouble you took. By the way, how are things going for you?"

"Jest fine, Miss. Don't ye worry 'bout nothin'," he said with a wink, and walked off whistling.

"What was that all about?" Ted asked with a frown.

"Nothing to concern yourself over," Kathleen replied casually. "Come on. We'd best head back to Chimera before Barbara and Mary start to panic."

The ride to Savannah and back had made her hot and dusty and wet with perspiration, and Kathleen was in no mood for what she found awaiting her at Chimera. Both Barbara and Mary were irritated with her, and Mammy's frown lines were clearly visible.

"How could you think of going off today, of all days!" Barbara scolded. "You'll never be ready in time!"

"Oh, posh! Barbara, I have plenty of time. If need be, I could be dressed and down in half an hour. Such a fuss over nothing!" With an exasperated sigh Kathleen mounted the stairs to her suite.

Yet another encounter awaited her when she entered her bedroom. There at the dresser stood Amy, calmly rifling through Kathleen's jewel case. Kathleen tossed her bundle on the bed. "Just what do you think you are doing?" she demanded, her cool voice laced with barely leashed fury.

"Why, cousin dear, I was hoping to borrow a few of your baubles for this evening, of course," Amy answered in a honeyed tone. "I was sure you wouldn't mind."

"I would not mind half so much if you had the decency to ask me first. Now, if you don't mind leaving, I'd like to start my bath." Kathleen was already unbuttoning her riding habit.

"Oh, so there it is," Amy announced, catching Kathleen's left hand. "I was looking for this ring. It would go so nicely with the emerald ear bobs."

Snatching her hand away, Kathleen headed for the water closet. "This, of all my jewels, is especially not for loan, Amy. It never leaves my hand. In fact, there are several of my pieces I am overly fond of and would not think of lending."

"My, my! I'd always heard it was the Scots who were so tight-fisted and stingy! Perhaps that goes for the Irish, too," Amy said sarcastically. "Oh, well, never mind. I will have all the jewels I'll ever need in a few months anyway. I suppose I can wait." Amy paused for the effect and added, "I'm sure Reed will buy me as many rings as I desire after we are married." Amy tossed her blond ringlets haughtily.

Kathleen's eyes narrowed slightly as she studied her cousin. "Married? I didn't realize the two of you were even engaged. Captain Taylor never mentioned it," she said evenly.

"Oh, yes! Of course it is not official, you understand, but there has been sort of an understanding between our families for years. I expect he'll give me a ring by Christmas and we'll announce the date."

"Really? I was under the impression that the captain had thus far shown no interest in any of the women around here. Are you so sure he's ready to settle down?" she said slyly, green eyes flashing.

"Now you sound like some of the others around here

who insist that Reed is not the marrying type!''

"Oh, no, Amy. I'm not suggesting that at all. In fact, it wouldn't surprise me if he has a wife stashed away somewhere already!'' Kathleen smiled inwardly.

"What a silly thing to say!'' Amy exploded.

"Well, he *is* a seafaring man, dear, and they are notorious for the way they carry on in ports around the world,'' Kathleen needled.

"That's ridiculous! Besides, Reed is mine. Do you hear? Mine! And you keep your grubby Irish hands away from him! You may have all the other men in the county charmed, but Reed would see right through you, Kathleen. He'd see right away that you are nothing more than a little Irish potato farmer, a red-haired baggage like your mother!''

Kathleen, who had wandered over and was leaning against her desk, reacted instantly and instinctively. In a flash of movement she had snatched up a letter opener that lay atop the desk, and now stood with the point pressed firmly against Amy's throat.

"Never—I repeat, never—say anything vile about my mother or her family again!'' Kathleen growled in a low, menacing tone. "She was more of a lady than you will ever think of becoming. You are nothing but a spoiled brat with a vicious tongue which you had better learn to control before someone else does it for you!''

As Kathleen was talking, Amy backed all the way to the door, eyes wide with fear. Kathleen lowered the letter opener and stepped aside. "Might I further suggest that you leave my rooms now and never enter again unless I request it.'' She motioned toward the door.

Amy felt behind her for the doorknob. "You are as crazy as a loon!'' she exclaimed excitedly. "You may be sure I will tell Reed what you have done!''

"You just do that, Amy,'' Kathleen said with a wry grin. "He'll probably wish I had at least removed your voice box. You screech like an old owl! I have no idea what he might see in you, except perhaps your pretty little face, which I shall ruin if you ever malign my relatives again.''

Without waiting for any comment, she continued, "And one more thing. Should I decide to set my cap for Reed Taylor, I shall do so, and neither you nor anyone else shall stop me!" Returning to the desk, she replaced the small weapon. "Do we understand each other, Amy, *dear?*"

"One thing I understand for sure is that you are looney! They should lock you away! I pity any man who marries you, Kathleen!"

Kathleen threw back her tawny head and laughed aloud as she thought of Reed and all the problems she had already brought him—and those she intended to cause in the near future. "There, at least, we agree, cousin!"

Amy gave her a queer look and departed hastily, muttering under her breath about insanity being hereditary, and hoping fervently if wasn't prevalant on her mother's side of the family.

While the servants filled her tub, Kathleen unwrapped her bundle from Dan. The first thing her eyes lit on was a pair of brand new black leather boots. With them lay a wide black belt with a huge gold-ornamental buckle. She ran her hand along the highly polished leather and let out a low whistle. They were beautiful! She pulled on the soft, supple boots, and they fit perfectly, coming to her knees.

She unfolded the note Dan had included and read quickly.

Cap'n. Found thees abored the Starbright *in yer quarters. They arrived at the estate after we sailed. Yer Pa ordered 'em months ago fer ye. He said ye needed some decent boots ta sail in so ye didn't get soked and wet or brake yer neck. Kirby had 'em sent on the next ship out. Dan*

"Papa. Dear, thoughtful Papa," Kathleen choked on the whispered words as the tears coursed down her face. "Always looking out for me. How I miss you!"

Hearing Mammy's heavy tread in the hallway, Kathleen scurried to the armoire. She pulled off the boots and hid them and the belt at the back of the highest shelf with her green material. Also in the bundle were needles and heavy-corded thread used in sewing up the sails. These she had requsted of Dan to sew up her costume. Reasonably sure they were safe from view, she hurried to her bath.

Chapter 15

THE ballroom sparkled, but so did Kathleen that evening. Dressed in a leaf-green taffeta gown with silver threads interwoven into the fabric, each movement brought a shimmer of light. The bodice was snug, cut very low and devoid of any lace or trim, and caught beneath the bosom in the high waisted style that was so popular in Europe. The gown seemed to flow to the floor from the raised waistline. There were no actual sleeves, as the dress was styled to be worn off the shoulder. She wore only an emerald pendant and ear bobs to set off the gown. Long kid gloves were her only other accessory, and the elegant simplicity of her ensemble was stunning.

Mammy had coaxed her hair into an intricate coiffure of braids and waves, leaving a few soft curls to frame her face, and one long curl to fall over her left shoulder.

True to her word, Kathleen was ready and standing next to Barbara in the receiving line as the first guests arrived. She was surprised at the number of people she had not previously met, and it embarrassed her slightly that Barbara insisted on introducing her as Lady Haley. Her dance card became an impossible jumble of names, and soon she had to resort to promises of half a dance if she could manage it. Barbara laughed and told her, "Your slippers won't last the evening, Kathleen."

"My feet probably won't either," Kathleen predicted.

Kathleen danced the opening cotillion with her Uncle

William and the next dance with Ted. Early in the evening she found herself rapidly becoming annoyed with Gerard Ainsley, who seemed to assume an attitude of superiority and possessiveness where she was concerned. By his actions he was openly telling everyone present that he was staking his claim on her. True, he was one of the most handsome and mature of her partners, but Kathleen was determined not to let him become too dominating. She found herself avoiding him when possible, and bestowing her most charming smiles on her other suitors. She remained very cool and aloof toward him, but it only served to aggravate the situation and make him more ardent.

"I've got to do something about this Don Juan before things get out of hand," she thought grimly as he claimed her for the next dance.

They were but a few steps into the dance, with Kathleen trying valiantly not to let Ainsley hold her as closely as he would have liked, when she felt compelled to glance toward the open veranda doors. Suddenly her breath caught in her throat, for there in the doorway, looking impossibly handsome and elegant in black evening attire, was Reed. His icy blue gaze rested on her for a moment before he flashed her a dazzling smile. Then he was gone from view as Ainsley whirled her about the floor.

Kathleen felt shaken, and her knees seemed to have turned to jelly. Her cheeks flushed with color as she struggled to regain her composure. Ainsley, taking advantage of her momentary lack of concentration, tightened his hold about her waist. In his conceit, he mistakenly contributed her confusion as a yielding to his charm. So confident was he, that by the time the music ended he had maneuvered her near the veranda. Taking her arm in his, Ainsley propelled her toward the doorway.

Kathleen came to herself with a jolt. At once she stiffened, withdrawing her arm. "Mr. Ainsley, I do not recall requesting to go outside," she said coldly, her eyes flashing in anger.

"Come now, Kathleen. It is overly warm in here and you are quite flushed. Some fresh air would do you no harm."

"Neither do I recall giving you leave to use my given name, sir, and I can do without the fresh air. You are entirely too fresh as it is," she continued sternly.

"Just for a moment, surely," he argued politely, taking a firm grip on her elbow.

"I said no!" Kathleen almost shouted at him.

"Release the lady's arm, Ainsley," came the terse command from behind them.

Turning to greet the intruder, Ainsley suddenly became very amiable. Thrusting out his hand, he said, "Hello, Reed. I haven't seen you for some time. When did you get home?"

"Just in time to rescue a damsel in distress, it seems," Reed stated with a slight bow to Kathleen. Their eyes met and held for a long second before Ainsley broke the spell.

"Oh! Forgive my lack of manners. Lady Haley, may I present Mr. Reed Taylor."

"We've met, thank you," Kathleen said, presenting her hand to Reed.

With an amused quirk of his dark eyebrow, Reed lifted her hand to his lips. "You look ravishing this evening, Kathleen." He grinned down at her.

"Thank you, Captain Taylor. You look very dashing yourself, as you well know," she countered with an easy smile.

"How is it you two know each other?" Ainsley interrupted, attempting unsuccessfully to place his arm around Kathleen's waist.

Reed's eyes narrowed slightly as he took in the movement, and Kathleen found herself holding her breath momentarily.

"I'd say we know each other rather well indeed, wouldn't you, Kathleen?" His gaze shifted to her, resting lightly on her exposed bosom, then traveling downward and gradually up again, silently undressing her with his eyes.

Kathleen blushed under his appraisal, but rallied quickly. Tossing her head, she looked coolly up at him and said, "Captain Taylor, you are much too bold, and you still have not answered Mr. Ainsley's question." Turning to Ainsley, she went on. "It is Captain Taylor's ship I took from Ireland. Now, if you two gentlemen will excuse me, I believe I promised this dance to Mr. Turnbill." With that, Kathleen walked calmly away.

Kate found her several minutes later in her sitting room. "Thought I might find ye here. Do ye intend on hiding up here all evening? Barbara noticed ye were missin' when Jerry Turnbill couldn't find his dance partner."

"He's here," Kathleen whispered, holding her head in her hands.

"If 'tis Reed ye're referring to, yes, I know he's here. 'Tis no great surprise, love, not to me, and surely not to ye. Ye knew he would be back sometime. After all, he lives here, and so, incidentally, does his wife."

"He's going to make things difficult, Kate. I don't think I'm up to handling it yet."

"If ye counted on Reed takin' this lying down, ye judged him wrong, lass. That's not his way."

"I know," Kathleen sighed. "I guess I just hoped he would stay gone longer." She stood and straightened her dress. "You're right, Kate. I can't avoid him forever. I'll just have to meet the devil on his own ground this time and hope I don't lose too badly."

"I'll walk down with ye," Kate offered. "Now smile and throw back yer shoulders. Don't admit defeat before the battle starts, lamb, or ye'll give him the edge."

"The sharp edge of my tongue is all he'll get willingly from me," Kathleen vowed with a wry smile.

Reed was waiting for her at the bottom of the staircase. "I was hoping for a dance with the belle of the ball," he said, catching her hand and leading her toward the dance floor.

"Oh, but I've promised them all already," Kathleen objected, pulling back.

Leaning his dark head down to hers, he whispered, "Ah, but I've never danced with my wife yet, and I'm not about to let this opportunity slip away." His hand tightened warningly on her arm as he guided her along.

"You are bruising my arm, Reed," she grated out be-

228

tween clenched teeth as she continued to smile for the benefit of those who might be watching.

"I'll bruise more than that if you don't behave," he answered tersely.

As they started to dance, he commented, "How nice. A waltz. Appropriate for our first dance together, don't you think?" When she made no comment, he drew her closer to him. Kathleen tried to draw away again, but his arm about his waist was like a steel vise. She turned stormy eyes up to his and he laughed aloud, then said quietly, "I am anxious to see if having you dancing with me is as interesting as having you dance for me."

Still she refused to speak. "Oh, surely you remember, my pet," he continued. "Surely you must recall how the evening ended—or should I say how the next day dawned."

"Must you remind me?" Kathleen asked in a low voice that she could not keep from quivering slightly.

"Could it be you miss me?" Reed went on in his taunting tone, bending close to her ear. "Do you miss having me to warm you bed and your body? Do you enjoy playing the innocent virgin, or does your body betray you when you are alone at night? Don't you long to be kissed and caressed and made love to till dawn?" His voice had itself become a caress by now and she trembled in his arms.

"Please stop it, Reed," she half pleaded, half demanded.

"Must I remind you to smile, my love? People are watching."

"Damn you!" she replied softly as she threw him the required smile. "If they are watching us, it is most likely because you are holding me more closely than is acceptable!"

As the music stopped, they were standing in the center of the ballroom. Keeping one long arm firmly around her waist, Reed reached into his pocket. "I brought you a gift, my love. A small token of remembrance, you might say." He produced a glimmering band of emeralds outlined in diamonds.

At first Kathleen thought it was a bracelet, and almost at once realized it was too long.

"Do you like it?" Reed was asking.

"It's beautiful, Reed," she breathed, "but what is it?"

He released his hold on her waist to place both hands behind her head and snap the band about her throat. "It is a necklace of sorts. The jeweler in New Orleans called it a collar."

"Please, Reed. Don't tease me like this!" Kathleen begged, looking up at him with wide green eyes. "You know I cannot accept it, no matter how much I love it. What will people say? What will they think? You are creating a scandal right now!"

"I have every right to give my wife jewels if I so desire," he countered, his eyes gleaming with malice.

"You are doing this on purpose, you beast!" she accused hotly. Her hands went up to unclasp the collar, but he grasped both wrists tightly.

"No! Leave it, Kat!" His look dared her to defy him.

"I can't accept it, and well you know it! And I will not back down and reveal our marriage just to keep a few jewels, Reed Taylor! You had better remember our agreement, too, if you know what is good for you!"

"Never fear, my precious. I promised not to tell anyone of our marriage. However, if you will recall, I never said I would keep quiet about our intimacies." At her quick gasp, he leaned forward and continued in a menacing tone, "If you do not keep the necklace and any other gifts I decide to bestow on you, I assure you that I will leave no doubts in anyone's mind as to how well I know your sweet body, including the shape and location of your birthmark."

"You—you cad!" she sputtered, and finally she snarled, "and just how am I supposed to explain this?"

"Well, sweetheart, they are your jewels. They are from the cache you recovered from that sunken ship in the bay. I merely had them set in the mounting for you. That should solve your dilemma a bit." He grinned down at her devilishly and added, "this time."

"Well why didn't you say so to start with, you Yankee devil!" she spouted, eyes flashing. "No! You have to put me through hell's own fire first, don't you?"

"Not so loud, my love," he cautioned. "Besides, this is just the first of many gifts to come. You see, Kat, I intend to court you openly, and God help the man who tries to stand in my way!"

"I don't want you to court me. All I want you to do is leave me alone," she groaned.

"There is really very little you can do about it, Kat, unless you set so little value on your reputation, so don't try refusing to see me or to accept my gifts."

"That's blackmail!"

"Precisely, my dear! How does it feel?" His eyes shot blue sparks as he finally released her wrists. "Now, we seem to have created quite a spectacle, so take my arm, smile warmly up at me, and we'll walk over to where Barbara is standing and looking very pale about now."

Kathleen felt as if she were running the gauntlet as they made their way off the dance floor. To her credit, she held her head high, smiling sweetly up at Reed. As he looked down at her, Reed thought her flashing green eyes nearly outshone her jewels in her anger. He leaned down and whispered, "You are really such a splendid actress, my love. You should have been on the stage."

"You are insufferable!" she hissed.

Mary, Barbara, and Kate were all standing together, Kate being the only one who did not look as if she were about to faint.

"Reed, dear, would you please explain what in the world is going on?" Mary insisted in an urgent whisper.

"I suppose, Mother, you are referring to the jewelry I just gave Kathleen?" he drawled.

"You suppose correctly, son."

"Does this mean you are engaged?" Ted interjected, popping up behind Kathleen.

"Heavens no!" Kathleen exploded.

Taking Kathleen's arm, Barbara admonished, "You really cannot mean to accept such a lavish gift, Kathleen.

And you, Reed," she continued, turning anxious eyes on him, "should know better than to create such a spectacle!"

"Please!" Kathleen cut in, holding up her hands for silence. "There is a logical explanation, I assure you. The jewels are mine, so of course I mean to keep them. Captain Taylor merely had them reset for me in a new mounting. He is acquainted with a jeweler in New Orleans, and was kind enough to take them there for me."

"See, Mother," Amy said in a disgusted tone. "I told you it was not as it appeared. Why would Reed give such an expensive item to a girl he barely knows?"

"Well, I must say this jeweler must be an artist in his field," Kate contributed. "I have never seen such a unique piece. What do ye call it?"

"It is a collar, my dear lady," Reed answered. "I must tell you that the design was my idea though the craftsmanship his."

Susan, who had come up quietly to join the group, asked, "It is magnificent, but why a collar? What put that idea into your head?"

"What could be more natural?" Reed answered with a laugh. "Here, have a look!" Before she had time to react, he placed his palm below Kathleen's jaw, tilting her head up. "What do you see here but the slanting green eyes of a feline?"

His smile widened as he continued, "Her name even fits, and it just struck me that every Cat needs a collar."

Jerking her face away, Kathleen whirled on him, her emerald eyes shooting flames. "That may be true, Captain Taylor, but you should be cautioned. Cats are very unpredictable creatures, and even the smallest kitten has sharp claws. Remember, sir, that no cat can ever truly be tamed."

"That is yet to be seen," Reed countered calmly.

"This conversation is beginning to bore me." Placing her hand on Ted's sleeve, Kathleen requested sweetly, "Teddy, dearest, would you mind taking me out for some air?"

"Certainly, Kathy. You know I am your most devoted slave," Ted replied, giving her a small bow.

They were halfway to the door when Reed called, "Kat!" Without thinking, Kathleen responded immediately to the accustomed nickname, only realizing her mistake after she had turned to face him. She felt the blood rush to her face as she met his triumphant smile and saw the wicked gleam in his ice-blue eyes. She sensed, rather than saw, several couples staring at them as Reed approached her. Calmly he handed her dance card to her, saying with a wry chuckle, "I believe you've lost something, Kat."

Pulling her tattered dignity about her, she replied with a nod. "The battle, perhaps, Reed, but not the war."

A little while later, Kathleen found herself holding court on the veranda. She was surrounded by several young men, each clamoring to be the one to escort her to the horse track the next day, or the barbecue at the Courtland plantation on Sunday. Inwardly, she was still smarting from her encounter with Reed, but no one seemed to notice. She laughed gaily at their antics as they coaxed and teased her. Ted, seeing himself outnumbered, went in search of Susan, chuckling to himself at his cousin's prowess. Kathleen had them eating from her palm as she turned her charm first on one and then another hopeful suitor. Fluttering her long lashes and smiling sweetly, she flirted outrageously with all of them, except poor Gerard Ainsley, whom she chose to ignore. It still rankled her that he would assume so much.

She stood with her back to the doorway, laughing up into the face of her nearest admirer. "Please, Mr. Godsey. Surely you realize that I am just one person. I cannot attend the races nor the barbecue with all of you. That would be entirely too selfish of me to deprive the other young ladies of their escorts."

"Yes, indeed, gentlemen," came a familiar deep voice from behind her. "Besides, Miss Haley has already agreed to go with me, haven't you, Kathleen?"

As she started to turn around, she felt a hand around

her ribcage, tightening in a silent warning. She hesitated for just a second and his long, strong fingers squeezed tighter until she felt her ribs would crack. She knew that to the others it looked as if he had simply placed his hand at her side; that they did not realize the hold he had on her. She gasped slightly, and Reed bent his dark head to hers.

"Did I hear you say something, sugar?" he prompted.

"Yes! I said yes, I have promised to go with you to the races," she said, trying to keep the pain from her voice.

"And to the barbecue on Sunday," Reed went on.

"That, too."

Releasing his grip on her ribs, he nevertheless let his hand remain at her side. "Well, gentlemen, I'm sorry. Perhaps another time." Firmly he guided her inside, putting an end to further conversation. He forced her to dance once more with him, and then seemed willing to ignore her for the remainder of the evening.

After that final dance with Reed, Kathleen developed a raging headache. Finally she had to make her excuses to her hostesses, and headed gratefully for her bedroom. Mammy helped her to undress and brought her a cold cloth for her brow.

"Shore do hope yoah not comin' down wif nuthin', Miss Kaffleen," she worried as she brushed out the copper tresses.

"No, no, Mammy. Don't worry. It's just a headache, I'm sure. I'll be fine after a good night's sleep."

"Yo wants me ta stay wif ya awhile, chile?"

"There's no need. I'm just going to try to sleep. You go on now. Amy and Barbara will need you later."

"Yo shore, honey?"

"I'm sure, Mammy. Go along now. I'm fine," Kathleen assured her tiredly. She closed her eyes, willing herself not to think of the disastrous evening now behind her, and fell into a restless sleep.

Kathleen did not know what had caused her to awaken. The house was quiet, and she realized thankfully that her headache was gone. The lamp still burned low outside her curtained bed. The night air was sultry, and her nightgown clung to her perspiring body. "It is going to storm soon," she thought groggily as she watched a flash of lightning streak across the sky.

Suddenly she froze, hardly daring to breathe. It was then she realized what had awakened her. Someone else was in her room. She felt his presence even though she could not see him. She sensed rather than heard a movement across the room, near her desk. So tense was she that she nearly screamed when he spoke.

"No need for alarm, Kat. It's only me," Reed drawled.

"Only you!" she cried out breathlessly. "What do you mean by stealing into my room in the middle of the night!"

"Come now, kitten. After all the other nights we've spent together, you are not going to object at this late date, are you?" He walked slowly to the end of the bed and drew back the sheer curtains. Vaguely her mind registered the fact that he had changed from his evening clothes into his dressing robe.

"Get out of here, you lousy conniving cur!" she stormed as she sprang from the bed. "Get out of my room and get out of my life!"

"Not on a bet, my love," he replied evenly. It was only then that she noticed the hard set of his jaw and the steel-blue gleam in his eyes.

"Wha—what do you want now?" she stammered, backing away from him as he stalked her.

"I think you know the answer to that."

"No. You can't! We made a pact!"

"This was not included in our agreement, Kat. I agreed to deposit you at your aunt's house and to say nothing of our marriage. I most assuredly did not agree to forfeit my conjugal rights." His hand flashed out and captured her wrist.

Kathleen's eyes glistened with spite as she snarled at him, "What a royal rogue you are, Reed Taylor! You are a blackguard and a filthy pirate, and a rutting jackass in the bargain! I despise you! If I never set eyes on you again, it would be too soon, you brute! You uncivilized barbarian!"

"Are you quite finished?" he growled as he pulled her

hard against him.

"I could go on into next week and still have not used all the epithets I could think of for you, you insidious rake!" she spat out hatefully.

"Then I can think of only one way to shut you up, you vicious-tongued Irish viper." Abruptly he tangled his hand in her hair and held her head motionless as his lips came down hard and demanding upon hers. She squirmed in his arms as he scooped her up and carried her to the bed. He threw her down on the mattress, her struggling body pinned beneath his, as he continued to silence her with his kiss. His lips and tongue ravaged her mouth even as his hands disposed of her nightgown and sought out her most sensitive areas.

She felt a coil of heat start in her stomach and fan out to encompass her entire being in a delirious burning desire. Suddenly it no longer mattered that this man was her enemy. She only knew she loved him and wanted him with a need bordering on desperation. Her muffled oaths became whimpered pleas as her arms found their way about his neck and drew him nearer. She arched her body to meet his caresses as his lips traced a burning path along her body. Her need overcame her pride, and with trembling fingers, she untied the belt of his robe and returned his kisses, his caresses.

As he mounted her, he bent to kiss her lips, and she saw her desire reflected in his smoky blue eyes. "Damn you, Reed," she whispered, and even to her ears it sounded like an endearment.

He took her with a savage urgency and she met him thrust for thrust. She stilled her outcries against his shoulder and raked his back with her nails as they reached a mind-shattering climax and the sweetest release she could recall.

The last thing she heard as she fell asleep, her head cradled on his shoulder, was his laughing retort, "I love the way you hate me, kitten."

Chapter 16

KATHLEEN awoke the next morning feeling rested and revived. She stretched languidly for a moment and breathed deeply of the morning air. It smelled fresh and clean, as it always does after a storm. Drawing back the bed hangings, she saw droplets of rain still dripping from the leaves and sparkling in the morning sun.

Recalling Reed's visit to her room the night before, she knew she should be angry with him, but in all fairness she could not deny that she had wanted him as badly as he had her. She smiled to herself as she wondered what Amy would say if she knew. Her smile disappeared as swiftly as it came when she thought of Mammy finding out. Hurriedly she leaped out of bed and fluffed up the pillows and made the bed. Scouring the room for signs of his having been there, she discovered his cigar in the ceramic tray on her desk and quickly discarded the remains out the window into the flower garden below. "The careless fool!" she grumbled to herself, and then giggled foolishly.

She washed and dressed quickly in a beige riding habit, and pulled her hair back into one long braid which she carelessly tossed across her shoulder. Studying herself in the mirror, she saw lively green eyes shining out of a face full of healthy color. Her cheeks looked flushed and her lips rosy, and she knew it was mostly caused by Reed's scratchy beard on her cheeks and his demanding lips on hers just a handful of hours ago.

She was in high spirits as she headed down to breakfast. She passed Mammy in the hall. "Hello, Mammy dear," she trilled. "No need to attend me this morning. As you see, I am already dressed and wanting my breakfast."

"Yoah headache mus' be bettah," Mammy replied with a questioning look. "Yo shoh are chipper dis mo'nin', miss."

"I feel just marvelous, thank you," Kathleen replied with a silly curtsy and a laugh, and nearly skipped down the stairs, resisting the crazy urge to slide down the banister.

As she entered the dining room, Kathleen saw Reed already seated at the head of the table talking with his mother and Ted and Uncle William. "Good morning, all," she chirped as she slipped quickly into the empty chair to Reed's right.

"My! You're in high spirits this morning, Kathy," Ted greeted her.

"I take it your headache is gone, Kathleen dear," Mary said with a smile.

"Oh, yes. I'd nearly forgotten it," Kathleen replied, reaching for a biscuit.

"Going riding this morning, Kat?" Reed asked, leaning back in his chair and eyeing her casually.

"Yes, I thought I would." She nodded to the young serving girl who poured her coffee. "I think I'll have an egg, sausage, potatoes, and a small glass of juice, please."

"You must be hungry this morning, Kathleen," William commented.

"Very," she answered, casting a quick glance at Reed who sat studying her.

"How do you like Chimera now that you've been here awhile?" he asked.

"It's beautiful! But of course you already know that," she said, sipping her hot coffee carefully.

"I'm glad you like it."

Amy entered the room with Susan just then and gave Kathleen a withering look when she saw her seated next to Reed. Kathleen smiled into the plate placed before her.

"Good morning, girls. Did you sleep well?" Mary inquired.

"Very well, thank you, Mother," Susan said, crossing to give her mother and brother a quick kiss. "Good morning, Kathleen."

"Good morning, Susan," Kathleen replied, and then leaned to look down the table at Amy. "Good morning, cousin."

"Is it? I hadn't noticed," Amy retorted, shooting her a spiteful look. "I see you are dressed for riding this morning. Which horse are you riding, Titan or your palomino?"

Kathleen nearly choked on her egg as William growled, "Amy! Order your breakfast!"

She shot a quick look toward Reed, who lifted his dark brows at her.

"I've already heard all about it, Amy," he commented, his steady blue gaze never leaving Kathleen's flushed face. "Congratulations, Kat. You never fail to amaze me."

"Are you very angry?" she asked in a small voice.

"Let us say I am slightly put out, but not overly surprised." Leaning closer to her, he added with a scowl, "It seems not even my animals are safe from your charms. That *is* how you did it, isn't it, Kat? A wee bit of your Irish witchcraft?" he taunted.

"Of course," she told him flippantly, refusing to be baited into anger. "How else could I have broken Zeus when you had already failed."

"Zeus, is it? Well, I should like to see how well you have trained your golden god. May I join you on your ride and we'll discuss this further? If—" he hesitated, giving her an icy glare, "my own horse will allow me on his back any longer."

"Be my guest, Captain Taylor," she smiled sweetly.

"Ah, so I'm back to being 'Captain Taylor,' I see," he said with a smirk. "How fickle of you, my dear, especially after last night."

Kathleen glared at him, her emerald eyes narrowed warningly. "Whatever do you mean by that, sir? What

239

about last night?''

He gave a brusque laugh and enumerated, ''Why, I gave you jewels and compliments, saved you from the lecherous Mr. Ainsley, and provided you with very interesting conversation and entertainment, and if Ted will be my witness, I am sure I heard you call me Reed last evening.''

Forgetting the others at the table, and rising to the bait, Kathleen proclaimed waspishly, ''Your conceit is showing, *Reed*. First of all, the jewels were mine to start with, and compliments come as easily to you as do barbed comments. I am capable of handling Mr. Ainsley without your aid, and your constant attentions gave me a horrendous headache!''

''All true, but you have once again forgotten yourself and called me Reed before all these rapt witnesses.'' Reed waved his hand to include the others in the dining room.

Angry, embarrassed, and thoroughly fed up, Kathleen pushed back her chair and stood. *''Touché,''* she said tersely. ''I'll meet you in the stableyard if you are there by the time I am ready to leave, *Reed.''* She stressed his name forcefully. Turning to the others, she said, ''Excuse me, please,'' and swiftly left the room.

As it happened, Reed was waiting for her when she finally entered the stableyard looking very pale and shaken.

''Kat. What's wrong? All the color has left your face!'' Reed came forward and gathered her into his strong arms, his own face full of concern.

''I know. It went at about the same time my breakfast left my stomach,'' she complained irritably. ''Perhaps you should not stand so near. I may be coming down with something.''

Reed placed his warm lips to her forehead. ''You have no fever, love,'' he said with relief.

''I could have told you that, had you asked me. However, this is the fourth time this week that this has happened. I've always been so healthy that it would worry me, except that once I vomit I am fine for the rest of the day,'' she explained.

"Oh?" He surveyed her speculatively for a moment, then smiled at her tenderly. "Would you rather not ride this morning?"

"I always ride when I can, and I'm fine now, so let's get going," she announced testily.

"By God, but you can be shrewish!" he exclaimed with a grin as he boosted her into the saddle. Deliberately he brushed his arm across her breast, and his smile widened knowingly as he saw her wince.

"Sorry. I didn't mean to hurt you. Are you sore, kitten?"

"Not that it's any of your business, Reed Taylor." As quickly as her bad mood had come, it left. She leaned down and kissed him full on the lips, a long, searching kiss. Then she straightened, wrinkled her nose at him impishly, and quipped, "I must be having growing pains."

"I'm sure of it," he grinned as he mounted his horse.

They rode companionably about the estate. Every so often Reed would dismount to inspect a field or check on the progress of his crops. The cotton fields were full of working slaves and the rice fields were maturing nicely. Later, he told her, he would sit down with the overseer and go over the books. Records were dutifully kept, not only for tax purposes, but to make sure the overseer was not led into temptation's way. As they were riding back, he handed her an envelope. When she opened it, she found it filled with money.

"Reed, what on earth is this?"

"It is money, of course. Haven't you ever seen any before?"

"Yes, of course, but why are you giving this to me?"

"Darling wife," he said with an exaggerated sigh, "I am responsible for your keep, and I do not relish the idea of my wife running around in public in rags for lack of a proper wardrobe. Use the money to buy whatever you find necessary. I thought you would appreciate the fact that I intend to continue to support you, and I did make a point of not giving you the envelope in front of anyone else."

She was instantly contrite. "Thank you, darling. I do appreciate your thoughtfulness." She touched his sleeve with her fingertips.

"Not enough to tell everyone we are married," he grumbled with a scowl.

"Stop it, Reed. We've been over all this before, and it's too beautiful a day to spend it arguing."

"I don't understand why you can't trust me, Kat." He turned his piercing blue eyes to hers, and just for a second she thought she saw pain in them. Then they were hard again.

"I've tried, Reed, but each time I do, something else happens. I can't afford the prices you extract," she said softly.

"Ah, well." He shrugged, "Perhaps someday." He smiled to himself as he thought of her ailments, "Someday soon she'll have to divulge our secret out of necessity if not out of love. I can wait."

They rode into the stable yard and he helped her dismount. His arm casually about her waist, they walked to the house. "I almost forgot to tell you, Kat. I saw Charles and Eleanore in New Orleans. He is going up to Boston to a doctor's seminar, so I invited them to stop here on the way. They should arrive Monday."

"Oh, Reed! How wonderful! I have missed Eleanore so."

"Charles can only stay a week, but if you like, Eleanore could stay on here while he goes to Boston." He smiled down on her glowing face and paused to brush a tuft of straying copper hair from her cheek.

"But how can I explain knowing Eleanore when it is you who have invited them?" She worried her lip with the tips of her small white teeth.

"We could say that I looked up my old friend Charles while I was in New Orleans and when I just happened to mention your name, Eleanore revealed that she knew you from England. Perhaps she could have met you while you

242

were in school there.''

"That's quite a shaky coincidence, wouldn't you say?" she asked skeptically.

"Perhaps, but who's to know for sure. You do want her company, don't you?" he inquired, peering down at her.

"Oh, yes! How kind of you to think of it." She smiled happily up at him, her emerald eyes glowing warmly.

"You can thank me properly later, wench," he suggested, his white teeth flashing in a roguish grin.

As she started up the stairs, his voice floated up to her. "Kat, change into something bright and cheery this afternoon, will you?" His voice carried a mysterious note to it, and she eyed him curiously before she nodded.

"If you wish."

The gown was lying on her bed when she entered the room. It was a beautiful muted orange, not quite a melon color, edged with antique ivory lace. Next to the gown lay a small pile of sheer, lacy French undergarments that brought a rosy blush to her cheeks. A delicate ivory fan, lace gloves, and matching slippers lay atop the gown. As she picked up the slippers to move them aside, her fingers touched something rough. Inside one of the slippers lay another collar, this one made entirely of large, delicately carved coral roses. The other slipper held a pair of matching ear bobs and a slim bracelet. "He never gives up, does he?" she mused tenderly, her eyes misting into emerald pools. "Incurable rogue!"

From the time they all piled into the open carriage, Kathleen could tell she was going to have trouble from Amy. Kate went along as a chaperone for the girls, so there were six of them including Ted and Susan. After one look at Amy, Kate whispered cattily to Kathleen, "I wouldn't miss this for the world!"

Kate maneuvered the seating arrangements so that Amy was between herself and Susan, and Kathleen between the two men. All the way to the track Reed deliberately cast his eyes from time to time down Kathleen's low-cut bodice, and Amy was becoming more livid with each mile.

"I hope you freckle, or better yet, burn, Kathleen."

She glowered nastily. "Then perhaps you'd learn to carry a parasol like the rest of us."

"But I'm *not* like the rest of you, cousin Amy," Kathleen replied, quite unruffled.

"I for one am glad you're different, Kathy," Ted soothed. "I admit, though, that I feared you'd be rather stuffy with your title and all."

"Even your jewelry is unusual," Susan stated. "You are wearing another collar today. It is very beautiful, but I'm slightly confused. I thought the collar ideas was yours, Reed."

"It is, Susan," Reed affirmed, moving his arm to lay it behind Kathleen's shoulders. At her baffled look, he said simply, "The same jeweler made this one up, too."

Something in his look transferred itself to his sister and her grey eyes took on a conspiratorial sparkle as she glanced from Reed to Kathleen. "Kathleen, you may be starting a new trend in jewelery fashion. On second thought, I think it is especially you, as personal as your own signature."

At Kathleen's amused look, she hurriedly said, "Please don't take offense. I meant it as a compliment. It makes you unique. Not everyone could carry it off successfully."

"Well said," Kate applauded.

"My sentiments exactly," Reed agreed, toying with a copper curl at the nape of Kathleen's neck.

His lightly caressing fingers sent a delightful chill down her spine, and Kathleen shivered. "Honestly!" she declared with a light laugh to cover her rising passion. "All this flattery could get me frightfully spoiled."

When they arrived at the race track, Amy finally managed to place herself at Reed's side, and stuck like flypaper. He devoted most of his attention toward Kathleen, but Amy managed to bend his ear quite a bit with her incessant chatter.

"Are ladies allowed to make wagers on the horses, Kate?" Kathleen inquired.

"The gentlemen usually place our bets for us."

"I'd be glad to do it for you, Kathy," Ted volunteered

eagerly.

"Would you like me to help you in your selections, Kat?" Reed offered.

"Really, Reed!" Kathleen erupted in temporary exasperation. "I'm not a child. There are some things I can do for myself, you know. Besides, who can judge horseflesh better than the Irish?"

Kate gave a sharp laugh. "She has ye there, Reed."

Out of five races, two of her choices came in first, two to place, and one to show, and Reed had to admit she had a sharp eye for a fast horse.

"Is there anything you don't do well?" Amy asked churlishly.

Considering this for a moment, Kathleen admitted, "I don't tat well. I don't take orders well, and I don't hold my temper well."

"Thank the Lord!" Amy exclaimed affectedly. "I was beginning to wonder if you were perfect!" In a lower tone she added, "You are disgusting, Kathleen."

Ted, who overheard Amy's last comment, added, "And *you* are jealous, Amy."

That evening, following a quiet supper and lively conversation in the parlor afterwards, Kathleen excused herself. She had Mammy prepare a bath for her, and help her undress and pin her hair up off her neck. "Ah'll be back up to he'p yo dry off when yoah ready, lamb."

"No need, Mammy. You go on now. You've done enough already."

"Well, holler when yoah done so's we can empty dat tub, den."

"I'd rather it waited til morning. I'm going to soak awhile, and the hot water will no doubt make me very sleepy. I'd prefer not to be disturbed after I get relaxed."

"Shoah, honey. See yo in de mo'nin' den."

Soaking until the water became uncomfortably chilly, Kathleen nearly fell asleep. She jerked herself awake, finished washing, and entered her bedroom with a large towel draped around her and tucked over her breasts.

"A very delightful outfit you have on, Mrs. Taylor."

The smoke from his cigar curled upward toward the ceiling from the bed.

"You again!" Kathleen exclaimed in mock anger. "You are like a persistent cough. I can't seem to shake you no matter what I try!" She removed the pins from her hair, and it tumbled down in a coppery cloud as she dropped her towel and stepped toward the bed.

He pulled her down next to him and buried his nose in her neck. "Mmm. You smell so good. Every time I smell lemons, I think of you. Do you know how truly beautiful you are, Kathleen? Do you realize how you fill a man's mind? You play havoc with his senses until everything he thinks, everything he feels, becomes a reflection of you. Then again, you can be so changeable. At times you are a playful kitten, at others a green-eyed spitfire. Other times you act the seductive cat, all sleek and graceful and soft, and a man can't help wanting to stroke you into submission."

His hands were already working their magic, and her own senses were swimming as she pulled him down to her quivering body. "Oh, Reed, what strange hold do you have on me that the mere touch of you can set me on fire? Your kisses drive all thought from my mind except my desperate need of you. Come to me, darling," she sighed tremulously. "Make love to me."

He stayed with her, held her throughout the night until she fell asleep at last, content enough to purr. Opening her eyes the next morning, the first thing she saw was a long-stemmed red rose lying on the pillow next to hers. She smiled sleepily to herself. "You court me, after you seduce me. Knowing you, that makes a strange kind of sense, somehow."

The day was bright and clear, promising to be hot. It was stuffy in the church pew, even with the breeze coming through the open window.

They all sat together, Reed's arm thrown carelessly across the back of the pew behind Kathleen, his fingertips lazily brushing her upper arm.

"Reed, please!" she pleaded in a low voice. "People are

noticing. What must they think?''

He gave her a mocking look, his blue eyes full of amusement. ''You certainly are an enigma, Kat. You don boy's breeches, ride astride, curse like a sailor, and perform the flamenco like a gypsy. You make love with the wild abandon of a courtesan, yet you are concerned now with what society will say?'' He chuckled deeply as he watched her blush.

To Kathleen it seemed that the good reverend directed the morning's sermon specifically toward her and Reed. He admonished the ladies against wearing alluring apparel that clung to their bodies and exposed a great deal of their breasts. He preached of the wickedness of men and women attempting to seduce one another. His entire message dealt with the sins of lust and seeking pleasures of the flesh. He railed at them and warned them that to fall low of temptation was to become the devil's playmate.

At this point, Kathleen, who was becoming more uncomfortable and angry with each passing moment, breathed into Reed's ear, ''I've always said you were a devil!''

''Yes,'' he whispered back, his eyes twinkling merrily, ''and I have found such a wonderfully wicked playmate!''

Reed continued to be charming and gallant throughout the afternoon. At the barbecue he brought her plate to her, and they ate on a bench beneath a shade tree where they could catch the breeze. They were not to be alone however, and Kathleen was relieved. ''He is worming his way into my heart bit by bit, and if I let down my guard completely I'll be lost for sure,'' she warned herself. ''Something has got to happen to prevent it—and soon.''

They chatted and ate, then strolled and played croquet on the rolling lawn. All the while, they were surrounded by admirers. Kathleen had collected her own entourage of young swains, while Reed seemed to draw the young ladies by droves. Amy had taken up her stance at Reed's side and was making a great show of proprietorship. She clung to his arm whenever possible, giving him long, loving looks and batting her lashes so swiftly that Kathleen thought she might fly away at any moment. She stroked his sleeve and patted him at every opportunity.

Kathleen became very weary of her own throng of beaus, and when the time finally came for the girls to rest before the evening's dancing, she was more than glad to retire.

The evening became another battle, as she was continually fending off advances. She began to wonder if it wouldn't be easier to announce her marriage and be able to relax her guard when she caught sight of Reed having similar problems. At one point, as she was dancing with Reed, he reminded her that they were indeed wed, and he would appreciate it if she would redouble her efforts to keep her would-be suitors at bay.

"You might do the same, Reed. What is sauce for the goose is sauce for the gander," she reminded him with a particularly suggestive smile. She spotted Amy hurrying in their direction as the dance ended, and her green eyes narrowed slightly as her long-nailed fingers curled unconsciously into talons.

"Your claws are showing, Kat," Reed drawled with a broad grin.

"That witch just rubs me the wrong way!" she declared, not at all aware of her pun until she heard Reed's loud hoot of laughter.

"Oh, hush, you big oaf!"

A while later, she was talking with Ted when she noticed Reed standing in a corner surrounded by a half dozen swooning young ladies who were hanging on his every word. He glanced up and caught her looking at him. With an impish smile, she flashed her wedding ring at him, reminding him to behave himself. He caught her message and quirked his brow at her before she turned away.

Still later, as Kathleen was dancing with a man she could have sworn had more than his share of arms and hands, she felt Reed's eyes on her. When she met his impudent look, he raised his champagne glass in salute, boldly flashing the onyx ring she had given him, its diamond glimmering brightly at her. She shrugged her shoulders helplessly and wrinkled her nose as she and her octopus whirled by.

It was late afternoon before Reed finally pulled up before the front entrance with Eleanore and Charles. Kathleen raced across the porch and down the steps to throw her arms about her friend.

"Good heavens!" Eleanore laughed. "You'd think it had been years since we'd last seen one another."

"It is supposed to have been," Kathleen said conspiratorially, and then gave Reed a dark look, "or didn't Reed tell you?"

"Yes, he filled us in. You are as stubborn as ever." Eleanore shook her head and clucked her tongue in a reprimand.

"Don't scold me, please. I get enough of that from Reed."

Charles chuckled as he kissed her cheek. "I would give you a good spanking if I were he, young lady."

"Heavens!" Kathleen threw up her arms in mock horror. "Don't give him any more ideas than he has already!"

"I still think it is poetic justice that you should end up staying here of all places!" Eleanore giggled.

"So do I. So do I," Reed declared with a laugh. "Come now, let's get you settled in."

Kathleen waited until Eleanore was settled into the rooms next to hers before she allowed herself to ask, "How is Dominique?"

"Fine. He asked me to be reminded to you. He seems distraught with the thought that you might forget him."

"I could never do that, Eleanore," Kathleen avowed adamantly, causing Eleanore to give her a queer look.

"Now don't you start, too!" Kathleen announced. "We are like brother and sister, and I do love him dearly, but not in any romantic sense."

"I'm glad to hear it, because whether you choose to acknowledge it or not, you are a married woman,"

Eleanore stated flatly.

Sinking into a chair, Kathleen retorted, "Please! Don't remind me!"

"Someone has to, dear." Leaning forward, Eleanore continued earnestly. "Don't you see how foolish all of this is, Kathleen? Here you are on this beautiful plantation in a marvelous mansion, with servants and horses and clothes and jewels. You have the most handsome and attentive of men for a husband. Most women would sell their souls for what you have. Reed can afford to provide well for you, and would probably grant your slightest wish. I fail to understand why you are being so obstinate, my friend. What more could you possibly want?"

"Love," Kathleen replied simply, her eyes resting on her friend's face. "Love or revenge, whichever the fates shall decree." Her emerald eyes shimmering with unshed tears, she went on in a strained voice, "Do you realize that Reed has not once told me he loves me?" Holding up her hand to ward off any interruptions, she choked out, "Sure. He has said I am beautiful, desirable. He has told me he desires my body, admires my charm and my talents, but never," her voice shattered, "never has he said 'I love you, Kathleen.'"

Tears coursed down her face as Eleanore gathered her into her arms. "I would give anything to hear him say those three small words. He comes to me at night now and shares my bed, and I cannot seem to find the courage to forbid it. I need him, Eleanore, as I need air to breathe. For now I take what affection he throws my way, but I cannot live this way forever. This is why I refuse to let him have it all. I will not be his chattel. I must have his heart and love or all this means nothing—less than nothing. All I am doing now is torturing myself, for soon he will be off to sea again, and perhaps with him gone I can begin to piece my life together."

"I am sorry I was so hard on you, petite. You are going through your own private hell, I know." Eleanore dried Kathleen's tears with her handkerchief. "Perhaps," she mused, "all is not as bleak as it seems. Have you told him how you feel, how much you love him?''

Straightening her shoulders and drying her eyes, Kathleen blurted, "I'll die before I crawl to him on my knees, Eleanore! I'll never give him the satisfaction of knowing that! He would only use it to hurt me as he has hurt me so many times in the few months we have known one another.''

"Ah, well, perhaps things will work out in time,'' Eleanore suggested. "I still cannot help feeling that he cares for you even if he does not say so.''

"I cannot share your optimism, even while I pray you are right. Let us say I will not hold my breath in waiting. Miracles are not as plentiful as they once were, I fear.'' Kathleen rose and checked her face in the mirror.

"Come. Let's find the men and chat awhile before supper. I'll introduce you to the others. I ask only one favor, besides keeping my secret.''

"What might that be?''

"Please, please don't like my cousin Amy, and if you do, don't tell me about it,'' Kathleen laughed.

It was nice having Eleanore to talk to once more. Early each morning the four friends rode together after a light breakfast. Afterward, Reed was usually occupied with overseeing the plantation and Charles would retire to his room and pore over his medical books, preparing for the seminar. If the two women felt so inclined, they would join the others, if not, they would go off by themselves to chat or take a ride over to see Kate.

When they were alone, Eleanore often helped Kathleen sew up her three sets of vests and breeches. They even found some fine lawn and made her three white shirts with long, billowing sleeves and tight cuffs. Eleanore tactfully never asked when or where Kathleen would wear her new clothes.

It became a ritual after supper to repair to the parlor, and, if they had no company or social functions to attend, usually started a lively game of poker, ignoring Amy's high-handed criticisms. Always, after everyone retired for the evening, Reed came to Kathleen's rooms.

It happened on Friday morning. As they were enjoying their ride, Eleanore, who was usually so tactful, related to Kathleen that Dominique had expressed concern over Kathleen's well-being. Upon discovering that Eleanore would be visiting her, he had asked her to remind Kathleen that he had promised her his aid if ever she had need of it.

Unfortunately, Reed was within hearing as Eleanore spoke, and it sparked his temper. "Damn that dirty scum!" he roared, causing Kathleen to jump at his voice. "He certainly takes a lot on himself where my wife is concerned! I can see that I am going to have to remind him who's responsibility you are." He glared at Kathleen, his eyes as icy as frosted crystal, his lips drawn tightly in a furious frown. "I'm beginning to wonder just what went on under my nose on that island that I wasn't aware of."

Kathleen attempted to calm him. "Reed, you know better than that. Dominique and I are just friends. He is your friend, too. Don't destroy that relationship. He cares a great deal about both of us."

"Don't try to placate me with that innocent act of yours, woman. It doesn't suit you. You forget. I know what you are like, and can only wonder how far you would actually go to extract your revenge on me. Tell me, Kat." He leaned toward her, his jaw muscle twitching in anger. "Is it my child or someone else's you carry within you, or perhaps even you are not sure?"

For a long moment she sat stunned, then looked from one person to the other as if she had never seen them before, her eyes enormous in her shock. "Child?" she repeated dumbly.

"Come now, Kat," Reed retorted sarcastically, "surely you must have suspected it by now."

The naked scorn in his voice sickened her, and she felt slightly dizzy and more than a little nauseated as she faced him. Tears formed in her eyes and she stared at him through a misty veil, the reality of his accusation hitting her like a blow. Her voice was shaking with hurt and anger. "How cruel you can be," she accused in a barely

252

audible voice. "Is it any wonder I despise ⟨...⟩ child, believe that it is yours and," she w⟨...⟩ wheeled her horse about, "believe that with a⟨...⟩ wish it were anyone else's!" Tears streaming dow⟨...⟩ cheeks, blinding her, she kicked Zeus into a gallop⟨...⟩ raced across the field. She let Zeus have his head in a headlong gallop as she clung desperately with one hand, wiping futilely at her tears and sobbing uncontrollably, her one thought to escape Reed forever.

A wave of dizziness threatened to overcome her, and she cried out in anguish as she felt herself slipping from the saddle. She landed with a sickening thud, and felt an agonizing pain shoot up her spine before everything went black.

Reed threw himself from Titan's back, kneeling at Kathleen's side, his face white with anxiety as he gently turned her over. He wiped the dirt from her forehead and tear-stained cheeks, and peered anxiously at her still face. "Dear God," he prayed, cradling her next to his chest, "let her be all right. Punish me as you will, but let her be all right."

Charles and Eleanore had dismounted, and Charles quickly bent to examine her unconscious body. He looked grimly at Reed. "She's bleeding badly. It is most likely she'll lose the baby. We have to get her to bed as quickly as possible and stop the bleeding or she may die."

Eleanore sat stroking Kathleen's forehead. "She's regaining consciousness, I think."

Swimming up to reality, Kathleen almost wished she had not. Through her pain, she opened her eyes and looked directly into Reed's worried face.

"Forgive me, Kat," he whispered hoarsely. "I didn't mean any of it."

She tried to lift her hand to stroke his cheek, but it fell limply to his chest. "I didn't know about the baby, Reed. Believe me," she pleaded weakly, biting back a groan. "And it *is* yours, I swear it!"

"I know, love. I know. I was just angry. Please don't try to talk. We've got to get you back to the house."

ne must know," she murmured into his shirt. romise me."

"No one but the four of us, kitten."

Her lids fluttered open again. "Take me to Kate's." At his confused look, she said, "It's closer, Reed. Hurry, darling. I hurt so badly!"

As he lifted her into his arms, she screamed out in agony and fainted.

Chapter 17

REED paced the hallway outside Kathleen's door. Charles had been in there an impossibly long time. Reed had carried her tenderly upstairs to Kate's guest room, and then had been promptly ushered out. Kate had scurried around gathering every item Charles requested, but she too was refused entry once the doctor had all he needed. Only Eleanore remained with him and Kathleen.

Kate, sitting in a chair near the bedroom door, tried to calm Reed as best she could, but she was extremely upset herself. "Tell me what happened, Reed."

"She fell from her horse," he said shortly.

"And Dr. DeBeaumont is a friend of yers? Was he riding as well?"

"Yes. We were all out riding together."

"Did her horse bolt?" Kate pried, and then declared, "I should have never let her have that animal! What have I done?"

Kneeling next to her chair, it was now Reed's turn to reassure her. "No, Kate. Her horse didn't bolt. I was angry and said some awful things to her, things I shall regret until my dying day. She was hurt and crying and spurred her horse into a gallop, trying, I suppose, to put as much distance between us as possible. Somehow she fell. We were closest to your place and she asked to be brought here." His face was lined with worry and guilt as he sat staring at the closed door.

Kate took his hand in both of hers. "What could ye have said to her? That spunky little lass is the type to stand up and throw yer words back in yer face tenfold."

"I can't tell you. I promised her no one would know."

"Know what, Reed?" Kate said irritably, clutching his hand. "That she was with child?" At his surprised look, she explained, "I have lived enough years on this earth to know a miscarriage when I see one, boy. Now, why don't ye tell me the rest? Ye are the father of the child, are ye not?"

Reed nodded. "Yes, but I was furious with her and stupidly asked her who had fathered her child when I knew damn well she hadn't even suspected her pregnancy yet. We are always at each other's throats, and now she has every right to hate me. It was unforgiveable of me, and I doubt she'll ever want to see me again. Kat does not forgive or forget all that easily."

" 'Tis done now. The words can't be unspoken," Kate said with a sigh. "I can't understand ye though, Reed. I've known ye since ye were in knee breeches. Ye've always been headstrong, but never deliberately cruel!"

"You have to know Kat to understand. She's such a temperamental, stubborn little vixen. From the moment she stepped aboard that ship in Ireland, I knew she meant trouble, but I was drawn to her as a bee to clover. I've never wanted another woman as I wanted this copper-haired beauty with those slanting green eyes. Right from the start we were as duelists, always clashing, but even as we battled there was an undercurrent pulling us toward one another."

"Do ye love her?" Kate asked when he paused.

"God help me, yes," Reed admitted, holding his head in his hands. "She's the one woman I can't live without, and she's the only one who resists me for all she's worth. Kat also seems to have a particular knack for sparking my anger faster than anyone else I've known. I don't understand it at all."

"I think I do," Kate offered, her eyes taking on a distant look. " 'Twas the same with Sean and me. Our love was like the ocean—wild and strong and constantly changing. When we weren't lovin' we were fighting, but our life together was never dull, and neither of us would have changed it. Through all the tiffs, our love continued to grow. Sean admitted he could never have stood being married to some quiet little milksop of a girl. He actually enjoyed some of our arguments, and deliberately baited me at times. He said I was beautiful when I was angry, and he admired my spirit.''

''It's the same with Kat and me, but this time I've gone too far. If she loses the baby she'll never forgive me. On the other hand, she told me she wished it was anyone's child but mine. She despises me that much. She may be glad to be rid of it.'' His clouded blue eyes showed plainly the heartache he was feeling.

''I'm sure she didn't mean that, lad. She was hurt and angry.''

''She'll be more than angry if she finds out you know all this.''

''I'll not be telling anyone about this. 'Twould cause quite a scandal, to be sure. Still, the lass is ruined whether anyone suspects or not. Ye are responsible for that, and I'm wondering what ye'll be doing now. Kathleen has lost all hope of making a good marriage. Unless the man is a complete moron, he'll realize something is amiss on his wedding night.''

''She can't marry, regardless,'' Reed blurted.

''Why not?'' Kate pressed, already knowing why but wanting to see if Reed would break his promise to Kathleen.

''I've said too much already. I made a vow and I cannot tell you more.''

Kate smiled sadly, proud that Reed would keep his word to her granddaughter. ''I think I can guess that too, Reed. She cannot marry because she's already wed to ye. Why all the mystery? That is what I really want to know.''

''Since you have seen through our charade, I will tell you if you do not reveal to Kathleen or anyone else that I have told you.''

"Ye have my word," she promised.

He related the tale to his friend, telling of Nanna's death, their marriage, and Kathleen's obsession with the fact that he had stolen her ship. Kate seemed surprised when he told her of Kathleen's strange alliance with the sea. He told her of their experiences on Grande Terre and why he had insisted on bringing her to Savannah. Finally he admitted that Kathleen had blackmailed him, and how their relationship had fared since he had returned to Chimera on the night of the ball.

When his story was done, Kate asked bluntly, "Does Kathleen know ye love her? Have ye ever told her in just those words?"

A puzzled expression crossed Reed's face as he contemplated her question. "I honestly can't tell you, but surely she knows that I do. I have told her how she affects me. I know I have been a beast at times, but she must know how I feel."

"Oh? Well I'll let ye in on a wee secret, my friend. Women thrive on compliments, but they need to be told they are loved. It is their life's blood. Tell her the first chance ye get."

"It's no use now, Kate. She has made it plain that she can't abide me. She can't deny that she enjoys our marriage bed, but now I am certain that even that aspect of our ill-willed marriage is a thing of the past. She won't want to become pregnant again, and I can't say I blame her. I wish I had never uttered those dreadful words to her today. Whatever slim chance our marriage had for succeeding, I have destroyed forever today."

"Only time will tell," Kate advised. She sat for a long time mulling over what both Kathleen and Reed had told her. Now she knew for certain that both of them loved each other very much. "They are at cross-purposes with one another," she thought sadly. "And here I sit, knowing both sides of the story and sworn to silence. I wish I could sit down with the two of them and help them straighten this mess out, but I think this may be one of those times when it is best to let them work their problems out in their own fashion. I could be a busybody and stick my nose in, but I think they would both resent it. It wouldn't cure Reed of his possessive jealousy or the need to dominate Kathleen. Nor would Kathleen learn to trust Reed and forget her desire for revenge. They must both learn that trust and love go hand in hand if they are ever to find true happiness in their marriage. Only when they are totally and blindly committed to one another will they come to know love as deep and strong as they are capable of, and they must do it in their own time and way. It is sad, but lessons learned the hard way are not so easily forgotten, and love earned the hard way is not so easily tossed aside. As much as I would like to save them the pain, I won't step in unless the situation becomes drastic or I see one of them in danger.''

Eleanore stepped quietly into the hallway, pulling the door shut after her. Reed, who had resumed his pacing, grabbed her by the shoulders. ''How is she?'' he demanded, fear causing her tanned face to appear pale.

Casting a quick glance at Kate, Eleanore pulled him farther down the hall before she spoke. ''Kathleen's going to be fine, with rest. She lost the baby. I'm sorry.''

''May I see her?''

''It is best that you don't, Reed. She needs to rest. Charles will be out soon to talk to you.''

Kate had risen from her chair. Now she too inquired, ''Will she be all right?''

''She'll be fine in a few weeks,'' Eleanore replied, retracing her steps. ''She took a pretty rough tumble. She'll need light foods and plenty of bedrest, but Charles could find no broken bones. She is very shaken, and will be dreadfully sore for a while as well as black and blue, but she's young and healthy and will recover quickly.''

''Thank you. It is a relief to hear you say so. Would you care for some tea now? You have been very busy these last few hours, and it is way past the lunch hour. Perhaps we could all use something to eat.''

''Tea sounds marvelous.''

''Reed, would you care for something to eat?'' Kate

offered.

"No. You ladies go on. I'll wait for Charles here."

The two women went down to the parlor without him. As they sat drinking and talking, Eleanore found herself studying Kate's features, the slightly upturned nose, the stubborn chin, and her arresting emerald eyes. The face was older, wiser, no longer as smooth as it was once, and the red-gold hair was liberally sprinkled with grey. Still, Eleanore was amazed that Reed had not noted the family resemblance. Once you knew the relationship, it was obvious, but perhaps if you were not aware of it you would overlook what was right before your eyes. Eleanore had been immediately struck by it the first time Kathleen had brought her to meet her grandmother.

Now, even as they shared Kathleen's secrets, they shared her loss and pain. Eleanore had told Kate that the baby could not be saved. Together they mourned the lost little life and the rocky marriage that seemed doomed if neither Kathleen nor Reed would make the first move toward the other out of faith.

Eleanore leaned back in her chair wearily. "What are we to do with those two? I know they love one another, but neither thinks the other feels the same way. They are both headstrong, and I doubt either would listen to reason. I could shake both of them until their teeth rattle!"

"And I'd help ye, but it would do no good. 'Tis best they settle this in their own way without our interference. 'Tis painful, but the only way."

"You are right, of course." Eleanore rose and started for the stairs.

"I see no reason why you can't sit with Kathleen if it would make you feel better. Reed can too, when the rest of her family is not here. We wouldn't want them to suspect anything."

"Oh, my!" Kate exclaimed excitedly. "I completely forgot to send any kind of word to Chimera!"

Eleanore chuckled, "I'm glad you did. It's been confusing enough around here today. I just hope I don't get all these confidences mixed up or start talking in my sleep. What a tangled mess these two have caught us up in!"

Charles emerged from Kathleen's room just as the ladies joined Reed. "She's resting comfortably," he said at their questioning looks.

"I'd like to see her, Charles. Just for a moment," Reed requested again.

Eleanore assured Charles, "It's all right. Mrs. O'Reilly knows all of it."

"All right," Charles conceded, "but don't wake her, and if she does stir, don't upset her."

"Is she in much pain?" Kate queried.

"A little, but the worst is over. She was only two months along, so it was fairly easy as miscarriages go. It was just that her fall was so hard and the bleeding so profuse. We've slowed the bleeding to normal, and now nature will start its healing process. She's young and active and healthy, and there is absolutely no reason why she couldn't be up and about in her usual fashion in a few weeks."

Reed breathed a tremulous sigh of relief. "Thank God!"

"Doctor," Kate asked hesitantly. "Is there any reason that she could not have more children later?"

Reed stopped with his hand on the doorknob, dreading what Charles might say next.

"Ease your mind, madam. There is no damage. Kathleen could have a dozen if she so desires. However," Charles turned to Reed, "I wouldn't recommend any activities of that nature for at least a month or more, if you please."

Reed nodded his assent. "Thank you, Charles," he said humbly, and entered the room.

He tiptoed to the side of the bed where Kathleen lay with her eyes closed. Her long dark lashes formed thick crescents against her pale cheeks. He took her hand in his and kissed her open palm.

"How are you feeling, sugar?"

"How do you think I feel?" she murmured irritably. "I feel as if I've been pulled through a knothole feet first. I

261

feel lousy! Does that answer your question?''

''Charles told me you lost the baby,'' he ventured again.

Hot tears gathered at the corners of her eyes and spilled silently down into her tangled hair. She stared up at him with pain-glazed eyes. ''That ought to make you very happy, Reed, since you obviously doubted its parentage!'' she declared vehemently.

''Kat, please, you must listen to me, darling. I didn't mean any of it. I knew the child was mine.'' He waited for her reply.

She sighed tiredly. ''Just get out, Reed. Go away and leave me alone. I really don't have the stamina to listen to more of your lies. It really doesn't matter much now anyway, does it? It's rather like closing the barn door after the horse has run off.'' She closed her eyes again, warding off further comments from him.

''I'll be back later,'' he said softly, and headed for the door.

''Don't bother,'' he heard her answer weakly. ''I'll get along just fine without you.''

After a good night's sleep, Kathleen was surprised at how much better she felt. Her bruises bothered her the most. She sat up for a breakfast of toast, tea, and eggnog, but then Kate and Eleanore both insisted she lie back down again. Reed had informed her family of her fall, and shortly before lunch they started invading her privacy. In a way, Kathleen was grateful, for it took her mind off of her loss.

Barbara and William visited first and were relieved to see her sitting up in bed, her hair combed neatly back. ''I told you that horse was dangerous,'' Barbara admonished. ''I hope now you will give him up and choose a more gentle mount.''

''Yes, my dear,'' William advised, patting her hand awkwardly. ''You should listen to your aunt and get rid of that animal before you get killed. You were very fortunate yesterday.''

Kathleen lifted her chin stubbornly. ''No! It wasn't

Zeus's fault. It was my own stupidity. However, I promise to be more careful in the future." She smiled ruefully at her own words and thought, "I most certainly must be more careful!"

Ted managed to wrangle visitation privileges he would ordinarily have been denied, and he brought her a stack of books and fashion periodicals.

"My heavens!" Kathleen exclaimed on seeing this, "I'm not going to be laid up that long, I'm sure!" Later in the day Mary Taylor visited with Susan, bringing a huge bouquet of flowers from the garden. Amy, thank goodness, stayed away.

Kathleen had engaged in an honest talk with her grandmother, confessing her own guilt over losing the baby. She deeply regretted the loss, and admitted to Kate that she would have loved to have born it. Kathleen cried until she had no more tears left to cry, and then silently brooded over what the child would have been like. She imagined a little boy or girl with black hair and startling blue eyes, and her depression deepened.

"Don't dwell on it, lass," her grandmother counseled. "I know it hurts, but ye are young, and the doctor says ye can have more children. There is plenty of time."

Kathleen shook her head determinedly. "Never! If you think I'll ever let that vermin-ridden skunk bed me again, you are sadly mistaken! It is my misfortune to be married to him, so I most likely will never have a child of my own. Ours is not your everyday marriage, in case you haven't noticed!" Her emerald eyes shone with spite.

"Aye. It has come to my attention," Kate commented with a short laugh. She added slyly, "However, if ye should find yerself truly desiring a child later, ye could always change yer mind and take him to yer bed until the seed were planted. It wouldn't have to be a meeting of the hearts, ye know, it wouldn't be any worse than him using yer body to ease his passions in the past."

"Kate!" Kathleen exploded in a shocked voice, and then giggled. "What a terribly wicked idea! I'll think on it, but I can already see it has its pitfalls."

"Oh?"

"I've told you before how Reed affects me. He's like a drug. Once I let him touch me, I'll be addicted again. It's strange how he can make me forget my anger. In fact, my mind nearly refuses to function at all when he's making love to me. Not only that, but should I have a child, it would only serve to tie me more closely to Reed. Then I should have to admit our marriage and abide by his rules whether I desired to or not. He'd have me right beneath his thumb, and once having gained that advantage, he'd move mountains to keep it. I'm not ready to be dominated that fully. Not now, perhaps never."

It was after supper that Reed came by. Kathleen had told Kate and Eleanore firmly that she did not wish to see him. Still, she was not unprepared when he strode into her room.

"As you can see, I made it past your guards," he growled. He stood at the foot of the bed, hands on his hips.

Kathleen's eyes glittered maliciously as she taunted, "That must have been a rare and difficult feat getting past two such muscular persons! What daring you possess!"

"Your fall didn't dull your sharp tongue at all, Kat."

"No, but it did shake some sense into me. I have come to the conclusion that the less I see of you, the happier I'll be, so please leave. I'm sure you know your way out."

"Not so fast, lovely wife," Reed sneered. "I have a few things to say before I go, and you are going to listen." He seated himself on the rail of the footboard and peered at her with icy blue eyes.

"How sweet! I am to be a captive audience to a raving madman!" she retorted, green eyes flashing. She waved her arm wide in a condescending gesture. "Talk away, sir. Have your say and then get out."

"Damn it all, if you aren't the most maddening woman I've ever come across!" he exploded. "I come with good intentions of apologizing and trying to salvage our marriage, and all you can do is provoke me to anger. If you were well, I'd be tempted to turn you over my knee and give your bottom a long overdue spanking!''

"If you have nothing more to say—" she began.

"I have plenty more to say!" he roared, his face dark with fury. He strode to the side of the bed. Leaning close, his hands resting on the mattress on either side of her, he continued in a barely constrained tone.

"To start with, I am leaving on a sea voyage tomorrow morning. I will be gone for a couple of weeks or more."

"The dear Lord does answer prayers after all!" she interjected acidly, meeting his steely gaze steadily.

His eyes narrowed dangerously as he clenched his fists tightly, striving to maintain control. "By the time I return, you should be back to normal again. At that time you are going to assume your position as my wife." He watched her face for her reaction.

Her features remained placid, but her eyes glittered like a spitting cat's. "Pray go on. This is becoming interesting. Shall I buy a new rope for your hanging or will an old one do?"

"There will be no hanging, Kat. You are going to forget all this nonsense and behave properly as a wife should. I don't know how I ever allowed you to bluff me in the first place. It is time I took a strong hand with you. Someone should have dealt firmly with you long ago!"

Kathleen started to laugh, a low evil laugh that sent chills up Reed's spine. He met her mocking look and stepped away from the bed before the temptation to strangle her became overpowering.

"Something I said amuses you?" he growled.

"You fool!" She sneered. "You bloody fool! If you think for one instant that I will bow to your insane demands, you are indeed crazed. I was never joking, Reed, when I said I'd see you hang for piracy. Don't try to out-maneuver me. It won't work. We are not on Grande Terre now, and your bullying tactics won't work here. Admit defeat gracefully and learn to live with it, and above all, don't ever make the mistake of pushing me too far, darling. It could be hazardous to your health!" Her eyes blazed like emerald fires as she glared at him defiantly.

Reed contemplated her words. "You really are a cold-hearted witch, aren't you? If I hadn't seen your blood running freely, I would swear you have ice water running through your veins! The only time you show any warmth is when you spread your legs for me in bed!"

Kathleen's temper flared. "Rest assured, Captain Taylor, skunks will smell sweet before you gain that privilege again!" Her hand flashed out, and before he had time to react, she threw the heavy porcelain pitcher from her nightstand at him. It hit him just above the knee, shattering on impact and sending shards flying across the floor.

With a roar of pain and anger, he lunged for her. Just as he reached the bed, the door flew open with a crash. Reed froze as Kate and Eleanore took in the scene before them.

"What in heaven's name is going on in here?" Kate exclaimed.

Kathleen, her breasts heaving with emotion, pushed Reed roughly from her bedside. "Captain Taylor was just about to leave. He has overstayed his welcome."

When he hesitated, she picked up the book nearest her and pitched it angrily at his head. He ducked the missile and glared at her, his eyes like two frozen pools. Executing a swift bow, he stomped toward the door, his jaw twitching madly.

"Have a nice trip, Reed, and don't expect any changes around here upon your return!" she shrieked after him, and threw her hairbrush at his retreating back for emphasis. She let out a strangled cry of anger as the brush fell short of its mark.

Reed turned, retrieved the brush, and tossed it carelessly onto the foot of the bed. "Don't feel too badly, Kat," he said. "Your aim is improving. One out of three is better than none at all. Practice while I am gone and you can try again in a couple of weeks." His jeering laughter lingered as he departed.

That next afternoon Kathleen had a long talk with her grandmother. Kate was especially curious about Kathleen's kindred spirit with the sea. "Ye know, Kathleen," she confided. "It is not as strange as it seems, especially in our family. Every third generation or so, this sort of thing emerges in a woman in our immediate family.

"There is a tale that long, long ago an ancestor of ours was out fishing on his boat and caught a mermaid. They fell in love and mated, and even though they had to live apart; he on the shore and she in the sea; they saw each other every day. When the girl-child was born, she was more human than mermaid, and having no fins, had to live ashore with her father. Even so, the sea was where she was born and was in her blood. Since that day it is not unknown for this sea blood to show up in one of her descendants, stronger in some than in others. When it does, the call of the sea is inevitable. You could not ignore it if you tried, but it is dangerous to one who tries to fight the allure or who does not understand her own limitations.

"I've heard tell of another female ancestor of ours who was also a sea pirate many years ago. She made quite a name for herself striking down Viking ships. Most folks think she is just part of a colorful legend, though."

"Such a pity!" Kathleen sighed.

"Yes, but better than another relative of ours who simply walked off in the middle of the night leaving a broken-hearted husband and three motherless children. They found her clothes neatly folded on a rock at the edge of the sea, but she was never seen again.

"Yet another beautiful young lass was captured in a raid long ago and taken prisoner aboard an enemy ship. She escaped by diving overboard, rather than submit her fair body to the lustful crew. It is said she sang a haunting ghost song long after she was supposedly drowned, and she lured the ship into unchartered seas in the nether world. Neither girl nor ship was ever found again. They say on a clear night the first of June every seven years just at the stroke of midnight, ye can hear her haunting song, and if ye look closely enough ye can catch a glimpse of the ghost ship still sailing on the dark waters.

"So ye see, lass, 'tis the special gift of the mermaid ye carry, and not so strange after all. Enjoy it and pass it on to later generations."

Kathleen smiled, not really ready to believe all Kate told

her, but relieved to know her gift had been shared by others of her family before her, and possibly generations to come. She felt honored in some special way to have been blessed with this unique affinity.

With Reed gone, she began to plan again, and almost immediately saw this as a golden opportunity. Charles came by to check on her after breakfast and was pleased with her condition.

"My ship sails for New York this evening, Kathleen, but if you wish me to stay, I will," he offered.

"You are a dear, Charles, to even consider it, but I'll be fine. Kate and Eleanore are excellent nurses. You go to your seminar and don't worry about me." She reached for his hand, her face suddenly thoughtful and sad. "Thank you for all you've done. I know you would have saved my child if it were possible."

Ted was to drive Charles to Savannah, and Eleanore was going along. By the time they left, Kathleen had entrusted a note to Dan into Eleanore's hands. "Ted will know where to locate him," she instructed. "Don't deliver it until Charles has sailed."

"This is all very mysterious, Kathleen. What are you up to now?"

"I'll tell you when the time is right. Please don't question me about it now."

In the note Kathleen had instructed Dan to make certain all preparations to sail were made ready. Everyone must be standing by and ready to move at a moment's notice.

Kate filled her in on Reed's plans. He was sailing down the coast to the Florida keys where he would lay in wait for a Spanish ship. There should be no problem in encountering either a galleon laden with gold and treasure sailing from Mexico, or one laden with goods from Spain heading west. They would be easy prey not only for Reed, but for any of Lafitte's seasoned privateers. After Reed captured the ship, he would take his booty to Grande Terre and return to Savannah.

"If all goes well, he will have nothing but sad tales to give to Jean." Kathleen's eyes shone like emeralds in her excitement. "I have sent a message to Dan to ready my frigate."

"Ye can't risk moving about so soon!" Kate exclaimed. "Surely ye cannot mean to take Reed's cargo while ye are so weak. 'Tis impossible, lass!" Her face was lined with worry as she studied Kathleen's determined look. "Wait until next time."

Kathleen explained eagerly. "Don't you see? Now is the perfect time. Even if Reed should see through my disguise, he would not believe it. I am here in Savannah recuperating. You must tell the others that I am severely homesick and depressed and wish to see no one for a while."

"It might work, but what of yerself?" Kate wanted to know.

"I will have nearly a week to recover aboard ship before we hit the Keys. Besides, I do not plan to leave until Tuesday night. Can you have my hair rinse ready by then?"

"I'm sure I can. There is another problem ye are overlooking. What of Eleanore?"

"Oh, dear!" Kathleen chewed her lower lip anxiously. "We'll just have to cross that bridge when we come to it, won't we?"

By late Monday afternoon her sailing gear was smuggled from Chimera and her hair rinse was finished. Her breeches fit snugly, but when she attempted to pull on the high boots, she found the material of the trousers too think to allow the boots to fit properly over them, and the legs of the breeches were too tight to fit on the outside of the boots.

"Drat!" Kathleen shouted in frustration. "What do I do now, Kate? We haven't time or material enough to start again."

"Ye'll jest have to wear yer old breeches this time," her grandmother commented before noticing the sly gleam in Kathleen's eye.

Reaching for her knife, Kathleen carefully sliced around the legs of her breeches. This complete, she donned her boots and breeches again. Her grandmother stared in amazement. Kathleen stood in black boots reaching to her knees and green breeches that barely covered her bottom. Her entire upper thigh was left bare!

"Ye Gods and little fishes!" Kate cried. "Ye don't dare wear that!"

"If I dare piracy, Gram, any outfit I appear in won't be more of a shock than that. At least I can move about easily this way." Studying her image in the mirror, Kathleen wondered to herself if she might further dare to wear the vest without the shirt beneath. After all, she was not supposed to present the picture of a proper lady, but that of a daring female pirate! It would be an extremely bold costume, but so were the actions she was preparing for.

"Why not be alluring as well as daring? Perhaps they'll be too stunned to fight!" Kathleen giggled to herself. The mirror showed that the small green eyemask covered her cheekbones just enough to hide the contours of her face and disguise her identity.

Kate had prepared the hair coloring herself. Flax seed had been cooked in water and strained to make a gel. Strong black tea had been brewed and the leaves strained from it. Then the tea was combined with vinegar in an iron pot and simmered slowly until only a small amount of dye liquor remained. When it had cooled, it was stirred into the gel. It would not be a permanent dye, either would it rinse out easily. It would take strong soap and baking soda to remove it, and a lemon rinse afterward, but it could be applied and removed time and again without causing damage to Kathleen's beautiful hair.

Kate helped Kathleen apply it with a small brush. Together they worked the dye through Kathleen's hair, then rinsed and brushed it dry. Kathleen exclaimed over the final result. It was as black as coal, and as shiny as satin. Even Kate was surprised at how much it changed her appearance. The effect was stunning! Her suntanned face suddenly seemed Latin, and her bottle-green eyes appeared even more brilliant in contrast to the ebony hair.

So involved in their examination were they that they failed to hear the bedroom door open until Eleanore let out a small gasp. "Kathleen! Is that really you? Whatever are you doing?"

Kathleen stepped forward and pulled Eleanore into the room as Kate hurriedly closed the door again, this time bolting it. Kathleen turned slowly in review, modeling before her friend.

"Don't you like it, Eleanore?"

"It's stunning, to be sure, but I am quite confused. Your own coloring is so beautiful! Why change it? This just isn't you! In fact, is took me a moment to realize it actually was you!"

"Good!" Kathleen stated satisfactorily.

Eleanore looked from one to the other of the ladies in bewilderment. "Are either of you going to tell me what this is all about?"

Kate answered first. "That depends on whether your loyalties lie with Kathleen or with Reed. After all, ye have known him longer."

"True, but so have you. Whatever mischief you are planning against him he deserves. I'll certainly not tell him." Eleanore turned to Kathleen with a sincere, slightly hurt look. "You should know that without asking, Kathleen. I was there. I saw how he treated you on the island, the way he allowed Rosita to taunt you, how he forced you to leave Grande Terre against your will."

Kathleen smiled sadly as she recalled the more loving scenes from that time. "Yes, but you also tried to tell me he loved me. You know of the passion that passes between us, and you've seen how generous he can be. The actions I am about to take against him may seem too drastic to you."

"Nothing short of murder would be too drastic, Kathleen!" Eleanore declared vehemently, her dark eyes smoldering. "I cannot honestly say I believe Reed married you for the *Kat-Ann* alone. I think he wanted you as well, but his methods were definitely underhanded. Perhaps I sympathized with him a little, even wished him well, but not when I recall the words he spoke to you the other morning." Tears filled her brown eyes. "How could he have said such awful, hateful things to you! I would not blame you if you chose never to forgive him. Whatever you intend to do, Kathleen, you can count on my help if you need it."

Kathleen quirked an eyebrow at her friend and asked quietly, "Even if Jean sees less of Reed's cargos in the future?"

Confusion registered on Eleanore's fair face. "What? How could that be unless you plan to have him caught by the authorities? You would be breaking your word if you did that."

"My word is true, Eleanore. I'll not dishonor it. Reed will simply be relieved of his booty before he reaches Barataria Bay."

"You are going to have him pirated!" Eleanore stood rooted on the floor, wearing a look of disbelief. Suddenly she began to laugh heartily, her slight frame quaking as her laughter filled the air. "Oh, Kathleen, it is perfect! Tell me, how did you find anyone daring enough to agree to this insane scheme?"

Kathleen flashed a look of deviltry toward Kate, both pairs of emerald eyes glowing with excitement. "I didn't have to look far, Eleanore. I will attend to the task myself. I have my own ship and crew standing by even now, thanks to the note you delivered for me Sunday evening." Kathleen laughed delightedly as Eleanore's mouth sagged open.

"Surely you are not going along on this venture!" Eleanore exclaimed when she had regained her speech.

"Sit down, dear, before your knees give out." Kathleen giggled, gently propelling her friend into a nearby chair. "In answer to your question, yes, I am going along. In fact, I am captaining the ship myself." At Eleanore's frantic look, she waved away her fears. "Please do not concern yourself. I am an excellent captain. I can handle a ship as well as I do a rapier."

A glimmer of admiration crept into Eleanore's eyes as

understanding dawned. "I'll be switched! Then it wasn't just a fluke when you sliced Pierre's arm! You actually are proficient!"

Giving an exaggerated bow, Kathleen smiled slyly. "At your service, ma'am. Now you see the need for my disguise. Reed must never suspect it is I who am pirating him. If you truly mean to help, you could assist Kate in covering up my absences."

At Eleanore's nod, she stated firmly, "Good. I sail tonight."

Kate voiced Eleanore's thoughts. "Are ye sure ye're well enough, lamb?"

"Don't you two worry about me. This voyage and the sea air will do me more good than anything Charles could think up. I draw my strength from the sea. It will provide all my needs and restore my spirit. It won't fail me when I need its power most."

Chapter 18

IT was nearly midnight by the time Kathleen pulled her horse up on the shore nearest the island. She had been forced to ride slowly, for she still had not regained all her strength and had recurring spells of weakness.

Two men stepped toward her from out of the shadows, and just for a moment Kathleen thought her heart had stopped. They came forward, and the taller of the two led her horse away while the other led her to the waiting dinghy. Not a word passed between them as they rowed to the ship.

Once aboard, she tossed her small bundle of clothes to a waiting sailor and told him to take them to her cabin. Dan waited for her on the quarterdeck.

"Are the stores in and everything in readiness?" she inquired. At his nod, she asked, "How many hands have we?"

"Forty aboard, and three at the warehouse I rented in town," came the immediate answer.

Kathleen frowned slightly. "Sounds a bit light, but it'll have to do. I suppose they are all good men."

"Aye, Cap'n. Good fighters and loyal."

"They know what we are about?"

"That they do, and they're eager fer it, let me tell ye. 'Specially if we come across a few of those devil-be-damned British."

Kathleen settled herself on the rail. "Tell me what

you've done since we last met."

"Well, let's see now." Dan scratched his beard. "We've painted her and dyed the sails and changed her name. She's real purty in the daylight, 'specially since we fixed the figurehead."

"Oh?"

"Aye, Cap'n. We took the stars out of her hands. She carries two cutlasses now. We darkened her hair a bit and put the stars in her hair. Call me superstitious, but it be bad luck to remove a ship's figurehead, and since we had to change her looks, we just rearranged some things a bit. Looks good, too, if I do say so meself."

"And the flag," Kathleen pressed. "Did the sailmaster finish the flag?"

"All done and ready to run up anytime ye give the order."

"Are the guns all in working order and plenty of balls and powder laid in?"

"All we should be needin'. And the hands are all armed to the teeth. Begorra! Some of the lads look like walkin' arsenals! They been practicin' every chance they git. Some of 'em are going to make ye right proud."

"That's a weight off my mind," Kathleen admitted. "I've been worrying a bit about that." She paused and went on. "That about covers all I need to know for now. Gather all hands on deck and I'll let them know what they can expect from me as their captain."

She drew herself up to her full height, hands on her hips, with one daintily booted foot propped on the rail. Eyeing them critically from the quarterdeck, she waited until all eyes turned her way and silence fell the length of the ship.

"Listen up, lads!" Her voice rang out commandingly. "We are about to set sail. The rules are simple enough. I am in full command here. Each and every order I give will be carried out immediately. Your lives may depend on it. Deserters will be shot. Mutineers will be hung. When we are engaged in battle, anyone who refuses to defend this ship will be keel-hauled—if he lives that long. There will be no gambling or drinking to excess as we will need clear heads at all times. You'll have ample time for celebration between runs. Disobey, and you'll feel the lash on your back.

"All booty will be divided into shares according to pirate law, although for the protection of your good names I'll not require you to sign an aritcle of piracy. Only Dan and I are in danger of being recognized, so as you see, we are disguising ourselves. You will call me Captain at all times, and never by my given name. Dan will simply be Mr. Bosun.

"Also, if you value your skin you will resist any temptations that may arise to take advantage of my feminine weaknesses. You may find my attire to be provocative in nature, but do not be so foolish as to be misled. The outfits will be solely for the purpose of disguising my identity from my husband and his men. It would prove fatal to underestimate my skills with pistol or rapier. All of you have sailed under my command when my father was alive. We go back many years together, and I know most of you well. I would hate to have to kill any one of you. I demand not only your allegiance, but also full respect as your captain."

Kathleen paused to let her words sink in, and to judge their reactions. They all seemed to be in agreement as far as she could tell. She scanned the faces below her until she found the one she sought. A smile played at the corners of her mouth.

"Hal Finley!" she called. "You will be our quartermaster, if you please, sir!" Her smile broke loose full force as she continued, "And please accept my apologies for taking your position as captain." She swept him a gallant bow.

The young man grinned and shouted over the laughter of his shipmates. "Thank you, Captain. The pay will probably amount to more now, anyway!"

Kathleen joined in their laughter. She felt self-confident and very sure of her crew. Finley would be an excellent man for quartermaster. He was experienced, and the men not only respected him, they liked him. He was in his mid-thirties, of medium height and build, with thick reddish-brown hair and beard. His face was pleasant, and there were laugh lines around his mouth and merry brown eyes. A man of good humor, he was nonetheless a skilled swordsmen and sailor. Kathleen recalled somewhat sheepishly that she'd had an enormous crush on him when she was twelve.

"All right, men! Let's get underway. Mr. Finley, take charge of your men while I man the wheel. Station a couple of lookouts to keep an eye peeled for navy cutters. No lamps and no noise!"

They stole silently out of the cove under Kathleen's delicate hand. It was a tricky feat maneuvering the shoals, but she managed it with ease. Under cover of darkness, they slipped past the shore patrol and headed south for the Florida keys. The sea breeze snapped the canvas to life, and as they sliced through the waves Kathleen felt a hidden part of her soul revitalize itself and spark to life. She felt new strength flow through her as she breathed deeply of the ocean air and felt the deck shifting beneath her feet. Again she realized she could never stay away from the sea for long. She must always return, for part of her soul rested deep within the bosom of the sea.

For four days the *Emerald Enchantress* skimmed across the ocean's surface. The weather remained fair and clear, and with each day Kathleen grew stronger. The ship ran smoothly under her command. Each day the men practiced their swordsmanship under her critical eye. They fought one on one and in pairs, repeatedly rehearsing a system of defense so that each man had his back covered in battle. At the end of each day, Kathleen fought the day's winner, sharpening her own skills, and securing the admiration of her crew.

During these sessions, Kathleen demonstrated technique, and showed her men the best means to take a man out of action with a wound rather than having to kill him. In her heart, she truly did not want any of the men she knew on either her ship or Reed's to have to die because of her thirst for revenge. She stressed this to her men.

If they engaged in battle with another ship and it was necessary for a few of the enemy to die, she would not grieve, but she did not want the blood of friends on her conscience, and neither did the crew, for they seemed sensitive to her feelings and agreed to her requests most readily.

On the morning of the fifth day they reached the keys. Most of the day was spent scouting the area until they found what looked to be a very strategic position where they could lay completely hidden by vines and hanging moss. *The Emerald Enchantress* could anchor between two islands and remain virtually unseen while a very large and undoubtedly well-traveled area was well within their view.

It was a day and a half before they spotted their first ship, but it was a slaver, and by mutual consent the crew declined to give chase. Nearly three hours later, the lookout spied the *Kat-Ann* approaching from the east and the alert went up. The hairs on the back of Kathleen's neck stood up as the *Kat-Ann* passed so closely that Kathleen could make out Reed's figure at the wheel with her naked eye. The *Kat-Ann* was riding too high in the water to be carrying cargo, however, so again they all settled down to wait.

While they waited, Kathleen passed a few hours helping Dan dye his frizzled beard and greying head with some of her black hair gel. The old sailor then wrapped a boldly printed scarf around his head, donned an eye-patch and a sash, and emerged as a fair representation of a pirate. Under Dan's lead, most of the others decided to dress up their appearances. Much to Kathleen's glee, she was soon the leader of as motley a crew as ever sailed the seas!

Their patience was rewarded just before dawn the next day. As the eastern sky was turning a pearly grey, the sharp eyes of the lookout picked out the shape of a Spanish galleon approaching from the west. In her wake rode her escort ship, her sole protection. Word passed silently and swiftly aboard the *Emerald Enchantress*.

Suddenly the sound of cannon fire split the still

morning air. Surprise registered on Kathleen's face as she looked questioningly at her lookout perched high in the rigging.

"Cap'n!" he shouted, the need for silence no longer pressing, "It's the *Kat-Ann!* She's attacking the escort ship, and she's lagging behind the galleon, with the gap wider by the minute!"

It took not a minute for Kathleen to make her decision. "To your stations, men! Captain Taylor has done us a good turn this day! We attack the galleon while he occupies the armed escort. To arms, my laddies!"

A great cheer roared forth from her crew as the *Emerald Enchantress* slipped swiftly from her hidden bower in pursuit of her quarry. The galleon was lumbering along laboriously, heavily laden and riding low in the water. In what seemed but a few seconds, the *Enchantress* was pulling alongside the Spanish ship, grabbling hooks linking the two ships.

"Board her!" came Kathleen's shout as she grabbed a line and swung herself aboard the galleon. Flexing her knees, she landed as lightly as a cat, her rapier ready in her hand. Hearing the light thud of boots behind her, she half turned to find Finley at her back. Her boots tapping a swift tattoo as she ran, Kathleen sprinted up the ladder to the quarterdeck in search of the captain.

As she cleared the final step, she had just enough time to shout a warning to Finley before she ducked the high black boot aimed for her midriff. Kathleen rolled quickly to her right and bounced to her feet to face the stout Spanish captain. Finley immediately covered her back and challenged the quartermaster. Astonishment was etched upon the captain's coarse face as he gaped open-mouthed at Kathleen.

"Surrender this ship at once, and I will call off my men!" Kathleen ordered in a gruff tone.

For a second he said nothing, and Kathleen was about to repeat the command in Spanish when he threw back his head and let out a loud laugh. *"Niña!* What kind of joke is this?" he asked. "Such a small *pirata* I have never seen before!"

Kathleen eyed him coldly. *"Señor,* 'tis no laughing matter! I intend to relieve you of this ship in short order. Defend yourself and your ship or surrender it! The choice is yours!" Taking her stance, she brought up her rapier. *"En garde,"* she challenged.

With an elaborate shrug of indifference, the Spaniard brought up his weapon. "So we must play out your charade, eh? So be it. I will try not to harm you too badly, *señorita.* It would be a pity not to be able to enjoy such a prize, no?"

She refused to be baited. Instead, she parried his first blow with a strength that made his eyebrows raise in surprise. Then she launched into her attack. The ease with which she outmaneuvered him almost disappointed her. In half a dozen swift blows she lured him to drop his guard, and neatly pierced his shoulder through. He dropped his sword at her feet, clutching at his wound in mute amazement.

Kathleen kicked his weapon beyond his reach. "Surrender the ship, sir, and your life will be spared." She watched as he called off his men, noting with great pride that her crew had obviously been besting the Spaniards anyway.

Stepping forward, her long legs splayed beneath her lithe frame, she called out her orders. At once all able-bodied seamen from both ships began transferring the goods from one hold to the other.

Stepping up beside her, Finley said quietly, "Beggin' yer pardon, Captain, but I'd say we have no more than a quarter hour to load and set sail before we have unwelcome company." He nodded toward the *Kat-Ann.*

"We'll take what we can and sink the rest," Kathleen decided cooly.

What little color remained in the Spanish captain's face drained slowly downward. "What of my men, *señorita?* Would you sink the ship with all aboard?"

"I am not completely heartless, Captain. We will allow you to lower the jolly boats before we fire the ship." Eyeing him amusedly, she gave a short laugh, and added with a toss of her head, "Of course, there is a price for my courtesy."

"Which is?" he asked doubtfully.

Pointing to the *Kat-Ann*, still engaged in battle but winning decisively now, Kathleen elaborated, "That ship you see engaging your escort will soon be wanting her rewards, which I am about to make off with. Either you head your jolly boats in her direction, thereby slowing her progress, or I shall direct cannon fire into your midst and you may join your mother ship in her descent. A simple choice, wouldn't you say?"

"Damn you for a sea devil!" the Spaniard swore.

From the corner of her eye, Kathleen caught Finley darting her a quick glance of admiration. "Order the jolly boats lowered, Finley, and prepare to fire the ship."

Two minutes later, Finley brought word that all prisoners but the captain were aboard the jolly boats and about to cast off. In the hold, a keg of blasting powder was ready to be set to the torch.

Kathleen gave the order to return to the *Enchantress*, and turning once again to the defeated captain, she said, "We depart with profound thanks and, of course, your cargo." With a tinkling laughter, she swept him a low bow. Then, graceful as a gazelle, she swung back to the deck of her own ship as Finley lit fire to the thin line of black powder leading into the hold.

"Cast off, and be quick about it!" she commanded from the quarterdeck. Several men were already severing the lines to the grappling hooks, while others raced to lengthen sail. They lurched away from the galleon, and catching the morning tide and trades, made haste to widen the gap between the two ships. The *Emerald Enchantress* responded to Kathleen's sense of urgency as she deftly manned the wheel. Even at a safe distance, the ensuing explosion rocked the *Enchantress*, and bits of flying, flaming debris littered her wake.

As they rounded the first island, Kathleen dared a look back, and the sight that met her eyes made her heart glad.

The *Kat-Ann*, wildly attempting to pursue her, was being effectively hampered by the jolly boats, three of which had capsized with the force of the explosion. Seamen were afloat and swimming everywhere in chaotic confusion. Kathleen could just make out the Spanish captain as he shouted at Reed and gestured frantically in the *Enchantress's* direction. Then they rounded the bend and entered the twisting maze of waterways that wound around the islands.

They were safely enclosed in their hideaway of greenery and snickering to themselves as they watched the Spanish sailors being rescued by the limping, but still seaworthy, escort ship. The *Kat-Ann* was gone from view, having made a vain attempt to catch the *Emerald Enchantress*. Lord only knew where in this labyrinth of channels she was now, but because she could well be close by, those aboard the *Enchantress* maintained a disciplined silence. Only the sparkle of mirth in each pair of eyes betrayed the smothered laughter.

Their silence soon proved prudent. As the last of the floundering Spaniards was hauled aboard the warship, new sounds, faint but distinct, reached their ears. Muffled voices, the creaking of lines and wood, became clearer minute by minutes.

Aboard the *Emerald Enchantress*, each man stood frozen in his stance as the *Kat-Ann* steadily approached. Hardly daring to breathe, they watched as the *Kat-Ann* passed their stern so closely that a man could have reached out and touched her planking. Kathleen clearly saw Reed standing at the wheel, and when it seemed he looked directly at her, she wondered that he had not heard her heart, it was pounding so hard in her chest. Blood hammered in her ears, and moisture gathered in her eyes from the strain of not blinking. She could feel rivulets of perspiration streaming down between her breasts and shoulder blades.

When finally the *Kat-Ann* had passed and gone, a collective sigh of relief rose from all aboard. Kathleen caught Dan's look of mixed relief and disbelief, and smiled broadly. "By the saints! I thought we were found out, old friend," she whispered. "I held my breath so long I nearly passed out!"

Dan grinned back at her. "Were thet yer heart or mine I heard poundin' so loud, Cap'n?" he chuckled low.

"Both, I'd imagine, and it won't surprise me if my boots slosh a bit when I walk, too."

"Yea, well I'll tell ye, it didn't feel none too good when I swallowed me tobacco neither, lass," he added with a grimace.

Tears of merriment danced in her eyes as she doubled over in an attempt to restrain an unladylike hoot of laughter. "Oh, dear heavens! You didn't!" she choked.

"Aye! And it burned like the fires of hell!" He eyed her with disdain. "Well, I couldn't very well spit just then, could I?" he exploded. "It would have sounded like a cannonball hittin' the water!"

At a strangled sound behind her, Kathleen turned to find Finley wiping tears from his eyes and struggling to regain his composure. "What are your orders now, Captain?" he managed between breaths. "Do we head for home with our take?"

"Nay. I'm thinking we'll cruise about a bit and see if we can't give Captain Taylor a good look at the lady pirate he's most likely very curious about just now."

Reed was more than merely curious. He was furious! Of all the absurd, idiotic things to have happen! To top it all, that imbecilic Spaniard telling that ridiculous tale about a daring lady pirate! Impossible! Especialy if he were to believe the Spaniard's description of her! He had said she was tall, with beautiful long legs encased in black knee boots. Reed smiled scornfully as he recalled the captain's words.

"She wore an outfit all of green and black, and her—uh—her breeches, well, they were cut off short to just below her buttocks. Such thighs! The shirt and vest did nothing to disguise her luscious breasts. When I think how they would fit into the palms of my hands!"

"Damn it, man!" Reed had stormed. "What did you

do, give her your entire cargo in exchange for a hop in the sack? You don't expect me to believe all this rot, do you?''

The captain had looked offended. ''The woman can fight like a man, I tell you! Do you think she did this to my shoulder with her fingernails? I still cannot believe that I was bested by a beautiful *pirata!* She is very bold, and so sure of herself. She is the captain of that band of cutthroats! She leads them!''

''And I suppose she has the face of an angel with a voice to match?'' Reed's voice was venomous.

''What I could see of her face, *sí.* She wore a mask, but I can tell you she has sea-green eyes and long hair as black as midnight.''

''I'll tell you what else she has!'' Reed countered. ''She has my booty!''

''Ha!'' The Spaniard laughed tersely. ''If I had to lose my ship, at least it was to her and not you!''

''Yes, well I intend to remedy that situation and see this *pirata* for myself. We'll see how good she is pitted against a real man!''

''Yes, indeed,'' Reed thought again. ''I know how to handle myself with a sword, and I'll be damned if some woman will best me! I hope we meet soon, lady pirate, and I won't be surprised to find you are as fat as a sow, with breasts like udders and thighs like tree trunks, and a face that would turn milk sour. These Spaniards have no taste—no taste at all!

The day was fine, clear and balmy with blue skies and a brisk wind to fill the sails. It had been two days since the *Emerald Enchantress* had attacked the galleon. Reed had searched relentlessly, constantly cruising in the hope of encountering the woman pirate and her crew. His anger had not abated, and his pride was still bruised. That, more than anything, made him continue to search.

When all at once they did come upon the sleek green vessel, Reed was taken aback. The *Emerald Enchantress* was not under sail, but resting quietly on the ocean waves, rocking slowly to and fro.

''What the devil?'' The frown lines deepened in his

broad forehead as Reed eyed the ship speculatively. She was flying no distress signal. A shiver ran up Reed's spine as he got the impression that this was no chance enounter. The witch was actually awaiting his arrival!

As if the thought of her had conjured her into being, Reed caught a flash of green as a slim figure detached itself from a group of men and walked across the deck. Her back was to him as she lightly bounded up the ladder to the bridge. Through his spyglass he got a quick look at long, well-shaped legs and a nicely rounded bottom below a shining curtain of long ebony hair. The short pants barely concealed the cheeks of her rump, which bounced tauntingly with each step she took.

"By heavens, I believe I owe that Spaniard an apology!" Reed whistled softly. The woman reached the bridge and turned to lean over the rail to speak to a man below her, and Reed sucked in his breath as he saw she wore only a vest with no shirt beneath. The manner in which she was leaning over was affording him a spectacular view of her breasts. He wished she was not wearing the mask so he could see her face, too. While he watched, she laughed at something the man said and, tossing her hair over her shoulder with a quick flip, she walked in smooth strides to the helm and strapped on her swordbelt.

The *Kat-Ann* approached rapidly, but with caution. They were well within firing range, yet no shots came from the *Enchantress's* cannon.

"Shall we fire on her, sir?" called Reed's quartermaster.

"Nay, Mr. Young. Have the men stand ready to board when I give the order. First, however, I should like to converse with the lady captain and find out her game."

Aboard the *Emerald Enchantress*, Kathleen was struggling to maintain her cool outer composure. Inside she was a quivering mass of jelly. "Oh, why did I ever start this venture? What if he recognizes me?" She fingered the hilt of her rapier nervously. It was the only outward sign of her anxiety.

She caught Dan's eye, and he came to stand beside her. "Don't ye be worryin' none, gal. Yer own daddy wouldn't recognize ye in thet outfit. Lor'! He must be spinnin' in his grave to see ye standin' here with all thet skin showin'!" He gave her a wicked wink, and said, "I should be tryin' to talk ye out of this, I know, but all I can do is wish I was twenty years younger."

"Dan!" she exclaimed. "You old rake! You really amaze me!"

"Yeah, well, ye got some amazin' to do yerself jest now. Buck up, lass, and remember what yer about. Yer not Mrs. Taylor nor Lady Haley today. Yer the bonniest pirate ever to sail the seas! Yer the best at sword and sail. Yer the Emerald Enchantress!"

Finley interrupted by pointing out that the *Kat-Ann* was nearly upon them. Kathleen pulled herself away from the men, squared her shoulders, and walked to the rail. She propped her dainty boot on the rail and pulled on her leather gloves.

"Ahoy! Aboard the *Kat-Ann!*" she called in an affected husky voice. "That is far enough! State your business!"

Reed stood in his black trousers and shirt, hands on hips, legs braced apart. She could see his arrogant sneer from where she stood.

"We're coming aboard!" he shouted back.

"We won't stop you, but I doubt you'll like the welcoming committee! What business have you with this vessel?"

"A little matter of settling accounts," Reed announced as his crew made fast to the *Enchantress*. He grabbed a line and swung himself across, landing just two feet from where Kathleen still lounged against the rail.

"Now, madam," he said, sweeping her a mocking bow, "if you would be so kind as to order your booty transferred to my ship, we can avoid any unnecessary bloodshed."

Kathleen drew herself to her full height, planting her feet firmly on the deck. "I think not, Captain."

"Come now, madam. It is not an old, slovenly Spaniard you are dealing with this time. I am an excellent swordsman, and as such I would hate to injure either your pride or your person." His eyes slid to her cleavage and her proud breasts displayed between the lacings of her vest.

Kathleen unsheathed her rapier with a smile, her pearly teeth flashing. " 'Tis your own pride you should fear for, sir. And just for the record, my title aboard this frigate is Captain, not madam."

Sounds of the fight rose to them from the deck below. Kathleen presented her weapon and shouted over the din, *"En garde."*

With a slight shrug of his broad shoulders, Reed did likewise.

At first Kathleen let him take the offensive while she parried his blows and studied his moves. Her eyes never left his as she danced away from his flashing rapier. Surprise registered momentarily on Reed's face, and then his face became a blank mask, except for his ice-blue eyes. Kathleen had counted on that, and found she could anticipate most of his moves by watching his eyes. Of course, Reed was applying the same technique with her, so the contest was near evenly matched.

"Damn!" Reed swore silently. "The wench is good! Where, I wonder, did she learn to fence so well? If she weren't such a slight little twit, I might start to worry, but I know that rapier is heavy and she'll tire soon."

As the match wore on, Kathleen showed no sign of slowing. In fact, if anything, she was countering better. A moment came when the tables turned, and both of them realized that Kathleen had placed Reed on the defensive. Her blows were strong and well-calculated. Her sword lashed out at him time and again, so swiftly that he had only time to deflect the blows, and none in which to attack. He side-stepped several of her lunges and suddenly he saw his opportunity. Throwing all his strength into a forward motion, his blade met hers and slid to the hilt, and he shoved her to the deck. As their swords parted, her blade swung up in an arc and slid up the inside of his leg. All at once he found himself in the precarious situation of having the point of her sword at his manhood. He drew in his breath sharply.

From where she had fallen to one knee, Kathleen looked

up at him. A smile dawned at the corners of her mouth, and a husky laugh broke forth to dance across the waves and echo back to them. Goose flesh worked its way down Reed's spine as he looked down into bright green eyes glittering with malice.

"Well now, Captain," she said softly. "It seems you have a decision to make. Tell me. Have you any children?"

"Not as yet." Somehow Reed managed to speak the words in a clear, unwavering voice.

"Then I suggest you hand over your weapon if you hope to father any in the future." She gave him a leering grin as he complied. "I applaud your wisdom."

She reached out and drew his knife from his boot top. "Now," she said, "very carefully remove your pistol with the fingertips of your left hand, and hand it to me." Her eyes searched him for other weapons. "Turn around slowly." She stood, and with her swordtip at his back, plucked another knife from his belt.

From her vantage point, she looked to the deck below. Proudly she noted that her men had fared as well as she. The battle was over, the day was won. All eyes followed them as she directed Reed down the ladder.

"The good captain and I are going aboard his ship," she announced with a chuckle that made Reed's blood boil. "I want three men to follow and check the hold for goods. She rides high, so I doubt she carries much," she ribbed.

Kathleen's guess was good. There was no booty to be had on the *Kat-Ann*. Finley reported this to her and asked, "Shall we fire the ship, Captain?"

Reed's face blanched white as he awaited her reply.

"Not this time, I'm thinking." Looking up at Reed, she added, "We'll meet again, Captain, and perhaps you will have more to offer next time."

The nerve in Reed's cheek twitched angrily, but he replied evenly, "Believe me, madam, we will indeed meet again. But do not count on the outcome of that day. You cannot be so outlandishly lucky twice running."

His eyes raked her body as he slowly undressed her in his mind. Then he looked deeply into her eyes and declared, "Rest assured, Captain Green Eyes, it will be more than booty I shall demand in payment for this day."

"We'll see, won't we?" Kathleen replied with a devilish grin. Turning her head, she instructed Finley, "Have the men remove the firing devices from all the cannons and place them in a gunny sack, then take them aboard the *Enchantress*."

When that was done, Kathleen returned to her own ship, but not before she swept Reed a gallant bow, winked wickedly, and commented, "Since you carry no cargo, I really should demand you pay me with your body, but I haven't the time to dally today. Perhaps when we meet again. I'll wager you surrender nicely."

They lifted anchor and Kathleen steered the *Emerald Enchantress* to the bow of the *Kat-Ann*. Then she had two of her crew tie the sack of firing devices to the *Kat-Ann's* bowsprit. By the time the crew of the *Kat-Ann* untied them and replaced them in the guns, the *Emerald Enchantress* would be long out of sight.

Chapter 19

KATHLEEN and Ted sat quietly in a corner of the library at Chimera playing chess. Eleanore sat reading nearby in an overstuffed chair, her feet curled up beneath her. Though it was early afternoon, lamps were lit already. The day was grey, with rain pouring in torrents against the window-panes. Periodically, lightning would slice through the clouds and thunder would growl ominously.

All three looked up simultaneously as Susan entered bearing a tea tray. "Momma thought you might like some spiced tea."

Eleanore uncurled like a lazy kitten. "Mmm, yes. That sounds lovely. What about you, Kathleen?"

Kathleen stood up and stretched. With a light chuckle, she walked to the divan. "It will give me something to do while Ted figures out his next move. Thank you, Susan."

Ted looked up indignantly. "Susie, bring me a cup over here, will you, please? Miss Smarty-Pants has me boxed in, and it looks like I'll be a while figuring a way out."

The three girls exchanged a look of amusement.

"That will teach you to play out of your league." Kathleen stuck out her tongue at him playfully.

"Well, it appears you are feeling much better," came the deep voice from the doorway.

Four heads swiveled. Kathleen's face immediately set into grim lines as she saw Reed standing there, his broad shoulders filling the doorway.

"Reed! You're home!" Susan exclaimed joyfully. She put down her tea and ran to embrace her brother.

"Glad you're back, Reed," Ted called from his seat near the window. "I've been hoping you would turn up soon. Eddie Newcomb's uncle has a horse I'm seriously considering, but I wanted you to see him first and advise me."

"Will it wait a day or two until I get some business around here in order?"

"Yeah, I suppose so, but Gerard Ainsley has his eye on him too, and I don't want to wait too long. I'll talk to Eddie and ask him to hold off for a few days."

Reed sauntered over and deposited his long frame on the divan next to Kathleen. She and Eleanore exchanged a quick look of anxiety.

Accepting a cup of steaming tea, Reed smiled lazily. "Hello, Eleanore. Have you heard anything from Charles yet?"

"As a matter of fact, I got a letter from him just today. He's staying with a fellow doctor and friend of his in Boston. He is very impressed with the caliber of doctors who are scheduled to speak at the seminar." Eleanore took a sip of her tea and asked innocently, "How as your venture?"

Reed gave a disgusted grunt. "You wouldn't believe me if I told you."

"Oh? That bad, huh? Well, now you've gotten me curious. Pray tell us what happened," Eleanore pressed.

Kathleen, who was having a devil of a time suppressing a grin, dared not look at either Eleanore or Reed. "By jiminy, what an actress you are, Eleanore," she thought as she mentally tipped her hat to her friend.

"Did you have some sort of trouble?" Susan added concernedly.

Now Ted was interested, coming to join the group around the tea table. "I say, did something unusual happen?"

"You might say that, thought after I tell you, that will seem an understatement. The fact of the matter is, I had

291

an encounter with a pirate ship, not once but twice.''

"A pirate ship!" Susan's hands went to her throat as she eyed her brother with both worry and awe.

"My! How exciting!" Eleanore sat up straighter in her chair, brown eyes twinkling. "What happened? Did they confiscate your cargo? Did you fight? Were many of your men killed?''

"Yes! Yes!" Ted urged. "Go on. Tell us all the gory and glorious details.''

"There isn't much to tell, actually. Yes, we fought. No, no one was killed, and yes, she has my cargo.''

"She!" Eleanore exclaimed, and then relaxed in her chair. "Oh, I keep forgetting you refer to all ships in the feminine gender.''

"You were right the first time, Eleanore. I wasn't referring to the ship, but the pirate captain. I still can't believe it myself!''

Reed gave Kathleen a long look. "You are remarkably quiet. Aren't there any questions you have about this unusual occurance, or are we boring you?''

"Not at all. I was just wondering what this woman pirate would look like. I suppose she was very mannish in appearance?''

"On the contrary. I have rarely seen a woman more beautiful!" He waited for her reaction, and got none.

"Pray go on, Reed!" Ted was all ears.

"You wouldn't have believed her, Ted. It is a good thing you were not along, or you may have joined up with her. I do believe most of my men were jealous of her crew. She has the longest legs I've ever seen on a woman, and they were in full view, as she wore breeches cut off so short that they barely covered her bottom.''

Ted's eyes widened, and Reed went on, "She wore a matching green vest with no shirt beneath, and the view between the lacings would have made you drool.''

Ted groaned aloud, and Susan's face flushed as she stared at her brother.

"I'm sorry, Sis," Reed apologized. "I shouldn't be saying these things before ladies.''

"That's quite all right," the three girls chorused together.

"Don't stop there. I'll die of curiosity!" Eleanore protested.

"She wore a mask, but still her face looked lovely, and she had eyes as green as Kathleen's and hair as black as Titan's coat, long and satiny." He glanced at Kathleen.

"An angel!" Ted sighed as he collapsed ecstatically into the nearest chair.

"Angel my foot! The devil's own daughter! She wields a rapier with a finesse most men would envy, and all the while she is tantalizing you with that voluptuous body. Her frigate is aptly named, for she calls it the *Emerald Enchantress*—after the color, which is entirely green, sails and all, or after her own tempting self, I am not sure which."

"Brother, dear, how can you go on so about this woman as if you admire her? She has stolen your cargo!" Susan was aghast.

"Quite right, Susie, but we shall meet again, I am sure. I know it as surely as I know the sun will rise tomorrow, and our next encounter will have an entirely different outcome."

"How can you be so sure?" Kathleen interjected. "She obviously bested you this time. It seems you have met your match at last, Reed." Her green eyes danced merrily.

"Are you jealous?" Reed's eyebrows raised curiously, a smug look on his face.

"Not particularly," Kathleen countered. "I'd have to be interested in you to be jealous, now wouldn't I? And we both know donkeys will sprout wings before that happens."

Amid the astonished expressions of her friends, Kathleen rose gracefully and exited the room. Reed was clenching his teeth in anger in an effort to hold back an equally scathing retort that would have made his gentle sister swoon.

Playing the recovering invalid, Kathleen had both her supper and breakfast the next morning served on a tray in her room. Unwilling to tip her hand, she had not ridden Zeus since her accident. Late morning found her leaning against the gate of the paddock coaxing him with carrots, when Reed sauntered up to stand beside her.

"How are you feeling, Kat?"

"I'm right as rain, thank you. Now go away. I don't want to talk to you." She flashed him a hateful look.

"I don't care!" He spun her around to face him, holding her chin in his palm so she could not turn away. "I want to speak to you, and you are going to cooperate," he told her firmly. "Have you given any thought to what I said to you before I left?"

"You said a great many things. To which are you referring?"

"Don't play dumb, Kat. It's not your style. I want you to tell everyone we are married and live with me as my wife."

"Look, you stubborn lop-eared mule!" she exploded, her face red with fury. "I told you no and I meant no!"

She, in turn, took his face between her palms, holding his head steady. "Now," she said, "watch my lips very carefully." Very slowly and distinctly, in a voice Charles could have heard in Boston, she repeated, "No!"

"Yes!" he yelled back.

"No, never!" she shouted, shaking her head back and forth on every syllable.

"Blast!" Reed expounded at the end of her tirade. "You are the most irascible, irritating, insensitive female God ever placed upon this earth!"

All at once he became contrite. He looked down at her, and his eyes were soft and warm. When next he spoke, his voice was tender. "Listen, kitten. I know you are angry with me, and you have every right to be. I said things I could cut my tongue out for!"

"Let me do it for you. I'll guarantee a better job," she broke in caustically.

"I'm trying to apologize to you. I never meant for this to happen. I've never wanted to hurt you, and I'm so sorry about the baby."

Kathleen had felt herself starting to soften toward him until he mentioned the baby. She pulled back as if stung by a hornet.

"Don't you ever mention the baby to me again, you demonic beast! After all you have put me through, all you have said and done, I'm not about to forgive and forget! You deserve everything you've got coming, and I hope you get your just deserts, you stinking barbarian!" With that, she spun on her heel and marched back to the house, and he never saw the flood of tears streaming down her cheeks.

Reed was at a loss as to how to deal with Kathleen, so he decided to start afresh and court her with all the loving tenderness he could command. In the days that followed, he sent her roses and books of sonnets. He seated her at the table, brought her refreshments when he thought she was thirsty, and generally treated her as if she were made of glass.

Eleanore was highly amused, and there were several pairs of raised eyebrows around the place, but Kathleen remained unmoved, at least on the surface. Inside, she felt as if she truly were made of glass, and all of it had shattered.

Though she showed him no encouragement, he persisted. He clamped a lid on his temper and was unfailingly polite, even when she railed at him or baited him. Many were the nights he sat undiscovered in her room watching her sleep, hoping for a miracle to soften her heart toward him. If nothing else, he knew he needed her forgiveness for the cruel words he had spoken before her fall.

Kate came by frequently, and just shook her head in despair at her granddaughter's stubbornness. One Thursday she popped in and announced that she was hosting the final picnic of the season the coming Sunday afternoon. They were all invitted to gather at Sandy Point, a lovely spot by the river on her plantation.

The same day, she pulled Kathleen aside for a little grandmotherly advice. "Little lass, don't ye think 'tis high time ye stopped feeling sorry for yerself?"

"Kate! How can you say that?"

"Ever since ye came back from yer little jaunt, ye've been sitting around playing the invalid. When are ye going to come to Emerald Hill and learn something about horse breeding? We made a bargain, if ye recall."

"You're right. I suppose I have been neglecting you, and I must make good my promise. I'll be by soon."

"That's me girl!" Kate gave her a big hug. "Besides, 'twill get you away from Reed, if that's what ye want." She gave Kathleen a wink and said, "I hear he's been very attentive lately. Perhaps ye're changing yer mind about him."

"Not a chance!" Kathleen declared.

"I also hear ye've been putting him through the paces. Are ye enjoying yerself, lass?" she asked critically.

"Yes! As a matter of fact, I am, and I don't need any lectures to that effect either." Kathleen's tone was belligerent. "He's a thoroughly rotten rogue, and he's getting his just desserts!"

"All right, lovely. Don't get yer dander up. Just step carefully and remember, ye get as good as ye give sometimes."

"I'll take it under advisement." Kathleen leaned over and kissed her grandmother's cheek. "And Gram," she whispered, "thank you for caring."

The weather Sunday was perfect for a fall picnic. The sky was as blue and clear as it could be. A light breeze tugged at Kathleen's coppery curls as she rode in the open carriage. Reed rode Titan at her side. She was finding it increasingly hard to ignore him. She had never seen him in this role before, and she had to admit he was charming.

They spread out their blankets beneath a large, shady oak near the water's edge, and left the unpacking of the lunchbaskets to the servants. A group of young men were trying to set up an improvised polo field nearby, and a gaggle of giggling girls were waiting while servants placed the wickets for a game of croquet elsewhere.

Reed courteously escorted Kathleen about, her hand tucked into the crook of his elbow. He partnered her in a game of croquet, and only when she agreed to rest and watch in the shade, did he join his neighbors in a polo match.

Kathleen sat quietly with Kate, and as hard as she tried not to, her eyes kept straying to Reed. She told herself that it was simply that she admired his steed and his horsemanship, but she knew it was untrue. He looked so strong and handsome and virile astride the giant black horse. His tan face was relaxed and smiling as he enjoyed the competitive sport, and her palm itched each time she saw him toss back the inky wave of hair that kept falling across his forehead. He had discarded coat and tie, and a shadowing of curly dark hair showed where he had unbuttoned the top of his shirt. Titan responded well to Reed's commands, and the two made a striking picture. Unconsciously, Kathleen's tongue slipped out to lick at her lips, and Kate burst into laughter.

"Ye should see yer face, lass!" she laughed. "No matter what ye say, ye love him still."

"I never said I didn't," Kathleen said softly.

Gerard Ainsley claimed her attention for a good portion of the afternoon, and it was mainly to avoid his ardent advances that after lunch Kathleen allowed Reed to talk her into a ride on the river. He placed her gently in the bow of the rowboat, and rolling up his sleeves, took up the oars.

Neither of them spoke until Reed finally asked, "Are you getting tired, Kat? You really must rest and not overdo, darling."

"I'm tired of resting," Kathleen complained. "I want to ride and run and play. I'm sick of being coddled. Just now there is nothing I would like more than to take off all my clothes and go swimming."

Reed chuckled at her petulant look. "I'm sure I wouldn't mind, but Barbara would probably faint dead away."

Kathleen grinned at the thought of her aunt's reaction. "Yes, I really must behave in a ladylike manner, I suppose, but sometimes it is so hard!"

"I still remember how you looked standing on the beach at Grande Terre with your hair all wet and streaming down your back, wearing nothing but my shirt clinging to you as if it were painted on." His tone was wistful, and his eyes had darkened to the hue that Kathleen recognized instantly. He wanted her in his arms with her lips moist and clinging to his.

Kathleen felt a familiar tug at her insides, and a warmth that traveled through her with shocking intensity. She turned her face from his with a shaky sigh. "Don't, Reed, please. Just row us back now."

He had read the look on her face, and for a moment felt hope. He shrugged at her words and thought, "Well, at least she's speaking to me, and I know now that she's not entirely immune to me. I've plenty of time to wear down her defenses. I'll just keep trying."

Talking to Susan on the way home, Kathleen was surprised to find that young girl had never been on a ship. "Do you mean to tell me that as long as Reed has been sailing he has never taken you aboard?"

Susan shook her head mutely.

"Reed Taylor! You should be ashamed of yourself!" Kathleen wagged an accusing finger at him.

"He's never invited me, either," Amy whined, tossing her blond curls. "He hasn't even extended the offer to his own mother!"

"This is an oversight that should be rectified immediately. Surely now that you have your own vessel you could see to it, Reed."

"Kat! You make me feel like a naughty boy in knee breeches caught sneaking cookies from the kitchen."

Her laughter bubbled. "You should feel naughty!"

"All right! You are all invited to come aboard the *Kat-Ann* before I sail next, but you must give me a few days to ready her for your tour."

"Tour!" Kathleen reproached. "Why not be magnanimous and give us a short cruise?"

"Next you'll want me to have champagne and finger cakes on board," he laughed.

"But, of course!" she nodded brightly. "Now you are getting the right idea!"

"Precisely," Amy concurred, for once in agreement with her cousin.

"Oh, yes, Reed. Please!" Susan begged prettily.

Mary smiled at Barbara and nodded.

"I'm outnumbered, it seems. All right, you win. Let's make it Thursday. We'll start out bright and early and I'll take you downriver to the ocean. We'll follow the coast for a short way and then return." Turning to Kathleen, he smiled. "Are you satisfied now?"

"Quite," she answered smugly.

The cruise was pleasant enough at the outset. The river was calm and the morning was mostly sunny, though a few clouds were gathered here and there.

"What an ungodly hour to be up and about!" Amy complained loudly.

Kathleen settled herself in a chair on deck and enjoyed the scenery and sun. Reed was busy at the helm, with Amy shadowing his every move. Ainsley, seeing his golden opportunity, glued himself to Kathleen's side. Most of the other young gentlemen eyed him ruefully and went on in pursuit of other game. Even Ted deserted her, courting Susan as if this were his last chance. Uncle William and Barbara were standing off to themselves and behaving like newlyweds, and Kate was busy convincing Mary that she was not seasick.

By lunchtime, Kathleen was considering tossing the persistantly pesky Gerard overboard. But Mother Nature relieved her of the necessity.

Once they cleared the mouth of the river and entered the open ocean, the waves were much choppier. The sky was becoming increasingly overcast, and the *Kat-Ann* was rocking more with the waves.

Kathleen, of course, was undeterred, and left a pale Ainsley in the dining area. On her way back on deck, she passed Amy, who was grumbling to herself about the wind ruining her coiffure. Kathleen grinned impishly. Amy did indeed look a sight! Beyond being windblown, she had a peculiar greenish cast to her face, and was clutching her arms about her waist.

A few of the guests suffered mal de mer, but most were gamely traversing the deck, determined to enjoy their outing despite a few inconveniences. Kathleen stood by the rail and enjoyed the feel of wind on her face. A few of her hair pins pulled loose. She removed the rest and let her hair tumble down her back in a coppery cascade.

Reed slipped up behind her. His arms snaked around her waist before she realized he was there. Placing his lips near her ear, he said softly, "Does this bring back pleasant memories, Kat?"

"Not all the memories are pleasant, Reed. By the way, I passed Amy and she looked miserable."

"Not much of a sailor," Reed agreed. "Where is your shadow?"

"Ainsley?"

"Who else?"

"He's below turning green about the gills." She giggled.

"Lord, I love it when you laugh," Reed vowed as he pulled her closer and pressed his face into her tumbled tresses. "Mmm, your hair smells good; just like lemons. It's been so long since I've seen you with your hair down. I love it!"

Kathleen squirmed in his arms. "Reed, please! People are starting to stare!"

"Let them," came the husky reply.

"I'm warning you, Reed Taylor!"

He dropped his arms and stepped back but a pace. "I'm sorry, but there is just so much a man can stand. Maybe you'd better go below and pin up your hair. You are much too tempting as it is."

"You and your obsession with my hair," she grumbled half to herself.

Reed's good behavior finally wore thin two days later. He'd gone up to Kathleen's room to ask if she would like for him to exercise Zeus for her. As he neared her veranda door, he heard voices. At the doorway he stopped to let his eyes adjust from the bright sunlight.

"Now, Nell, quit fussing about and let's get this done." Kathleen's voice floated out to him.

"But, ma'am, yoah haih be so purty! Ah doan know why yoah wants ta cut it. 'Sides, ah ain't nevah cut haih lak yoahs befor'."

"For heavens sake, Nell! Just pick up those scissors and get started. I'd do it myself if I could reach the back, but I can't."

Reed had heard all he needed to. He stepped over the threshold, and in three long strides reached Nell just as she held the scissors toward the first of Kathleen's curls. He knocked the scissors out of Nell's hands.

"Pick up those shears and get out of here, Nell! Don't ever let me catch you even thinking of assisting Miss Kathleen in cutting her hair, or so help me I'll whip your black hide!" he roared.

Nell had never seen her master so furious, and it frightened her so that her hand shook as she reached for the scissors. Her eyes were enormous with fear.

Kathleen whirled on Reed. "How dare you barge into my room and order people around! I asked her to help me, and she's going to. It's none of your business what I do, Captain Taylor. You don't own me!"

"You think not?"

"I know not!"

"That's debatable. At any rate, I do own Nell and I am ordering her not to assist you." He regarded her stormy face. "Furthermore, I'll make sure no one else comes to your aid."

Nell scurried from the room.

"Damn you, Reed! Who do you think you are?" Kathleen screamed.

"I know who I am. It is you who seems to find it convenient to forget your place. I'm here to remind you that, like it or not, I am still your husband. I still have some say in your life, and I'm ordering you not to cut your hair."

"It's my hair!"

"True, but you are my wife and I am warning you, if you cut so much as one lock of it I'll turn you over my knee and give you as spanking you'll never forget."

"You despicable bully!" Kathleen faced him, hands on her hips. "You don't scare me!"

"You may not be scared now, but defy me on this, Kat, and you certainly will be. I promise you! I want your word that you will not cut your hair."

"I'll think about it!"

"That's not good enough. I want your word on it."

"All right!" she relented with a sigh. She gave him a disgusted look. "You have my word, you ogre. Saints, but you can be a beast when you want to! I see you've dropped the 'adoring suitor' facade. I wondered how long it would last."

"I can see now that is not the way to handle you." Reed glared down at her.

"And this is?" she queried hatefully.

"This, at least, has some effect," he answered stiffly.

"Don't push your luck, dear."

"Don't worry your pretty little head about it. After Sunday I won't be here to pester you."

"Are you going in search of your lady pirate?"

"As a matter of fact, yes. I have a score to settle with her."

"Well, she has my vote. I, for one, hope she skins you royally!"

At breakfast Monday morning Kathleen told everyone that Kate wanted to teach her something of horse-breeding, and that she would be spending a great deal of time there. Eleanore decided to go with her.

After the meal, she had Mammy pack up some of their clothes and send them to Emerald Hill.

Kate was waiting for them when they arrived. "I've been expecting ye," she said with a rueful smile. "I heard Reed left this morning. I don't suppose ye'll be learning much about horses in the next few days."

"I'm sorry, Kate, really I am," Kathleen apologized. "I'll be sailing this evening. I've already sent word to Dan.

We can't disappoint Reed by having his pirate lady fail to show, can we?''

Kate rolled her eyes and chuckled. ''Heaven forbid!''

They used their previous hiding spot between the two islands. Lady Luck was with them. They waited but six hours before they swooped down upon an English brigantine, confiscated her entire cargo of milled goods, and sent her to the ocean floor. They set her men adrift on the seagoing tide in dinghys.

Two days passed and they hit upon a Spanish bark directly from Mexico. Chests of gold, silver, and jewels were brought aboard, and she too was sunk. The green-eyed Enchantress of the gulf was alive and well, and her reputation was spreading like wildfire.

Going through the chests, Kathleen came across two beautifully worked snake-shaped gold arm bands. An emerald was set in place for each eye. These Kathleen took for her own, along with a pair of gold hoop earrings and a large gold medallion with an emerald center-stone. She added these to her pirate attire, and the effect was stunning.

The following day passed uneventfully, but dawn the next day brought an alert from the eagle-eyed lookout. The *Kat-Ann* had been spotted. Kathleen manned the wheel, and they threaded their way in and out of the islands and slipped silently up behind the *Kat-Ann* before anyone realized it. When they boarded her, most of the crew was still below in their bunks. Reed was in a rage.

Kathleen stood poised before him on the *Kat-Ann's* bridge, her rapier drawn. ''It seems we meet again, Captain Taylor. I must say it was very rude of you to come along so early and disrupt my beauty rest.'' She faked a delicate yawn.

''My apologies,'' Reed remarked mockingly.

''As retribution you might hand over your plunder. I see you ride low in the water this trip.''

He glowered at her, his eyes glittering like cool blue diamonds. ''You brazen sea-witch!''

She laughed up at him, her eyes a perfect match to her

emeralds. "I'm sure that is not what you would like to call me."

"How right you are! Where in Hades did you come from so suddenly?"

"Why, Captain, surely you don't expect me to answer that question. Let's imagine that I materialized out of the mist. Just possibly I am but a spirit, a figment of your imagination."

"Somehow I doubt that. I can feel the heat of you from here."

"How perceptive of you! You must also realize you have no recourse but to surrender to me once again."

"You'll forgive me if I do so with regret."

She nodded, and her earrings tinkled, their musical notes irritating him all the more.

He called to his men, ordering the transfer of goods to her ship. "Reports of your pirating ventures are reaching a good many ears, yet no one knows a thing about you," he stated.

"What is there to know?"

"Such things as where you hail from, why we've not heard of you before, your name," he pressed.

She gave him an amused look. "That is my private business. Next you will be asking that I send engraved announcements of my next attack. I come from home. I had parents like everyone else did, and if you must call me by some name, let it be simply Captain or Emerald."

"We'll meet again you know, Emerald," he grated from between clenched teeth.

"You repeat yourself, Captain," she chuckled softly. "Now I must say *adieu*. It has been very lucrative running into you."

She swung herself aboard her own ship. "Say hello to your friend Lafitte for me and tender my apologies for cutting into his profits." She shouted her parting remark to him as they sailed swiftly away.

Her hold full, the *Emerald Enchantress* headed for home.

Once in Savannah, Kathleen decided to appease Kate

by spending a few days actually studying breeding. On her fourth day back she, Kate, and Eleanore were sitting on the veranda enjoying the sunset when Reed rode up on Titan.

"My, you're back soon!" Kathleen remarked.

"Yes, I just couldn't stay gone from your radiant face and loving ways," he grumbled churlishly. He plopped himself into a wicker porch chair.

Kate chuckled. "Ye two really should try to be nicer to one another."

"Reed, what really brings you back so quickly?" Eleanore inquired. "Did you run into your pirate lady again? Kathleen and I were into Savannah the other day for fittings. The town is buzzing with talk of her."

Reed gave Eleanore a look that clearly stated he wished she would mind her own business. "If you must know, yes, I did encounter her again."

"And?" Kathleen prompted, her eyes glowing with mischief.

"She was lucky once again. She was on us before we were aware of it. I could swear she appeared from nowhere!" He gave Kathleen a disgusted look. "Yes, my love, she took my entire cargo. Does that make you happy?"

"Satisfied would be a more appropriate term." She held his gaze unwaveringly. "Did she sink your ship as she has the others we've heard about?"

"No, and I am quite mystified over it. I cannot understand it at all. I think the vixen wants us to meet again. It is as if she takes personal satisfaction in harassing me time and again. Once is not enough in my case."

"Perhaps she is enamoured of you, darling," Kathleen purred.

"Would that be so amazing?" Reed asked sourly.

"No. Not until she got to know you, love."

Reed glared at her. "I've come to take you home. Mother says you've been here since I left."

"I don't need a keeper, Reed."

He ignored her comment. "I'll only be home for a few

days. Surely you can stand my presence that long."

Kathleen heaved a sigh and pushed herself from her chair. "I suppose I can try. Come, Eleanore. We return to the lion's lair."

In the few days that he was home, Reed escorted her to several functions and around Savannah. Kathleen would almost have enjoyed seeing him try to dodge Amy's blatant flirtations, were she not so busy trying to fend off the ever-ardent Ainsley.

As it had done the first Sunday in October every year, the church held its annual bazaar. Eleanore worked a booth for the children, and Mary and Barbara held court in the quilt booth. Amy lost her usual position to Kathleen and wound up with Susan in the embroidery booth, while Kathleen reigned supreme in the kissing booth. She was having the time of her life and collecting a good deal of money for the church, knowing full well that each kiss she bestowed upon some besotted man was driving Reed mad with jealousy.

"You really don't have to be so generous with your favors, Mrs. Taylor," he complained.

Ainsley, through some fluke, won Kathleen's hamper in the lottery drawing, and she was compelled to dine with him at supper. He found a quiet spot at one of the more secluded tables that were set up in a grove next to the church. The evening was balmy and starlit, and Gerard was determined to make the most of it.

When they had eaten their fill, he led her a short distance away, out of the lantern light. Without warning she found herself drawn into his arms, locked in a tight embrace. He brought his lips down on hers in a possessive kiss that left her cold, but obviously had the reverse effect on him. As his moist, full lips pressed down on hers, she felt the swelling in his breeches against her thigh.

When he finally came up for air, she squirmed in his embrace, trying to unlock his arms from about her waist. "Gerard!" she cried out. "Release me this instant!"

"I never want to let you go!" he delcared adoringly. "Oh, my sweet love, I knew it would be like this! That

chaste kiss I bought from you this afternoon merely served
to whet my appetite for more.''

"Ainsley! Control yourself!" she screeched. "Let me
go!''

"Not until I hear you say you feel for me as I do for you!
I know you have just been playing the coy female. I know
you have merely been trying to arouse my jealousy by pretending interest in Reed Taylor.''

"Have you lost your mind?" she demanded. "I've been
pretending nothing of the sort!''

"Surely you share my feelings," he protested. "How
could you not?''

"It is quite easy, let me assure you!" She wriggled in his
grasp, trying to free herself, but exciting him more.

"Perhaps I've been too delicate with you, my beloved.
Perhaps actions speak louder than words." He tightened
his hold on her and tried to recapture her mouth.

"Quite right!" she agreed as she brought her knee
sharply up into his groin.

He released her abruptly and doubled over with a loud
groan. "Dearest girl, we shall never have children if you
continue in this manner," he groaned.

"Now you're beginning to get the right idea!" she
retorted huffily. She whirled about and started to storm
away to find herself face to face with Reed.

"Oh!" she squeaked.

"A very understated speech, Kat," he commented
acidly.

"Reed, you don't understand!" she protested.

He ignored her, concentrating totally on Ainsley.

"Ainsley, I can't say it would bother me greatly to have
to kill you, but I hope it won't come to that. It needn't if
you heed my warning and never come near Miss Haley
again.''

Ainsley drew himself up as best he could in his pain and
protested, "Taylor, you presume too much upon yourself.
What passed between the lady and myself does not
concern you!''

Kathleen looked anxiously from one man to the other.

"I heard the lady tell you she does not care for you. I also heard her ask you to release her. With my own eyes I saw you attempt to force yourself on her," Reed continued tersely. Kathleen sensed the fury behind his evenly intoned words.

"It is no business of yours," Ainsley repeated.

"I'm making it my business. The lady is mine, Ainsley. She belongs to me. Get that through your head and don't ever forget it!" Reed stalked up to Ainsley and grabbed him by the lapels. He brought his face close to Ainsley's and snarled, "This is your first and last warning. The next time, I'll kill you!" He shoved Ainsley unceremoniously to the ground, grabbed Kathleen by the arm, and marched away with her in tow.

He said very little to her, except to stress one point. "Don't ever put me in the position of having to defend my rights or your honor again. You started this little game, so you'd better learn how to hold your suitors at bay if you don't want the streets littered with their bodies. Do you understand me, Kat?"

In the face of his anger she merely nodded and said meekly, "Yes, Reed."

He left the next morning and she sailed after him on the following morning's tide. They reached their hidden bower and lay in wait like a spider in a web. It took two days of patiently waiting before the *Kat-Ann* came into view. She was heading east from Barataria, so Kathleen knew she was in pursuit of quarry and her holds were still empty.

She gave Reed good lead and then the *Emerald Enchantress* sneaked from her cover and followed stealthily, careful to keep from view. Around mid-afternoon the *Kat-Ann* attacked another ship, a sleek little sloop flying no flag and most probably a pirate vessel. Through her glass, Kathleen watched the battle, ready to go to Reed's rescue if his life became endangered.

The two crews seemed fairly evenly matched, but soon it became evident that the tide of battle was shifting in favor of the strange pirates. Kathleen instructed her men, and with full sails they flew into the fray.

As she told Finley, "If anyone kills the bloody black-guard, it will be me!"

She swung herself aboard the sloop and fought her way to Reed's side, her sword flashing in the late afternoon sun. From the corner of his eye, Reed caught sight of her. "Come you as friend or foe?" he challenged.

"Both!" she shouted back, beating back a charging pirate. "Just now I'll aid you in putting down this vermin who call themselves men!"

He flashed her a broad grin as he turned away his opponent's blade. She stationed herself at his back, and together they cut a swath across the bridge. He leaped to the lower quarterdeck and caught her in his arms as she followed suit.

Her crewmen were pairing up with Reed's men, fighting the enemy together as comrades. Together they made short work of the enemy. As the last pirate took to the water, she turned to him. On a prearranged signal, each of her men now turned his weapon on one of Reed's men. Kathleen knew that the men of the *Kat-Ann* had fought longer and were more tired than her own men. They could take them easily.

She looked up into eyes as blue as the sea they rode upon. "Captain Taylor, I now challenge you for possession of this sloop and all her booty."

"I can't say I'm surprised. You are a heartless wench, a beautiful barracuda, all teeth and no tenderness."

"Then we make a fine pair, for you are about as helpless as a shark and twice as shifty."

"Perhaps we should team up rather than wasting our energies fighting one another."

She shook her head. "Thank you for the offer, but I must refuse, Captain."

"Oh, well, I tried," he sighed as he took up his stance.

She matched his move. Just for a second something passed between them as their eyes held. Then the spell was broken as she whispered, *"En garde."*

They were both tired, but she met him blow for blow.

They were like dancers in a ballet. First he led and she followed, and then she led and he came after her. Theirs was a symmetry of motion, beautiful but deadly.

On deck, Reed's men had been defeated. Now all watched in silence as the two captains dueled alone, oblivious to anyone but each other.

Dan Shanahan blanched as he saw Reed strike a blow that severed the drawstring of Kathleen's vest. Her vest parted to reveal most of her snowy round breasts. Reed hadn't even pricked her skin.

Reed smiled wickedly into her blazing green eyes.

Kathleen never faltered, though she felt the blood rush to her face in anger and embarrassment. To do so would mean defeat, and she could not afford to let that happen. Carefully she watched his eyes, and when she saw her chance, she struck with the speed of a cobra. With one blow she sliced upward and out, striking him a slicing blow to his right side and at the same time knocking the rapier from his hand.

Blood seeped from his torn side, but he did not cry out or turn from her as he awaited her next move. He stood tall and straight, boldly meeting her gaze as he wondered if she would kill him now.

Kathleen glanced at his wounded side and back to his face. "Captain, I have drawn first blood and will be satisfied with that. The sloop is mine."

Reed let out his breath slowly and winced as his side pained him.

"Have you a doctor on board your ship, Captain Taylor?"

"I'll tend to my wound myself, thank you," Reed ground out.

"I repeat. Have you a doctor?" she said firmly.

"No!"

"Then I will have mine tend to you." As he opened his mouth to refuse, she said, "You are in no position to argue, my handsome fellow."

She pulled her vest together and called over the rail, "Dr. Bishop! Bring your bag to Captain Taylor's quarters at once. Finley, lend a hand here!"

Together they got him to his cabin, though he refused to lie down on the bunk. When the doctor arrived, Kathleen sent Finley to oversee the transfer of cargo. "Wait until we are ready to sail before you fire the sloop," she instructed.

As she watched the doctor cut away Reed's shirt and unveil the wound, Kathleen almost felt faint. She felt his pain as if it were her own.

Reed eyed her quizzically. "What kind of corsair are you who can't stand the sight of blood? You were standing in it on deck just moments ago. It didn't seem to bother you then, or when you were carving up those pirates one by one. You don't have to stand there and watch, you know. I won't harm your precious doctor."

"I didn't think you would," she commented lamely as she watched the doctor examine and clean the wound. It wasn't as bad as she had feared. She had sliced a flap of skin loose along his ribs, but it was a surface wound, and properly dressed and cared for would leave only a thin scar as a reminder of this day.

Reed gritted his teeth as the doctor stitched up the flesh and poured a powder on it. Then he wound strips of cloth about Reed's chest. "Change the dressing twice daily, and apply some of this powder. Keep the wound and dressing clean, and in about a week get someone to remove the stitches."

"Thank you, sir." Shooting Kathleen a spiteful glance, he asked, "Doctor, you wouldn't care to join my crew, would you? We could use a man like you."

The doctor grinned and replied, "As long as I'm well paid, I'll remain loyal to the lady."

"Lady! She may be a wildcat, but she's no lady!" Reed retorted.

On her way out of his cabin, Kathleen chastised, "Captain Taylor, there is no reason for you to malign my character. The fight was fairly fought." She turned to him. "And another thing. You needn't try to solicit my crew. These men are all quite loyal to me."

"I see." He smirked knowingly as he raked her lithe form with his eyes.

"No, I don't think you do, Captain. I shall leave you now. I'll have the good doctor see to any of your men who are wounded before I set sail. *Bon voyage, mon cher!*"

Chapter 20

RATHER than head north to Savannah, the *Emerald Enchantress* took a southern tack, heading for Matanzas, Cuba. Kathleen told herself it was because her crew needed some time to enjoy themselves. The truth was that she could not resist a visit to the pirate stronghold to see how nortorious she actually was among the brethren. Besides, she had plenty of time before she had to be back to Savannah.

It was three o'clock on a Sunday morning when they slipped into Matanzas Bay. Savannah would have been to bed long before, but not Matanzas. When they had reefed the sails and anchored securely, Kathleen released half her men for shore leave. The rest she kept with her to protect the ship. She warned them to stay close to their mates and not wander off alone. Matanzas was a den of pirates, and safety was a thing unknown here.

After one glimpse of the rowdy little harbor town, Kathleen delayed her own entrance until late Sunday morning. She, Dan, Finley, and a big burly Irishman by the name of Kenigan set out together along the narrow rutted road that served as the town's main street. Kathleen noted thankfully that few persons were about. Most were still sleeping off their Saturday night rum. The few that did see them stared as though they couldn't believe their eyes.

The four of them turned into a tavern called the Red Bull, or so the weather-beaten sign declared. They seated themselves at a table near the door, careful to keep a wall to their backs.

A skinny little man with grey wispy hair and a filthy apron strutted over, looked at Kathleen, and announced, "We don't serve ladies."

"Fine," she countered with a glare. "I'll have a powder and rum, and the same for my lads." She gave him a level, steady gaze, her smoky green eyes daring him to offend her.

He looked from her to her companions and gulped. "Be you the captain of that green frigate we heard so much about?"

"I am."

He fidgeted nervously with his apron. "I'll get yer drinks directly." He scurried off like a scared rabbit.

"Saints and salvation!" Kathleen swore under her breath. "What must that man have heard about us?"

"I don't care what he heard, lass," Dan spoke up. "Do ye realize what ye ordered? Thet drink is rum and real gunpowder! How are ye goin' to drink thet?"

"I'll manage." Kathleen grinned at him. "The question is can you handle it? I really should have let you order your own, but I doubt they serve goat's milk here," she teased.

"You rascally female!" Dan exploded. "Why, I've knowed ye since ye was knee high to a leprechaun! Don't ye get lippy with me!"

The other three collapsed in laughter at Dan's ire.

"Ye're laughin' now, but wait till ye have to drink that rot! 'Twon't be so funny then."

The bartender returned with their drinks. He stood there watching and the challenge was plain to see on his face. Several dirty-looking pirates entered the tavern just then and also stopped to watch.

Kathleen picked up her glass. In one movement she tossed the shot to the back of her throat, clamped her teeth shut, and swallowed. It took her breath away, and it was all she could do not to let it show on her face. Her stomach felt on fire.

Evidently she was successful, for Finley gave her a wink and raised his glass. "Here's mud in yer eye," he intoned.

They sat talking quietly for a while, watching the tavern gradually fill up as more and more of the hungover raiders of the night before felt the need for a bit of the hair of the dog. They wandered in by twos and threes. A few stared openly at Kathleen, but most slid her a glance from the corner of their eye and went on to question a shipmate or the bartender, who would nod his head excitedly and whisper fervently into their ears. All in all, she was creating quite a stir.

In a little while, the whores started trickling down from the upstairs rooms. Some of the younger ones were rather pretty, while others looked like a hundred miles of bad road.

A smile tugged at the corners of Kathleen's mouth as she watched Kenigan. He was shifting around in his seat, and the more he tried not to notice the girls, the more uncomfortable he became. He was as fidgety as an old maid at a bachelor's convention.

"See something you like, Kenigan?" she taunted with a grin.

He looked at her sheepishly.

"Go on," she said, "but be careful. I'd hate to lose a good man over a few moments of pleasure. Try to pick one free of disease, if they have one in this fleatrap."

Kenigan scooted out of his chair like a hound after a hare.

"Oh, and Kenigan!" she called after him. "If I were you, I'd watch my money and keep my back to the mattress. Let her do the work, and keep your blind side covered."

Her voice had carried, and several loud guffaws and hoots of laughter erupted at neighboring tables. Finley's dark eyes twinkled in merriment, and Dan looked flabbergasted.

"How about you, Finley?" she asked. "Do you feel any of the baser urges building up?"

He gave her a lopsided smile and shook his head. "I've

got me a little piece o' fluff waiting in Savannah. At least there I'm sure of what I'm getting. With her I can relax and enjoy myself without keeping one eye open."

They both turned to stare expectantly at Dan.

"Hmph!" he grunted disgustedly. "I suppose ye think I'm too long in the tooth."

"I never said that," Kathleen placated with a smile.

"Ye didn't need to. 'Tis all over yer face. I'm surprised at the way ye're actin', lass. What happened to thet wisp of a gal thet I used to know, all rosy cheeked and innocent?"

"She grew up," Kathleen stated flatly.

"I'll stay with the captain if you want to seek out some amusement," Finley offered.

"I jest might do that," Dan retorted.

"We'll be here when you get back, old-timer, if you don't take too long about it." Finley chuckled. At Dan's nasty look, he said, "If you're not back in two days, we'll come and claim the body."

" 'Tis real cute ye are, you two. Well, just behave yerselves while I'm gone, and try to stay out o' trouble." As he rose from his chair, he glanced toward the door. "Uh-oh. Look at who jest walked in," he commented in a low voice, nudging Kathleen.

She looked toward the door. There stood Vincent Gambie and the ugly Cut Nose. The door hadn't swung shut when in came Pierre Lafitte followed by Dominique You.

Kathleen let out a low whistle and eyed Dan apprehensively. "Sweet dancing dolphins! It looks like old home week at Barataria Bay! It's a good thing we're in disguise, Dan, or I wouldn't give us a snowball's chance in Hades of getting out of here with our skins."

"Maybe I'll jest stay with ye after all."

"Nay, Dan. The two of us together may get noticed faster. You go on. If I need help, I'll whistle. Some of our boys are starting to filter in now. I'll have plenty of support. Besides, Finley is here."

"Anything ye say, Cap'n." Dan shrugged and shuffled off.

He had no sooner disappeared than Pierre swaggered up with Dominique. He pulled out a chair, turned it around, and sat down with his arms crossed over the back of it. He starerd at Kathleen, letting his eyes travel over her body.

"Well, well, so this is the famous *pirata* who has been giving us so much trouble." He smirked.

Dominique quietly took the chair next to Kathleen.

"Can't see that you are much different from any other woman I've seen," Pierre went on. "Maybe a little better put together, but otherwise . . ." his voice trailed off. "Take off the mask and let us see the rest of your face. You are among friends here."

"Honor among thieves?" Kathleen asked sarcastically.

Pierre shrugged and gave her a lewd grin.

"Sorry. The mask remains," Kathleen stated firmly.

"You wound me." Pierre groaned in fake dismay. "You do not take Pierre at his word."

"Mother didn't raise a fool," Kathleen said off-handedly.

Dominique guffawed and slapped his knee. "Neatly put, small one. Allow me to introduce myself. I am Dominique You, and this loud friend of mine is Pierre Lafitte."

Kathleen's small hand disappeared into his large one. "I am the captain of the *Emerald Enchantress*, commonly known as Emerald."

"What brings you to these parts? The last we heard you were up around Florida, harassing Reed Taylor. That was really very nasty of you to cut up his side like that."

"He is a friend of yours?" Finley cut in.

Dominique nodded. "A very good friend. He works for Jean Lafitte."

"Word travels fast in these waters," Kathleen commented.

"Yes, well, we ran into him on his way back to Grande Terre. He was quite upset with you. He's out for your blood, you know."

"I don't see why. It was a fair fight, and I was gracious enough to have my doctor sew him up and see to his wounded crewmen."

"My brother is none too happy either, *cherie*," Pierre stated. "In fact, he told us if we were to run into you, we were to invite you back to Grande Terre." He leered at her over his mug.

"Invite or escort forcibly?" she inquired coolly.

"Whichever way you prefer."

"Emerald, Jean would much rather have you join up with us than work against us. By raiding Reed, you are cutting into our profits," Dominique explained.

"I see."

"Then you will come willingly?"

"I'll think about it," she lied.

"Perhaps I can aid in convincing you. One night alone with me, and you would never want to leave my side again, my dove," Pierre assured her with a smug look.

"I doubt that," she returned, her green eyes flashing over the rim of her whiskey glass.

"I will prove it," Pierre said as he reached for her over the table.

Her hand automatically flew to the hilt of her rapier, only to find Dominique's already there.

"Easy does it," he soothed. He pried her fingers loose from his as he looked her directly in the face. He had been speculating on her features since he had first sat down, trying to convince himself he was wrong. Now, as he held her hand in his, he turned it palm up and uncurled her fingers one by one. There on the tip of her index finger he found the small scar he sought.

His dark eyes sought hers, his full of accusations and bewilderment, hers with guilt and pleading. "*Cherie*, you will walk with me outside, no?"

She nodded, unable to speak.

"Unfair!" Pierre objected hotly. "I saw her first!"

Finley didn't know what to do. "Captain, are you sure?"

"It's fine, Finley," she assured him. "I'll meet you all back at the ship soon."

This was not in their plans, as they were not to sail until dawn, so Finley figured he'd better round up his men dis-

318

cretly and head back. Something was in the wind.

Dominique led Kathleen outside. Wordlessly they walked down the road toward the shore. In a secluded spot, Dominique picked out a large rock. Finally removing his large hand from her upper arm, he pointed to the rock.

"Sit," he directed.

"No, thank you," she said politely enough, still wondering if he had truly seen through her, or was just guessing.

"Sit, I said," he commanded in a no-nonsense tone.

Docilely, she complied.

"All right, Kathleen, you can remove your mask now." He stood shaking his head at her, much like a father correcting his truant child.

She did so reluctantly, wondering just how angry he was with her. "What are you going to do?" She looked up at him questioningly with wide green eyes.

"That depends on how good your explanations are, little sister. You can dispense with the wide-eyed innocent looks, too. They don't suit that outrageous outfit you are wearing."

"I thought the disguise was quite appropriate to the role, myself," she pouted. "If it were not for the scar on my finger, you would never have had any idea it was me, especially with my hair dyed black. I've clashed with Reed three times now face to face, and he hasn't seen through me."

"You are lucky, and you are wrong. I had my suspicions about you since I first sat down next to you."

"How? Why?"

"What really started it was your immediate dislike of Pierre, yet you seemed amiable enough toward me. We were supposedly both strangers and pirates. You should have been wary of me, too."

"Oh," she said in a small voice.

"I realize, of course, what you are doing. What I don't know is why, or how you have managed it. Obviously you are the lady pirate who had been giving Reed all his problems. A personal revenge as I see it, but why? You are

319

supposed to be with your aunt in Savannah. Where did you come by your ship?"

"It's a long story," she evaded.

"Then you had better start telling it right away." He was firm.

She gave a heavy sigh. "I didn't tell you all of it on Grande Terre."

"Oh?" Dominique sat down on the ground in front of her.

"Reed tricked me into marrying him. He made me believe he had stolen my virtues while I was in a stupor over Nanna's death. In order to salvage my tattered reputation, or so I thought, I married him, though I didn't really want to. Then I came to realize all he wanted was the *Kat-Ann*. I was just sort of a bonus. The *Kat-Ann* was mine, bequeathed to me by my father. I decided then and there that if the right opportunity ever arose, I would pay him back for his treachery."

Dominique nodded. "Go on."

"For a while I almost thought our marriage would work out. I tried, Dom. Really I did. But Reed is just impossible! I know deep down I love him, but I just can't live with the oaf! We don't get along at all. If arguing were a prerequisite for success in a marriage, we'd be the happiest couple alive."

"And?" he prompted.

"Well, you know how it was when we left Grande Terre. I had told Reed that if he thought he was going to dump me with his mother like an old suit, he was crazy. I blackmailed him into taking me to my aunt's and pretending I was still a single, virtuous maiden."

"How did you manage that?"

She gave him an abashed look and hung her head. "I told him that if he dared to expose our marriage I would tell the world he was associated with pirates and watch him hang," she said in a low voice.

"Kathleen!" Dominique was astonished.

"Oh, I'm so sorry, Dom!" Her eyes filled with tears of remorse. "I know how it must sound. I would proudly tell the world I know you and Jean. I'm truly not ashamed of you. I adore you! I would not hurt you for the world! It's just that Reed had me backed into a corner, and it was the only opening I could find that would work.''

Dominique folded his long arm about her quaking shoulders. ''It is all right, *cherie*. I understand. I believe you. But where did you get your ship?''

She brightened at his words and continued. ''It, too, belonged to my father's shipping line, along with six others. You see, Reed knows I have an estate in Ireland, though he has not yet inquired about it much, and he knew the *Kat-Ann* belonged to me. He doesn't realize my father had a shipping line with eight frigates, and I never volunteered the information. I thought I should keep that ace up my sleeve as long as I could. I felt bad enough about the *Kat-Ann*.''

''You are sneakier than you look,'' he complimented her. ''Continue.''

''When the *Emerald Enchantress*—she was the *Starbright* before we renamed and painted her—pulled into port in Savannah, I commandeered her. Every time Reed pulls out of port, I follow. I get away by saying I'm staying at my grandmother's plantation for a while. She lives near Chimera. Reed knows we are friends, but he doesn't know she is my Irish grandmother.''

''You are confusing me, little one. What has this grandmother to do with anything?''

''Well, you see, her name is O'Reilly. She's my mother's mother. The Bakers of Savannah are my father's English relations. They don't admit any family ties. Reed confides in Kate a lot. She knows all about his privateering for Jean. As long as he doesn't know she is related to me, he keeps confiding in her. She passes along the information of where he'll be and when. She's my spy and my alibi when I'm gone.''

''Holy catfish!'' He was frankly admiring. ''That explains the why and the how, but where is all this leading in the end?''

''Well, I figured if I could successfully pirate Reed often

enough, I could drive him to just enough financial ruin to force him to sell the *Kat-Ann*. Then I would have an agent from my own company buy it back for me. You know Reed's family situation. He can never use the proceeds from the plantation to buy a ship. It is stipulated so in his father's will."

"I know. So you don't want him killed. You don't want his head, just his ship."

"Precisely. If I can force him to give up the *Kat-Ann*, my revenge will be complete. However, I also find I am having the time of my life! I can take the *Kat-Ann* swiftly or slowly, and I choose to do it slowly and prolong Reed's agony."

"He may still sail for Jean," Dominique pointed out.

"I know, but he won't have my ship."

"And what then?"

"What do you mean?"

"After you get your revenge what do you do? Do you continue on as a Savannah belle? Do you agree to live as Reed's wife then? Do you go back to Ireland? What?"

"I hadn't really thought that far ahead," she admitted. "He's constantly trying to get me to admit our marriage publicly. I just don't know what I'll do yet."

"What if you are caught? I'm not just talking about Reed, either. You are playing a game of chance with the shore patrol too. After all, you are running their blockade."

"Then the fat will be in the fire, won't it?" For the first time since they sat down, the twinkle was back in her dark-fringed eyes. "You won't tell on me, will you?" she cajoled.

"You finagling female!" he laughed. "You know I won't, even though you are cutting into Jean's profits. He's not too happy about that."

"No doubt! I wish you could explain to him—but we daren't. Anyway, my booty is going into a warehouse in Savannah. When this is all over, I'll send it all to Jean to make up for now."

"You amaze me! You've thought of almost everything."

"I hope so, but I hadn't planned on Jean insisting on seeing me at Grande Terre. I can't show up there, Dom."

"Obviously," he agreed dryly.

"Then you'll help me get away tonight?"

"Better than that. I think from now on I'll sail as gunner for Reed. He can't turn down an offer like that. I'm the best there is." Dominique grinned at her devilishly. "That way I can keep an eye on you both, in case you get caught."

Kathleen threw her arms about his neck and kissed him on his rough cheek. "Oh, Dom! You're the best brother a girl could have! I'll feel better knowing you are there," she added sheepishly.

"You really do cover yourself well. You make quite a pirate after all!" He hugged her affectionately. "Come. Let's get you aboard your ship and away before Pierre realizes you are gone."

"So you think I cover myself well, eh?" she teased with a toss of her head. "It is obvious to me that you are blind in one eye and can't see from the other then, my big friend." She danced around in front of him, pivoting about, letting him see all sides of her.

"You minx! You had just better watch yourself! You are quite a tempting dish as it is, without displaying yourself like that. You are safe with me, yes, but what of all the others?"

"The outfit comes complete with rapier, pistol, and knife," she mimicked as though modeling the lastest Paris fashion. "Besides," she said, pulling a face at him, "can you see me captaining a frigate in all those skirts? I'd trip myself up twelve times a day, and what a sight I'd be climbing the rigging! Talk about revealing! That would certainly give the sailors an eyeful!"

"You could wear long trousers and a shirt!" he argued with a frown.

"Now, Dom. If I'm to play a pirate, I'll do it all the way. I've yet to meet one who is prim, proper, and prissy!"

He laughed and shook his head in mock dismay. "Oh, Kathleen," he sighed. "You are one in a million. Reed certainly has his work cut out for him, and I'm not sure I envy him. You may turn out to be more woman than any man can handle."

"I agree," she concurred, tossing her head arrogantly. She walked tall and proud beside him to her ship.

They had no trouble leaving the darkened harbor, and no one followed in pursuit. Kathleen chuckled to herself as she imagined Dominique getting Pierre roaring drunk and fobbing one of those horse-faced tarts off on him for the evening. To herself she admitted what a close call she had had, and determined not to indulge herself in such idiotic ventures again. She would stay away from Matanzas and other such places. Her men would have to content themselves in Savannah from now on, and she would have to keep a firm hold on her avid curiosity.

They were quite a way up the Florida coastline when Kathleen noticed a change in the air. Dan noticed it too.

"We're in for a blow, Cap'n," he announced in his casual way.

"Aye, Dan, and it's going to be a good one. Look at the murky color of the sea, and there is an odd color to the sky and an eerie stillness in the air. Unless I miss my guess, we're in for a hurricane once this cuts loose." Excitement tinged her voice.

"How soon ye figure?"

"Sometime early tomorrow is my guess."

"If we put her to full sail we might make home port," he calculated.

"What, and miss all the fun?" she teased with a pretty pout.

"Are ye daft, lass?" he demanded.

"Probably," she concluded. "Still," she said dreamily, "I'd like to ride her out. I want to feel the wind and rain in my face. I want to face the waves head on and meet the challenge of the sea."

"Ye're young and foolish, and crazy as a loon," he intoned under his breath, "but ye're the captain."

A few hours later the lookout called down. "Ship

ahoy!''

"Where away?" she yelled up at him.

"Two points off the port bow," came the answer.

"Can you identify her?"

"Maybe if I had the glass!"

Kathleen turned the helm over to Finley and climbed up the rigging with the glass. Perched in the topmost yard, clinging to the mast with both legs and one arm, she raised the glass in the direction the lookout indicated.

"Well, blow me down, if it isn't the *Kat-Ann!*" A smile found its way across her face. "Here, Timmy. Take a look and see if I am right." She handed the glass to the youth.

"Yep! That's her all right!"

"Talk about the luck o' the Irish!" she said to herself as she climbed nimbly down.

She took the wheel from Finley. "It's the *Kat-Ann,*" she informed him. "I want full sail. We're going after her."

Slowly, little by little, because of the swells, they crept up on the *Kat-Ann.* Finally, when they were within hailing distance, Kathleen picked up the horn.

"Ahoy, the *Kat-Ann!*" she called.

She watched as Reed reached for his own horn. "I haven't time for your silly games now, you daftie!" he shouted back. "There's a storm brewing in case you haven't noticed!"

"I know! Where are you heading, Taylor?"

"Savannah port! You'd better head for shelter of your own! 'Tis brewing into one helluva blow!"

"Not just a blow, you novice!" she called in return. "A hurricane! Can't you see the signs?"

"You're probably right. Catch me when I'm not in such a hurry, sugar," he shot back.

"Why all the rush? Come play a while!" she baited, her laughter ringing more clearly in his ears than her words.

"Your brains have turned to seaweed!" he exclaimed.

She shook her head negatively, and her raven locks

broke free of their restraining ribbon and tumbled about her shoulders, blowing wildly in the wind. Now more than ever, she resembled a sea temptress, the enchantress of some old sea tale.

"Come ride out the hurricane with me!" she shouted. "Let your men take your ship into port if you fear for them, and come ride with me on the *Emerald Enchantress*."

"You can't mean to stay out in this!" he retorted.

"I do indeed. Are you afraid of me or the storm? Which is it?" she taunted. "I promise no harm will come to you. I'll deliver you safe and sound on Georgian soil as soon as the storm is over. You have my word on it."

"The word of a pirate!" he spat out.

"My word as a lady, then."

His laughter grated on her.

"If you have guts enough to accept my challenge and meet it well, you shall have a reward!" she yelled over the rising wind.

"What might that be?"

"If you can weather this storm as well as I, if you survive and keep your stomach, you shall ride more than wind and waves this day." She threw the final challenge. "You shall ride me as well. What say you, brave, strong Captain Taylor?" She was close enough to see the surprise register on his face.

"And what do you get, fair lady?" he shot back.

"Why, you, of course!" Her tinkling laughter rang out again. "One evening with the captain of your choice! Say you yea or nay? This is a one-time offer!"

"I hope the sheets are clean. I'm allergic to bedbugs! Hold fast while I issue orders to my crew, you sea witch!"

He swung himself aboard with a self-satisfied smile. "I'll hold you to your bargain, you know."

"My word is as good as your gold." She couldn't resist the dig. "I'll handle the helm. You help out with the rigging. You're not here just for the ride, my friend," she said.

He saluted smartly. Over his shoulder he said, "I'll see

you later, sugar.''

The storm worsened gradually throughout the evening. Kathleen figured the brunt of it would hit in the wee hours of the morning. By midnight the swells were gigantic, washing over the decks with such force that Kathleen ordered all sails reefed and tied down securely. She lashed herself to the wheel and ordered her crew below decks. The wind buffeted the frigate about like a toy ship on a lake.

When the rain started, it came in blinding sheets that nearly took her breath away. By feel alone, Kathleen kept the heading she had set before dusk. She relied completely on her inborn sense of the sea's movements, praying it would hold her in good stead. She tacked in the wind, riding the huge troughs of the sea. The planking creaked and groaned in agony as the frigate reached fearful heights and then plunged straight to the heart of the trough, only to rise miraculously to the crest of the next wave. The ship and all aboard were at the mercy of the elements.

Kathleen was buffeted about as she fought the wheel. If not for the lines about her waist, she would have been tossed into the boiling sea. The rain beat down on her face with such force that it felt like a thousand needles plunging into her skin. Her hair whipped back and forth in great wet strands, nearly blinding her at times, but there was not much to see anyway. She dared not let loose of the wheel to tie it back, as it took all the strength in her arms to control the wheel as it was.

Her arms were tiring, her back ached terribly, her face stung, and she was soaked to the skin and beyond, but still she was enjoying herself. It was more exilarating than anything she had ever experienced. She stood with her feet braced beneath her and let her soul fly free. She felt released, free and clean and pure. She stood with her head bowed beneath the rain and felt herself at one with the wind and rain and sea. A feeling that could only be described as holy settled within her breast and started to grow. She was reborn, a creature of the sea, Venus rising from the waves. She was filled with wonder and awe. A power filled her being, and she gloried in it. She felt invincible; defeat was a thought that never occurred to her.

This was her hour of glory, and she reveled in it. She was a goddess reborn, heeding only the call of the sea. She felt its might, and it became her own. Fear had no place in her. She was in her element and at peace within herself. Her soul had found its haven in the midst of the turbulent storm. She was home!

Reed stood on the deck below her, clinging to the mast. The rest of the crew had long ago gone below deck. He watched Emerald and marveled at her strength and endurance. Who would guess that someone so slight could withstand the force of such a storm for so long? From his chosen post he saw her head both beneath the wind and rain. From this position he caught a glimpse of her rain-washed face and was amazed to see her smiling. Her expression was one of complete confidence, and something more. With a start he realized it was exhilaration and unbound joy and unleashed power all wrapped in one. Even all wet, with her hair straggling in her face, she was magnificent to behold. She resembled some high priestess with all the powers of the unknown at her command, competent and completely in charge. He felt drawn to her as the tides were drawn by the moon.

Kathleen felt Reed's arms close about her waist. "I think it's abating!" he yelled into her ear. "The eye of the storm should be coming soon!"

She nodded her agreement.

He reached around her and untied the line, winding it about both of them, lashing the two of them together to the wheel. "Lean into me!" he screamed over the storm. He felt her resistance. "Don't be stubborn! Let me help you!"

She relaxed against him. She could feel the steamy heat of him coming through his wet clothes and hers. Instantly she was aware of his hard muscular contours pushing against her from shoulder to knee. His warm breath on her neck raised goose flesh along her spine.

With one hand he gathered her streaming hair from her

face. Then he placed his own strong hands upon the wheel, and with his aid she found it easier to handle by far.

For perhaps half an hour they silently battled the elements together. Then it was as if some great hand from above had suddenly shut off the storm. All at once the rain stopped and the sea became as calm and smooth as glass. The wind stopped and everything was so unearthly still that Kathleen could hear Reed's heart beating as clearly as she could her own.

She sighed audibly in relief. Releasing the wheel, she flexed her numb hands and aching arms. "The eye," she said softly. With stiff fingers she untied the line binding them to the wheel.

The hatch to the forecastle slammed open and Finley staggered onto the slippery deck. His face was wan and drawn. He looked up at Kathleen and Reed. They looked wet and bedraggled, but damn if they didn't look as if they'd enjoyed themselves. How dare they! He gave them a dark scowl.

"How is it below?" Kathleen called down to him. "How did the men fare? Is anyone hurt?"

"Other than being half drowned and thoroughly sick, everyone fared well enough," Finley grouched. "There'll be one helluva mess to clean up later, though."

Kathleen chuckled sympathetically. "Did we take on much water?"

"Not as much as I feared, but more than enough for my taste."

"We are in the eye of the hurricane now. It won't be long before the backside catches us. Put some men to the pumps and others busy bailing with pails. Get rid of what water you can," she ordered. "Also, take a few men and inspect the ship. See what damage was done and repair what you can in the time allotted us. Report back to me."

She turned to Reed. "I'm going below to towel off. I'll bring a towel up to you."

"Fine. I'll stay here and man the helm if you wish."

She hurried to her cabin. Hastily she lit a lamp and

viewed herself in the small mirror. "I look like a drowned rat," she grumbled. "At least my hair dye isn't washing out. That is something to be thankful for." Quickly she stripped off her wet clothes, dreading the thought of putting them back on. There was no sense in exchanging them for dry ones, however, since they would soon be soaked too. As she toweled off her trembling body, she looked down and immediately dropped the towel.

"How stupid of me!" she exclaimed as she eyed her body. "Anyone would think I wanted to get caught!" She dashed to her trunk, rummaging until she found her hair dye. With the small brush she gently, but thoroughly, applied the dye to the small triangle of tightly curled red-blonde hair between her legs. Then she scrubbed and dried until she was sure it was set.

As she pulled her wet clothes back on, she cursed herself for a fool. "I'm going to have to be more attentive to small details. This makes the second close call in two days!" She tied her hair back with a leather thong and grabbed a towel for Reed. Blowing out the lamp, she scurried to the bridge.

Finley has assessed the damage and reported to her. "Nothing too bad, Captain. A couple of bow beams seemed weakened, so we braced them. No tears or splits along her seams. She's sound as a bell. Anything we missed can't be that important, and we'll go over her real good once we hit home port. Most of the water's been pumped out, too."

"Thank you, Finley. You'd better get below and batten down that hatch. It's starting to get rough again. The reprieve is over."

Once more she took her heading and altered course into the wind. When she started to lash herself to the wheel, Reed stepped up behind her. She gave him a level look and handed him the line, and he strapped the two of them again to the wheel.

As the wind and waves rose, so did Kathleen's spirits. She leaned comfortably into Reed's chest as the ship was buffeted about. The rain drenched them, but could not dampen their devil-may-care mood. Her head bowed beneath the onslaught, and his lowered over hers. Moments later she felt his lips nibbling at the nape of her neck. A shiver that had nothing to do with the cold rain traveled through her. He pulled her tightly against him, and together they rode out the storm.

As the storm lessened, he caught at her earlobe with his teeth. "I want you," he murmured huskily.

When it was safe to do so, Kathleen turned the helm over to Finley, giving him his heading and instructions. She led Reed to her cabin. Carefully she lit but one lantern in the far corner of the cabin. It left the bunk in shadow.

She walked back to Reed. He reached up to untie her mask, and she stepped hastily backward. "No! The mask stays!"

"I want to see your face," he insisted.

"That wasn't part of the bargain. Satisfy yourself with seeing the rest of me."

A devilish grin lit up his features. He reached for her again, this time to undo the lacing on her vest.

Silently she removed her gloves and unbuttoned his clinging shirt. She looked up at him from beneath her thick lashes, laughter dancing in her brilliant green eyes.

He released the buckle of her belt and her weapons fell to the floor with a clatter. She smiled seductively at him as she did likewise with his weapons. Then her nimble fingers were working at the buttons on his trousers. She motioned for him to sit, then bent and pulled off his boots.

"What about your side?" she asked, eyeing his bandage.

"Let me worry about that."

She stood, as did he. He pulled her close, running his hands up her bare back to her neck. Holding her head between his hands, he lowered his lips to a fraction above hers. She arched toward him, bringing her lips fully against his in a long, tantalizing kiss. Her tongue snaked into his mouth, entwining with his, searching, exploring, arousing.

His hands traveled to her waist, releasing her breeches. He stepped back and let his eyes roam over her appreciatively. She stood proudly beneath his frank appraisal. She was completely naked except for her boots and mask. Her hair hung in damp strands, one dark tendril curling around her right breast as if in a caress.

She tossed the errant curl over her shoulder. Taking up a towel, she fluffed her hair dry, all the while edging her way across the room, careful to keep the telltale mole on her left buttock out of sight. Stopping below the lamp, she said, "If you have looked your fill, we can douse the light. Then, and only then, will I remove my mask."

He walked to the bunk and sat down. One last time he let his eyes devour her. Beneath the lamplight her skin had a golden glow. At last he nodded.

She blew out the light, removed the mask and made her way to the bunk. His arms enclosed her waist as he tumbled her onto the bed. On his knees, he reached down and pulled off her boots. His hands searched their way up her legs, slowly, endlessly, until he found where they met. He heard her gasp, and the sound led him to her lips. They were full and moist beneath his kiss, tasting of sea spray. His fingers threaded through her hair, and it was silky to his touch.

He continued his blind exploration, kissing her nose, her eyes, her cheeks. Nuzzling his nose into her neck, he found her ear.

"I want you, my proud beauty," he whispered huskily. His hands continued in their travels, first finding her breasts. They were firm and high, and eager for his touch. The nipples rose immediately against his palms. He chuckled knowingly. His hands roamed her body, seeking, searching out her most vulnerable spots.

She moaned and thrashed beneath him, and then suddenly she wriggled out from under him. He felt the mattress give with her movements, but in the dark he could not tell what she was doing. For one brief moment he imagined she might be preparing to plunge a knife into his breast.

Then he felt her hair tickling the inside of his leg. Her

tongue flicked a path of fire along his inner thigh, heading ever higher. He drew in his beath in anticipation. For long, agonizing moments she toyed with him with tongue and teeth and fingers. Then her mouth sheathed his aroused member and he moaned aloud. She worked with him until he thought he would explode, but she allowed him no release. Each time he was about to climax, she eased up and changed her tactics, finding new ways to drive him mindless.

"Enough, Emerald," he groaned at last. "Give over!"

She laughed huskily. "Not this time, Reed Taylor. In my bed I give no quarter. Neither do I expect any."

She swung herself up to straddle him. She eased onto his tortured shaft, enclosing him in moist velvet. "You can ride next," she promised in a sultry whisper. "Right now it is my turn."

His hands found her hips as she lowered herself until the tips of her breasts just lightly brushed the curling hairs of his chest. Her lips found his in a demanding, almost angry kiss. Her trim young thighs rose and fell, and he found her rhythm and matched it with his powerful thrusts. Her hair fell around them like a curtain, shutting out the world and enclosing them in their own pleasures. Their passions built to heights unbound until they both reached a release so ecstatic that it hurled them beyond reason, into a time and space of their own making.

Both of them cried out and Kathleen clung to him, shaking, until the wonder subsided and she could once again control herself. Finally she sat up, brushed her hair back, and said, "Now it is your turn, for what is left of the night. Until dawn I am yours to do with as you will."

For the rest of the night he made love to her in every conceivable fashion he could imagine. He teased and tantalized until she was drunk on his kisses and his touch. She murmured love words to him in a provocative mixture of Spanish, French, and English that awakened his senses time and again. Her voice was deep and throaty with emotion.

He pushed his body and hers to the limits of their endurance, and just when he thought he was thoroughly exhausted, he found himself wanting her again. She had said that she was his to do with as he wished and would ask no quarter, and he put her word to the test. He took her both roughly and gently, and granted her no reprieve when he knew she was weary and sleepy.

Toward dawn, he took her one last time, drowning her senses and his in a rapture so complete that at the explosive moment of their stupendous climax she cried out, "Oh, Reed, I love you! I love you so!"

As he floated back to reality, his mind caught at her words. Not only her words, but her voice. Just for a moment he had thought it was Kathleen's voice crying out to him in the dark. But no, that could not be. Emerald had uttered the words he so longed to hear Kathleen say, and she had said them in a moment of extreme passion. His mind was tired and playing tricks on him. Kathleen was back in Savannah, and Kathleen desperately despised him . . .

At the thought of Kathleen, guilt stabbed him. Through his exhaustion, he acknowledged to himself that he had just committed adultery. Worse, he had thoroughly enjoyed it. It bothered him that he should be so drawn to another woman when he loved Kathleen so completely. He didn't understand his strange attraction to Emerald.

Reed winced to think what Kat would do if she ever found out about this. But, damn it all, if she had been behaving like a proper wife and meeting his male needs, he would probably never have strayed. With this to salve his conscience, and a firm resolve to straighten things out with Kathleen soon, he drifted into a deep sleep.

Kathleen lay beside Reed until she was sure he was asleep. As she had drifted down from her cloud of ecstasy, she realized what she had cried out in her passion. She prayed that Reed had not caught her words, or if he had, that he would not ponder them. She hoped that if he had heard them he would consider them spoken only in and for the moment, and not question her. Above all, she hoped she had not exposed her cover. When he said nothing and fell asleep, she breathed esier.

When she was sure he was sound asleep, she eased herself from the bunk. Quietly she dressed, taking a fresh mask from the drawer rather than risk waking him by trying to find the other in the bed. She slipped silently from the room and went on deck.

She had Timmy bring her coffee and peaches from the galley, and feeling the need for privacy, climbed high into the rigging and perched there. As she sat there thinking, Kathleen realized that she was, in a way she had never anticipated, the victim of her own scheme. As Emerald, she had been free to entice and seduce Reed in a manner far more bold than ever before. The encounter had been daring and glorious, but now Kathleen had to deal with an odd sense of betrayal and jealousy. It was ridiculous, of course to feel jealous of Emerald—she *was* Emerald.

Still the sense of betrayal lingered. Reed had been unfaithful to her! Technically, he had not committed adultery, but he had no way of knowing that, and it was this that both saddened and angered her. How could he do this to her! Then, on a more honest note, she conceded that she'd brought it on herself. Besides, if truth be told, the night just past had been nothing less than spectacular! A delicious smile curved her lips as she recalled their passionate lovemaking.

For some time, Kathleen wrestled with the problem of Reed's infidelity and her own muddled reactions of it. Finally she resolved it somewhat by adding it to her list of grievances against him and renewing her pledge of piracy and revenge. With a confused sigh, she settled back to forget herself in the beauty of the morning.

She was still there an hour and a half later, among the taut shrouds, when Reed came on deck. He eyed her for a moment, then climbed up to join her.

"Good morning," he grinned. "Where are we?"

"About ten minutes from sighting the Georgian shore," she told him. "If you don't mind, we'll drop you at the mouth of the river. I really can't afford to get caught in a trap of my own making." She gave him an easy smile, though in another sense, she had done just that.

"I understand. I'll have no problem hitching a ride from there."

"How is your side?"

"Your good doctor just removed the stitches."

Kathleen laughed. "He does better needlework than I would ever hope to," she joked.

When they parted, she offered him her hand. "Well, Captain Taylor, this has been a memorable encounter!" She gave him a sly look.

"I should hope so!" he retorted with an unfathomable gleam in his eyes.

CHAPTER 21

THE social event of the season was St. Theresa's Ball held in mid-October every year. This was the moment every debutante awaited, her moment of glory. The ball was held in a huge ballroom with an encircling gallery, and made a perfect setting for debuts. Each girl of sixteen would walk down the wide staircase on the arm of her escort to be formally introduced into society. She would be dressed in traditional white, while girls previously presented would wear pastels.

This year Amy, who had turned sixteen in the spring, would make her social debut. So would Kathleen who, although she had turned seventeen in April, was new to Savannah. Since Susan's birthday was but a few days after the ball, she too would be presented.

Havoc reigned at Chimera as preparations were underway for all three girls. Kate had alerted Kathleen a month previous, and only a final trip to the dressmaker for a fitting of her gown was required. Since all the gowns were to be white, competition for originality was furious, and Kathleen allowed no one to see her gown but Eleanore. The design was Kathleen's own idea and she paid Mrs. Fitz extra to keep it secret, especially from the ever curious Amy.

Reed, of course, was being a problem. "But, darling," he told Kathleen in his cynical way, "you are supposed to be an unwed, untouched maiden when you are presented.

That is why all the debutantes wear white, don't you know? You are flaunting tradition by doing this." He smiled crookedly.

"Go to the devil, Reed!" she snarled.

"Only if I can take you along."

"I can't very well come out and say I'm not a virgin, Reed. Barbara would have fits."

"Can't or won't? You could always tell them the whole story if you wanted."

"You don't have to escort me, you know. I'm sure I can find someone else to do so."

"Never fear, fair lady. I shall be only too glad to do so. It will seem a bit odd, however. I'll wager I'm the first husband ever to escort his wife to a coming out ball."

"What they don't know won't hurt them," she replied stubbornly.

The day of the ball Kathleen rode into Savannah and collected her dress personally. That afternoon she closeted herself with Eleanore refusing entry to anyone else. She took a light supper in her room, giving Amy no opportunity to sneak in and see her gown.

Eleanore dressed her hair for her, sweeping it up into a soft coiffure with gentle waves framing her delicate face. Just above each ear she pinned a jeweled butterfly. When the time came to depart for Savannah, she wrapped herself in a hooded white satin cape that covered all but her face. Even once they had arrived at the ball she refused to show even a glimpse of her gown.

"Not until we are announced," she said firmly.

"This had better be worth all the suspense," Reed grumbled.

Kathleen merely smiled up at him sweetly. Then she let her eyes rove over him. Once again she was struck by his handsomeness. He looked so debonaire in his formal attire.

He caught her look and grinned knowingly. "Your thoughts are showing, Kat," he whispered. "Beware! One of these days someone is going to see you eyeing me in that fashion and know immediately that here is one maiden who is not as virtuous as she would lead others to believe.''

Kathleen favored him with a scathing look and said nothing.

Soon they were at the head of the stairs hearing their name announced, ''The Lady Kathleen Haley and her escort, Captain Reed Taylor.'' All eyes focused on them as Kathleen slipped the cloak from her shoulders and handed it to the attendant. A collective sigh rose from the crowd below. Reed turned to offer her arm and nearly lost his balance as he beheld her.

She stood regally in a dress of shining pure white satin with a plunging vee neckline and an empire waist. It clung sleekly to her figure. The sleeves were long and tightly fitted, but overlaying the sleeves and draping beneath her arm like angel's wings was the sheerest white pleated chiffon. The gown was embroidered with scores of brightly jeweled butterflies that twinkled and shone against the snowy satin. About her neck a choker of perfectly matched pearls clung, their mates kissing her dainty earlobes. Kathleen herself looked like the rarest of winged creatures newly emerged from its cocoon, all shiny and new and beautiful.

Reed could not believe his eyes. As many gowns as he had seen her in, she had never looked more beautiful and radiant than now. Almost reverently, but certainly proudly, he led her down the staircase. At the foot of the stairs they faced the committee of dowagers and town leaders whose acceptance she was symbolically seeking. He held her arm firmly as she gave them a low, graceful curtsy. She rose and they took their places to await the rest of the announcements.

Amy was livid! She had chosen a dress with rows of delicate ruffles edged in fine lace. A beautiful old cameo rested at the base of her throat, a gift from her parents. To offset the white gown she had a wreath of pink roses about her head. She really looked quite lovely, but next to Kathleen's smoothly draped sophistication she felt like a little girl at her first party.

Susan delighted everyone, and was probably the second most beautiful young lady there. She appeared in a charming gown with off-the-shoulder sleeves and a peasant ruffle of Irish lace flowing gently over her shoulders and bosom. The bodic was just low enough to reveal a glimpse of her maturing breasts. The waistline was fitted and the skirt flowed gently to the floor with an overlay of Irish lace. Tiny miniature orange blossoms of coral studded her ears, with a larger matching pendant hanging from a gold chain around her neck. A sprinkling of genuine orange blossoms set off her dark hair to perfection. She was charming as she was escorted by a bedazzled Ted.

Somehow word had gotten around that Kathleen had rejected Gerard Ainsley. The attention Reed was lavishing on her this evening left no doubt in anyone's mind that Kathleen and Reed were now an item. By some unwritten code, and in some intangible way, Reed had officially staked his claim to her this evening and everyone knew it.

Amy was not the only one angered by this turn of events. Kathleen was a confused mass of conflicting emotions. While she enjoyed tantalizing Reed and reigning supreme at the ball, she was upset that, in some way she could not pinpoint, she had given Reed the upper hand. She was proud to be at his side, knowing what a sensational pair they presented, yet angry that Reed had taken advantage and publicly earmarked her as his alone. She was highly irritated both at herself and at him.

When he held her close against him as they danced, she was alarmed at her own reaction. She felt her breathing become shallow and her heart pound faster in her chest. Reed felt it too, pressing his advantage. He held her even closer and whirled her about until she was breathless and nearly helpless in his arms.

"Reed, please," she complained faintly.

"Please what, my sweet?" he taunted. "Please take you home and make mad, passionate love to you as you know you want me to?"

Kathleen nearly choked keeping the word yes from her lips. He was so handsome and strong this evening, his arms so familiar and powerful, his breath warm and tingling on her neck. His lips were too close and inviting, the masculine scent of him made her head spin. The timbre of his voice loosed butterflies in her stomach, and the sensuous look in his incredibly blue eyes set her pulse pounding crazily.

How easy it would be to surrender to him this evening. There was magic in the air and the promise of heaven in his arms. Her body ached for the feel of him and her lips tingled to feel his upon them. Her heart yearned for him even as her head called her seven kinds of fool. The memory of that incredible night aboard the *Enchantress* lingered and served to fuel her hunger.

She lowered her gaze from his searching look and murmured fervently, "Damn you, Reed! Damn you to hell and back again and myself as well!"

He chuckled warmly in her ear. "Give in, Kat. You know you want to. Stop tormenting yourself."

Kathleen was both disappointed and relieved when the evening ended. The ride back to Chimera seemed endless. The three men rode beside the carriage, Uncle William nearly asleep in his saddle. Inside, Susan sat with a dreamy look on her face, while Amy was purely petulant. Kathleen sat quietly staring at her folded hands in her lap and fighting the rising tide of need for Reed. From time to time she glanced up to find Mary or Barbara eyeing her expectantly, or to meet Eleanore's understanding look. More irritating than anything else was Reed's cheerful whistling as he rode near her door.

"He knows and he's gloating," she thought. "Well, go ahead and enjoy yourself for all the good it will do you, Reed Taylor!"

Once undressed and tucked into her own bed, Kathleen found sleep elusive. She tossed and turned and argued with herself until she finally gave up. Rising, she donned a light robe over her sheer nightgown and tiptoed barefoot onto the veranda. Her feet led her unerringly along the veranda toward Reed's room as if drawn by a magnet. Her good sense told her to flee back to her room, but she could not.

She found him outside his room quietly smoking and evidently expecting her. Tossing his cheroot over the rail, he took her arm and steered her inside his door. Without a word he closed the door and, turning to her, slipped the robe from her shoulders. Words were not necessary as he threw off his own robe, pulled her gown over her head, and led her to the bed.

Shielded by the exotic dragon-festooned black bed drapes, she felt wrapped in his love. As he caressed and kissed her, bringing her body to an agony of yearning, she felt her anger and frustration melt under his expert touch. She yielded to his mastery and as her passion mounted, she forgot everything but her need of him and reveled in the desires he awoke.

Reed could barely believe it when he saw Kathleen coming to his room. He had hoped and prayed, but it seemed a miracle. She had actually come to him for the first time in months. If this was a turning point in his life, it could only be for the good.

The glow of her body as he undressed her in the lamplight excited him. The sight and scent of her assaulted his senses, and he could barely keep control as he led her to the bed. With effort he controlled himself and took his time exciting her body. Her passions flared to match his. He felt her hands gently explore his sweat-filmed body and he groaned in anticipation. She brought her lips to his and he crushed her to him in a long, passionate kiss. Her lips were warm and pliant, her tongue a sweet, twisting snake entwined with his.

Their breath mingled and their hearts beat in rhythm. He lowered his mouth to her rosy peaked breasts, and she moaned out her desire. Her hand caught the object of her need, and gently she led him inside her. Urgency took over and soon they were beyond themselves, taking and giving in return. Together they reached the summit and plummeted into rapture, clinging and crying out softly to one another. Kathleen felt as if she were drowning in ecstasy and her only solid safety was Reed.

Their passion spent, they lay still entwined, her head

pillowed on his shoulder. She traced her fingers along his ribs and encountered the still-red scar of his wound. "You've been hurt. This scar is new."

She felt him nod. "It's healing well," he said.

"How did you get it?" She kept her face carefully hidden on his chest.

"That sea-witch of a pirate left me a token of her abilities," he growled softly.

"She bested you?" Kathleen did her best to sound incredulous.

"Yes," he said curtly. "Now, let it rest at that."

Tactfully she changed the subject. "You knew I'd come tonight, didn't you?" she whispered.

He shook his head. "I waited and hoped you would come."

"This doesn't change anything, Reed."

"I hadn't expected it would," he lied. "Will you stay the night?"

"I'll stay as long as I dare."

The rest of the night was heavenly, filled with tender loving and whispered words in the dark. No promises, no recriminations, no ill will or words, just love, pure and sweet and warm.

As Kathleen walked back to her room the next morning, she did not berate herself for going to Reed. She only felt a bittersweet sorrow that things could not always be as good between them, but that was wishful thinking and better left alone.

She had gotten up quietly, hoping not to awaken Reed. For a moment she stood looking down at him, wishing she could stay. Reed's eyes suddenly opened, catching her by surprise. He noted her wistful expression and said quietly, "Anytime you want, things can change, kitten. Just say the word, and you need never leave my bed at dawn alone."

"Don't Reed," she answered softly, shaking her head. "Don't make me hurt any more than I do already."

A couple of days later Ted came dashing up the stairs shouting, "Kathy! Kathy! Where are you?" .

"What in heaven's name is going on?" Kathleen dashed from her room.

"The greatest thing has happened!" he babbled excitedly. "There is a band of gypsies camped about five miles upstream. They're dancing and horseracing and selling things . . . just everything! They are trading horses and selling the most beautifully crafted silver and gold jewelry. You simply must come see! They are even telling fortunes!"

"And robbing people blind, no doubt," she put in.

"No doubt," he agreed with a broad, boyish grin, "but it's great fun! Say you'll come."

"Of course I'll come. Aren't you going to tell Susan and the others?"

"I already saw Amy downstairs. She just sniffed at me and threw her nose in the air." Ted imitated his sister perfectly.

Kathleen laughed at his antics. "You go get Susan and I'll tell Eleanore. We'll meet you at the stables."

Ted stopped in mid-stride. "You aren't going to ride Zeus, are you?"

"Why not?"

"They might steal him." Ted, for once, was serious.

"They might try!" Kathleen retorted with a sly grin.

She and Eleanore were dashing down the back stairs when they encountered Reed. "Where are you two going in such a hurry?" he inquired.

"To the gypsy encampment up river with Ted and Susan," Eleanore called back over her shoulder.

"Yes! If we don't come back, you'll know we joined up and ran off!" Kathleen teased with a cheerful laugh.

"If you don't come back, it will probably be because you've been kidnapped by some gypsy," he informed them as he reversed directions and followed them.

Kathleen's green eyes widened in mock wonder. "Oh, my! That sounds intriguing! I wonder if they are good lovers?"

Eleanore giggled excitedly.

"That settles it!" Reed stormed. "If you are set on

going, I am coming along."

"Spoil sport!" Kathleen stuck out her lip in a pretended pout.

"Spoiled brat!" Reed shot back.

The gypsy encampment was a kaleidoscope of brilliant colors. The wagons were all outrageously painted, and their clothing was bright and cheerful. Susan hung back in awe of the gayly dressed, dark-eyed gypsies. Eleanore and Kathleen were simply delighted and eager to explore. They immediately gravitated toward the displays of jewelry.

They were still browsing when Kathleen felt someone staring at her. She looked up into a pair of ardent dark eyes. The owner flashed her a daring smile, showing rows of startlingly white teeth. Instinctively she knew not to smile back, for that would only encourage the young rogue. Instead, she ignored him. From the corner of her eye she saw disappointment cross his face momentarily. Then he shrugged good-naturedly and transferred his gaze to Eleanore.

"Fickle lad!" Kathleen chuckled inwardly.

"If you think that passed my notice, think again," Reed said smoothly in her ear.

"You are going to be the nosiest old busybody in your dotage!" she grumbled playfully.

They wandered down to a flat stretch of beach where the horse racing was being organized. Several of the neighboring planters were already there.

"Hey, Reed! You going to enter Titan?" shouted one fellow.

Reed shrugged.

Dave Murdock came ambling up, followed by a middle-aged gypsy. "This fellow says he has a horse that can outrun anything we run against him."

The gypsy man shook Reed's extended hand, bowing acknowledgement of the ladies.

"Is this true?" Reed asked.

"It is so." The man pointed to a magnificent white stallion tied beneath a tree. "He is as fast as the wind."

"What is the wager?"

"We are simple people, but we love horses. I will wager my horse against yours. The golden stallion you brought with you for mine."

"The black stallion is mine. The golden one you speak of belongs to the lady." Reed gestured toward Kathleen. "I cannot wager what is not mine."

"I understand. Perhaps your lady wishes to wager." He eyed her hopefully.

Kathleen walked slowly to where the snowy stallion was tethered. The walked around him, examining the horse thoroughly. Finally she spoke, "Your animal is indeed magnificent. He might truly be able to beat either my stallion or Reed's, but I love my horse and could never part with him. If you would care to wager something else, perhaps."

The man grinned, showing a gold tooth in front. "Your man will race his black against mine." He looked at Reed. "If I win, you will send two sheep to our camp, yes?"

"Yes. And if I win?" Reed inquired.

"Then you shall all join us and share our supper." He grinned broadly. "Since it was probably your sheep that wandered into our camp and dropped over dead, how can I lose?"

Reed threw back his dark head and laughed heartily. "How indeed? It is agreed."

The race was set. Just for sport, Kathleen decided to race Zeus with Titan and the white stallion to see how he would fare.

"You can't race him yourself, Kat." Reed was firm. "You can't ride astride in a skirt."

"Then I'll ride sidesaddle."

He took her aside and spoke more gently to her than was his usual style. "Kitten, another time perhaps. I am asking you not to race today. Your fall is still too fresh in my mind. All I can envision is you falling again. Please wait awhile." He looked down pleadingly at her.

Kathleen closed her eyes against his look. The pain of

346

her recent loss flooded her. When she opened her eyes again, she answered, "I won't race Zeus myself, Reed, but I would like to see how he would do."

"I will ride him for you for a small price," piped up an unknown voice.

Kathleen turned to find a small ragged boy with enormous dark eyes and a head of curly black hair.

"I ride very well." He puffed out his chest proudly.

"I don't doubt if for a moment," Kathleen told him earnestly, holding back a smile. "Who might you be?"

"I am Raul. My grandmother is the seeress."

"Well, Raul, what price did you have in mind?"

"Five dollars," came the prompt answer.

Reed stepped up beside Kathleen. "That sounds a bit steep to me."

Raul thought a moment. "Four will do—in gold of course."

Kathleen smiled, her eyes twinkling merrily. "You will be riding a rare animal, Raul. It is you who should pay me for the privilege."

"Three then, and not a penny less." Raul's chin jutted out.

"Two and not a cent more!" Kathleen countered.

"All right, two!" Raul grinned.

"I think he would have settled for one if you had kept going, Kat," Reed whispered.

"Oh well, easy come, easy go!" she quipped.

"That is fine for you to say," Reed grumbled, "especially since it is I who foot your bills!"

"Should I feel like a kept woman?" she teased.

"Just keep in mind that a kept woman always pays for her keep one way or another, my dear." He grinned at her rakishly.

The horses raced with Raul astride Zeus. Kathleen watched from the sidelines with the others. Several other planters also entered horses, and there was a final field of ten animals. Reed, on Titan, came in a half length ahead of the others, but to Kathleen's delight, Zeus and the gypsy's white came in tied for second place a good two lengths ahead of the field. Now she finally had some idea of Zeus's speed, and since it was his first race, he was sure to get better. He might even beat Titan before long. She was bursting with pride.

As per terms of the wager, the five from Chimera stayed for supper. Reed sent word to Chimera and had another sheep sent to the camp anyway. While they waited, Kathleen let Ted and Eleanore talk her into having her fortune told by Raul's grandmother. She followed the old woman into the darkened interior of the woman's wagon. A few candles glowed dimly and the smell of incense was strong.

She seated Kathleen opposite herself at a small, low table and took the girl's hands. For long, silent moments she stared into her palms, then gazed deeply into Kathleen's emerald eyes until Kathleen almost felt herself drowning in the fathomless dark eyes opposite her own. She felt herself start to sway.

The wizened old woman began to quiver and shake as if in a fit of palsy. Still holding Kathleen's hands, she closed her eyes and her head fell back. Her voice, when it began, was strange, seeming to belong to another person altogether.

Caught up in the strangeness of the moment, Kathleen listened in awed silence as the woman spoke.

"You have come from a far land by the sea. More than that, you come from the sea. You belong to it, and will return to it time and again, until you at last rest forever beneath its depths. But fear not, for that will be many years hence. I see many years of life for you, but you must take heed. Never stay long from the sea, for from it you gain your strength from the trials life brings you.

"I see one great love in your life. He is tall and dark and commanding, with eyes like the morning sky. Tread carefully here. You are often at cross-purposes with this man. His domination threatens your independent nature. I see you in combat against him. I see you in conflict within yourself.

"I see a young woman who resembles you—no, she *is* you. She has hair as black as the raven's wing, and wears an outfit of green. She is wild and bold. She is your other freer self when your spirit needs to fly free. She wields a sword and sails the seas. I see her in grave danger from another man, a short man with an accent and a scar along his arm. Avoid this man. He wishes your life's blood.

"I see a deep grief, an unborn son lost recently. Share your grief with one who also bears it and you will feel your heart lighten. Grieve not. There will be another son and more children in your future.

"I see a marriage that is not as it should be. It brings distress to its partners where there should be great joy. Your pride keeps you from this joy. When you can put aside your vengeance and pride, only then will your life be complete. You fight your love for this man, but he is your only love. You are not whole without him. Cut him away from yourself, and you cut out your reason for being.

"There will come a great crisis. I do not see it clearly, but it is a battle of wills and swords. Weather it with love and you will survive it unscathed. Deal with it with revenge and hate, and you will always live with regret. The paths are both there. The choice is yours. Walk one path and sorrow abounds. Choose the other and joy will light your life and that of those you love.

"You have two great gifts—that of your sea spirit, and that of true love. Do not waste them. Nurture them and they will serve you well, with rewards beyond your dreams. Heed my words."

The old woman's voice trailed off. For long seconds neither moved. Finally the gypsy opened her eyes and breathed deeply. "I can see no more. I can add only this. Because of your great gifts you will save many lives. Your personal destiny rests more firmly in your own hands than most. This can be a burden and trial to you, but it can also be your salvation."

She stood and led Kathleen outside. "Go with God and follow your heart, girl. Your happiness awaits you."

Kathleen was very quiet when she rejoined the others. From time to time as they ate, she would glance up to see Reed eyeing her with concern. She thought with great care about the old woman's words. The more she mulled it over in her mind, the more confused she felt. The knowledge that her unborn baby was a boy weighed heavily on her.

Finally Reed broke into her thoughts. "I don't mean to pry, Kat, but something has obviously upset you. Was it something that old fortuneteller said?"

Kathleen blinked back her tears. "I don't want to talk about it."

"Kat! I'm surprised at you! That is all a bunch of hocus-pocus! There isn't a grain of truth in any of it, darling."

Kathleen gave him a decidedly hateful look. "That's how much you know, Reed Taylor! You should have heard! Oh, God, you should have heard!" She leaped from the ground and ran off along the river's bank.

Reed started to go after her and swiftly changed his mind. Instead, he went off in search of the old woman. He found her outside her wagon.

"Old woman, what did you say to my wife that upset her so?"

She looked up at him tiredly. "If she wishes you to know, she will tell you herself. I told her only the truth."

"Most palmists tell their clients nice things, don't they? That they will have long life and much wealth and happiness? What kind of gloom and doom did you forecast for her to have such a downcast reaction?" Reed was trying to control his anger.

"Life is not like that, is it?" the old woman answered. She stood slowly and took his palms.

"I will tell you this. Your young wife is more complex than you can now imagine. You and she will have your trials, but if you cling firmly to your love, you can win her. Even should you discover she has deceived you in some way, hold fast to her. It is never wise to try to cage such a bird. By letting her feel free, you will tame her.

"You will reach a crisis in your marriage. You will know it when it comes. Beware! At this time do not flee from her, for if you do you may lose her forever. You must curb your anger and use understanding and forgiveness as your truest weapons. Your life and marriage will be resolved if you can learn to temper your anger with love and kindness."

Reed jerked his hands from hers. "You have told me nothing!"

The old woman shook her head sadly. "I have told you much. You have not listened. I only hope you will recall my words when you should, that they should serve you well." She turned to enter her wagon. "Go now. Your wife awaits you at the bend in the river by the big rock. She will be ready to talk to you."

Indeed, he found her exactly where the old woman had said. She was sitting on the rock, crying softly into her hands. He walked up and touched her gently on the shoulder.

Without hesitation she threw herself into his arms, sobbing wildly. Surprised beyond words, he could only hold her until she calmed down enough to talk.

"Tell me about it, kitten."

"Oh, Reed!" she sobbed. "All these weeks I've been blaming you! But it was my fault! I killed our baby, Reed! I killed your son by my own stupid actions!"

"You didn't fall deliberately, Kat. I was right behind you. I know. It was an accident that took our child, love. An accident." He stroked her hair soothingly.

"It was a *boy,* Reed, a son! Your heir! She told me so, and even if she hadn't, I knew deep inside anyway. Oh, dear God! How can you bear to look at me?" she cried.

"It wasn't your fault. You were upset and I had said some foul things. Can you forgive me, Kat?"

"Me forgive *you?* Of course! But, Reed, I have to tell you this. I would have wanted that baby had I known. I know I told you differently, but the words were said in anger and I didn't really mean them, darling. I would have loved him dearly, not just for himself, but because he was yours. Please believe me, Reed. I would have loved your child!" She nearly screamed the words at him. He held her tightly for a moment, feeling the tears stinging behind his eyelids.

"Thank you, kitten. I believe you. Someday perhaps

there will be other children.''

He expected her to pull away from him angrily but she only hiccuped and nodded her head as she snuggled into his chest. The sweet sorrow and growing joy he felt at that moment were indescribable. They clung to each other for a long while before they started walking slowly back to the camp.

He handed her his handkerchief. "Here. Wipe your face and blow your nose.''

"Thank you.'' She accepted the offer. "I already feel better. Besides, I forgot to tell you the good part.''

"The good part?'' He looked at her questioningly.

"Yes. She told me we would have another son and more children besides.''

He grinned down at her radiant face. "I could have told you that, you adorable ninny!''

By the time Reed and Kathleen reached the gypsy camp, the festivities were in full swing. Singing and playing and dancing were being enjoyed by one and all. As the only outsiders present, they had expected to leave, but when they found Susan and Ted, they could not immediately locate Eleanore. Then Ted pointed into the center of a group of dancers. There was Eleanore, being swung about by the handsome young man who had flirted with her earlier.

"Do you suppose she'll invite him to the Christmas cotillion?'' Ted joked.

"Knowing Eleanore, she just might at that!'' Kathleen retorted. Looking up at Reed, she said in a low, slightly husky voice, "Doesn't this remind you of another evening, a similar group of people, guitars, dancing?''

He gazed at her adoringly. "I recall one dancer in particular who held me spellbound that evening.'' He held out his arms. "Would you care to dance, señora?''

She wasn't surprised that he picked up the dance steps of the gypsy music so well. She was finding her husband to be a very versatile man. They danced for a while and finally found themselves near Eleanore and her dashing partner.

"Kathleen, Reed, I would like you to meet Pietre. Isn't he magnificent?"

"Indeed," Kathleen muttered as the two men shook hands.

"I was just about to tell Pietre how marvelously you can dance the flamenco, Kathleen," Eleanore said.

"Eleanore!" Kathleen hissed in annoyance.

Pietre looked at Eleanore confusedly. "You are joking, yes?"

Eleanore giggled. "I am joking, no! She really can!"

Kathleen surveyed her friend critically. "Eleanore, are you tipsy?" she asked incredulously.

"Maybe just a bit tiddly. I had a wee tad of wine, but, my lands, it was strong!"

"We should head for home now," Reed suggested diplomatically.

"Just a little while longer," Pietre pleaded elegantly. "I have just found this delightful angel. Don't take her away from me so soon."

Then he turned to Kathleen. "You must dance for us. Are you as good as Eleanore suggests?"

"Our friend exaggerates at times," Reed explained.

"Especially when she's in her cups," Kathleen added in an undertone.

"You will not try?" Pietre asked.

"Thank you, but no. I must refuse," Kathleen said politely, all the time glaring daggers at the grinning Eleanore.

"As you wish." Pietre shrugged philosophically, leading Eleanore back into the dance.

Kathleen sighed in relief, silently promising to strangle her friend at the first opportunity.

After dancing a while longer, they sat about the campfire listening to tales retold and songs and stories invented on the spot. As one young girl took her place with her guitar, Raul stepped up to Kathleen and Reed and whispered, "Atalia asks that you listen well to the words of her song. She says you will recognize the man and woman in the verses."

Kathleen and Reed exchanged a confused look. Then they shrugged and settled back to listen as Atalia began to strum the beginning chords. Her voice rang out soft and clear as she sang her ballad to them.

> There once was a god and a goddess
> Different as day and night,
> And when they came together
> 'Twas the meeting of fire and ice.
>
> She was the sunshine, he the darkness;
> Enchantress and devil were they.
> A flash of lightning, a roll of thunder,
> As they clashed throughout the day.
>
> She as a small orange kitten,
> He as the panther dark.
> She as the playful dolphin,
> He as the deadly shark.
>
> She as soft as the clouds above,
> He as hard as tempered steel.
> And spirits say when they made love
> It made the universe reel.
>
> His eyes were blue as a cloudless sky,
> His hair was black as midnight.
> Her emerald eyes could weave a spell,
> Her tresses long and spun of sunlight.
>
> She was the honey; he the bee,
> She the rose and he the thorn.
> She the turbulent, mystic sea,
> He the heavens racked with storm.
>
> He swore to tame the fiery maiden
> With all his power and might.
> She said she'd see him first to fall;
> They battled through endless night.
>
> His lips so warm, demanding,
> Hers so soft and tender;
> He used his strength, she her wiles,
> Testing who would first surrender.

But when the storm clouds parted
And the newborn day had dawned,
They both lay sleeping, still embracing,
Conquered by the love they'd spawned.

As Atalia's voice faded away, Kathleen and Reed sat staring into one another's eyes. Kathleen blinked back the tears that were sparkling in her eyes. Reed grasped her small hand in his, fingering the emerald ring he had given her.

"It's uncanny!" Kathleen murmured.

"How many times have I called you my kitten?" he said wonderingly.

"The comparisons fit so well."

"At least she has given us a happy ending in her ballad," he whispered wistfully. "Do you think perhaps she could be right?"

Kathleen sighed sadly, "Perhaps in time, Reed." To herself she added, "Dear God, I hope so."

They rode home thoughtfully. Reed took the tipsy Eleanore before him in his saddle as Kathleen led her horse. Kathleen was acutely aware of the speculative looks she got from Ted and Susan, though neither said anything.

"Oh, dear! What a day this has turned out to be!" she groaned to herself.

When they at last reached Chimera, she fled to her room. She would certainly have a few words for Eleanore in the morning!

Exhaustion, emotional as well as physical, took its toll, and she fell into a deep sleep. She awoke to the feel of warm lips parting hers, and calloused hands stroking her limbs.

"Reed?" she questioned sleepily.

"In the flesh!" came his chuckled reply.

Her hands reached out to search his body. "And not much else it seems," she giggled. He massaged her stomach and she made small sounds in her throat. "You don't even cheat fairly, Reed," she purred.

"I know," he returned cheerfully.

Kathleen stifled a yawn. "What time is it?"

"Nearly one."

"I'm half asleep still."

"Part of my plan," he admitted. "This way you are less apt to throw me out and much easier to manage."

"You rascal!"

They made slow, langourous love, drawing out their pleasures. Afterward they slept soundly, awakening only when they heard Mammy enter the sitting room.

Kathleen sat up with a start and gave the smiling Reed a great shove. "Damn it, Reed! Don't just lie there grinning! Get up and out of here!" she hissed.

They heard Mammy humming as she straightened the other room.

Slowly Reed climbed out of bed. Kathleen tossed his robe at him and pushed him to the veranda doors. "Hurry up! If you are caught here, I swear I will never speak to you again!"

"Is that a threat or a promise?" He winked at her.

"Both, you big lout! Now go!"

She shoved him onto the veranda and leaped back into bed just as Mammy opened the bedroom door.

"Mornin', chile," Mammy said cheerfully. "Ah thought ah'd bring up yoah coffee befoah breakfas'. Ya went to baid so early las' night. How's dat haidache o' yoah's?"

"It just left," Kathleen said ironically, rolling her eyes.

"Whut? Sometimes ah just cain't figure these white folks."

"Nothing, Mammy. Thank you for the coffee." Kathleen sat up, fluffing the pillows before Mammy could round the bed. She swung her feet to the floor—and smack into Reed's slippers! Quickly she kicked them beneath the bed.

"I'll kill him!" she thought furiously. "I'll simply murder him!"

"Whut ya'll gwine to wear to town t'day?" Mammy was asking.

"To town?" Kathleen's brain felt as if it were in a fog.

Mammy eyed her curiously. "Ta dat Octoberfest thing."

"Is that today?"

"Shore is."

"I suppose the green brocade with the cream lapels."

"An' thet little cream cape whut matches? Jest in case it rains?"

Over the breakfast table Reed grinned at her. She glared back, her eyes blazing. As he assisted her into the carriage later, she hissed at him, "You horse's rear!"

His eyebrows shot up. "What brought that on?"

"Let's just say the gardener will find your slippers in the bushes below your veranda. I tossed them on my way down to breakfast!"

His deep laughter rumbled.

"Stop laughing! It's not funny!" She stamped her small foot at him. "I believe you did that deliberately."

He shook his head. "No, kitten. Next time I'll be more careful."

"Pigs will hatch from eggs before I take that chance again!"

He changed the subject. "Tell me, Kat, I'm curious. It struck me this morning that your legs are tan all the way up. How is that so?"

For a moment she couldn't breathe. All she could think of was that damned short pirate outfit. Her brain raced frantically. "Promise you won't tell a soul," she whispered at last.

He nodded.

"I've been slipping out by myself whenever I can. I swim in my chemise and pantalettes."

He stared at her in disbelief. "Kat! You go too far! Suppose someone saw you?"

"No one will. I've very careful."

"No more, Kathleen. I forbid it."

"Oh you do, do you? And who's to stop me?" She glared at him belligerently.

"I am, Miss Sassy-pants!"

"You and whose army, Captain?" She turned her back on him and climbed into the carriage. Before she knew what he was about, he had slapped her smartly on the backside.

"Ouch!" She wheeled on him, but fortunately for him, Barbara chose that exact moment to start out the door toward the carriage. "You'll pay for that, you scurvy weasel!" she muttered.

"Don't I always?" he returned dryly.

Chapter 22

FOR a week Reed continued to escort her to harvest feasts, hayrides and fall parties. He and Kathleen were decidedly cool toward one another. Much to Kate's sorrow, their war was on again. Eleanore would gladly have gotten drunk again if she could have healed the breach. She knew beyond a doubt that Reed and Kathleen had made love recently. Her room was next to Kathleen's and the walls were not very thick. She had heard them as she was nursing her sick stomach and aching head.

Reed left for Grande Terre on the first of November. According to Kate, he would not return until his birthday celebration on the thirteenth.

Kathleen, of course, followed, and in the next week and a half attacked his ship twice, relieving him of sizable cargos both times. The first time she merely outmaneuvered him smartly, having the advantage of the *Emerald Enchantress* being lighter and swifter. She caught Dominique's amused gaze as she squared off with Reed. They sparred heavily for nearly ten minutes, neither gaining the advantage until Kathleen, with one furious uphand stroke, snapped Reed's blade in two near the hilt.

For a moment they both stood mutely eyeing the broken blade. Then Reed raised his gaze to meet Kathleen's gleefully victorious look.

"Damn!" he muttered.

"It seems if you don't have bad luck, you have none at

all, Captain Taylor," she mused mirthfully. "Do you concede the contest, sir?"

"Aye," he snapped. "What choice do I have?" If looks alone could kill, she would have fallen to the deck.

"Well then, I await my cargo." She gave him a smug smile.

On the second encounter, they exchanged cannon fire for a time, until the *Emerald Enchantress* landed a blow to the *Kat-Ann's* rudder. Kathleen was sure Dominique had a hand in making sure that the *Emerald Enchantress* suffered no hits. Just how he had managed it she could not figure, but she felt sure he had helped her.

When she once more faced Reed on the bridge, she heaved a sigh and said, "Captain Taylor, this is really becoming monotonous! Let us this time make it a contest between the two of us. The good doctor is weary with patching up my wounded men as well as yours. What say you?"

"For the good doctor's sake then," he nodded. "Will first blood satisfy you, or are you after my head today?" His blue eyes shot angry sparks at her.

"First blood it is," she agreed.

They squared off and silently began the battle. Not a sound was heard but steel kissing steel as they parried and counter-parried one another's blows. As was her habit by training, Kathleen held strict eye contact with her opponent. Finally, her arm becoming weary beneath the weight of her rapier, she took a calculated risk. Deliberately she left herself open, her blade lowered. Immediately Reed swung in close, his blade reaching for her bare midsection. Kathleen twisted lithely aside and brought her own blade up swiftly, barely scraping the inside of Reed's right wrist, but it was enough to draw a thin line of blood beneath the slice in his shirt sleeve.

She backed off, her breasts heaving beneath her vest. "Give off, Reed. I have drawn first blood. I am tired and you have a rudder to repair before you head for port this day."

"Once more I concede to you it seems," he ground out.

He swept an arm toward the hold. "Take your rewards and be gone." With that he turned and stalked to his cabin.

Kathleen looked at Dominique and arched her brows. "Surly sort, isn't he?"

"It would seem so," he commented with a sly wink.

They were following the Georgia coastline and were nearing the inlets to the cove. Kathleen and her crew were weary, and she was looking forward to getting home to a steaming tub. All lanterns were doused, and silence reigned on board, but Kathleen cursed the full moon that still made them so visible. She felt uneasy, as did most of the crew. They carried a full hold and could ill afford to get caught by the ever-diligent shore patrol.

Kathleen's unease soon proved prophetic. Word passed to her that two ships were sighted, one to her bow and one to her stern, both heading in her direction. Through her spyglass Kathleen quickly calculated the speed of the ship before her. To her dismay she recognized it as a U.S. Navy patrol ship, and also realized she could not make the first of several inlets in time to avoid being seen. Swinging her glass about, she searched out the other ship. It appeared to be a merchant heading most likely for Savannah.

"Well now, lass, what's to be done fer it?" Dan whispered near her ear.

"Tell me. With the *Enchantress* between them, do you suppose either of them will see the other first, Dan?"

"My guess is they won't be looking fer another ship so close."

"My thoughts exactly. We are about to be spotted by both at any second, I expect."

"Aye, Cap'n, and then what?" Finley inquired softly.

"Well, the patrol will give chase or return fire for sure. The merchant may even fire on us, as our reputation has become so well known. If we act before we are spotted, swiftly and silently, with no lights, there is a plan which might work."

Quickly Kathleen outlined her course of action, and Dan and Finley scurried off to ready the crew. Within seconds Kathleen had swung the *Emerald Enchantress* about to face the open sea. The sails were trimmed to catch the maximum breeze. Just as the wind snapped the sails into service, the *Enchantress's* port and starboard guns simultaneously roared out once, breaking the silence of the night. Then the *Enchantress* slipped speedily out from between the two approaching ships and broke for the open sea.

It was as if Kathleen's guardian angel had chosen that exact moment to awaken from her slumber, for the moon suddenly ducked beneath a bank of clouds, covering the *Enchantress's* flight. The sleek dark ship blended with the darkness of the sky and sea and was no more.

Aboard the naval vessel, the captain was issuing swift orders to return fire on the unfortunate merchant ship before him. Not having seen the darkened *Enchantress,* he assumed the shots had come from the trading vessel.

At the same time, similar thoughts were racing through the other captain's brain. Having received fire from the shore patrol, and not wanting to get caught with his hold full of contriband, the merchant had to decide whether to run or fight. Figuring that discretion was the better part of valor, they turned tail and ran, the shore patrol close behind.

Meanwhile, aboard the *Enchantress,* Kathleen cut a wide circle around the two fleeing ships, came up well behind, and furtively scooted into her inlet. Once anchored safely in the hidden cove, all aboard heaved a collective sigh of relief.

"I'll be hanged if I didn't think we'd bought it thet time, lass!" Dan declared.

"Please! Don't say the word hanged within my hearing for a while. That was much too close for comfort." Kathleen sighed shakily. "It is one thing to pirate Reed or another vessel one on one, and quite another matter to tangle with the American government. It would please me if I never encounter another situation such as this again in my lifetime!"

"Aye, Cap'n. Me, too!" Dan said fervently, as Finley nodded his head in full agreement.

"We must be more careful in the future."

As soon as she had bathed and changed her clothes at Kate's, Kathleen headed for Chimera. Reed was due home at any time, and she wanted him to find her there for a change. Also, she needed to wrap and hide his birthday gift before he arrived.

She had no doubt that he would like it, and it was a perfect gift since personal items were unseemly for a young lady to give a gentleman. Dan had grumbled loudly, but finally agreed to sell her the detailed scale model of the *Kat-Ann* he had carved out. He had worked on it for over a year, and finally finished it not a month earlier.

"What ye be wantin' to give it to him fer?" he grouched.

"It's his birthday, Dan. Besides, since I plan to relieve him of the *Kat-Ann* in the near future, he will at least have its likeness to remind him of his pirate love."

At this Dan chuckled. "Oh, lass, ye're a devious one, ye are!"

Kathleen had been in bed for a couple of hours when she heard Reed's tread outside her veranda door. Carefully regulating her breathing, she feigned sleep as she heard him approach her bedside. She felt his gaze upon her for long moments and finally he bent to kiss her tenderly at her temple. Then he turned and left.

The next day dawned clear and bright, a beautiful day for Reed's birthday. Kathleen took her place next to Reed at the breakfast table.

"Good morning, all," she said, glancing around the table. Turning to Reed she smiled, "Good morning, birthday boy!"

He caught her teasing look. His deep voice was amused as he replied dryly, "Good morning, Kat. Did you sleep well?"

"Fine, thank you."

"Reed, dear." Mary claimed her son's attention. "You know we've planned a supper party for your birthday. Is there anything special you would like today, or anything you want to do?''

"Darling mother.'' Reed took Mary's hand in his and kissed it, his eyes twinkling in amusement. "I'm a bit old for a pony or sugar plum pudding, and I think the party is more than enough.'' He looked up, and his gaze locked with Kathleen's for a long moment. "No,'' he sighed, "there is nothing more you can do for me, Mother.''

"Well, what are you going to do?'' Susan prompted. "It's such a lovely day.''

"It is just another work day, Susie, love. I have a lot of business to attend to around here. It is difficult sometimes to keep up on everything here and tend to my ship, too.''

Amy leveled her baby-blue eyes at him, batting her lashes blatantly. "Surely you aren't going to lock yourself in that stuffy office of yours all day,'' she crooned sweetly. "After all, all work and no play . . .''

Eleanore caught Kathleen's look and rolled her eyes in mock dismay. "I wondered where all the hotcake syrup went this morning,'' she mocked innocently.

Kathleen choked back a chuckle. "Well, I don't know what the rest of you intend to do today, but Zeus needs exercise.'' Kathleen pushed her chair back from the table. "Excuse me, please.''

"Just a second, young lady.'' Reed's imposing tone stopped her before she could rise.

She looked at him with raised brows over surprised green eyes.

He gestured toward her plate. "You haven't eaten much this morning. Are you feeling well?''

"I'm perfectly well, Reed, thank you. No need to worry yourself. I'm merely watching my figure.'' Kathleen rose from her chair.

"If you need any help, let me know.'' His blue eyes danced merrily as he saw her start to blush.

"Reed Taylor! You are a—''

"Yes?'' he drawled, his eyes roving her body.

"Oh! You're an uncouth cad!'' she declared as she stormed out of the room.

"I say! Breakfast around here gets more interesting all

the time," Ted commented. A second later he yelled, "Ouch!" as his mother kicked his ankle beneath the table.

Tears of merriment danced in Eleanore's brown eyes as she tried desperately to swallow a mouthful of orange juice.

"Really Reed! I thought I brought you up better!" Mary reproved gently. "Stop baiting the poor girl."

"I really can't help myself sometimes, Mother." At her frown he said, "All right, I'll go apologize. Titan needs to stretch his legs anyway."

He found her later atop a knoll overlooking the river, a frown creasing her forehead.

"Why the thoughtful look, Kat? Surely you are not still angry with me."

"Why shouldn't I be?" she responded. "You are altogether too forward sometimes, Reed, especially in front of other people. I can imagine what they must think!"

"The worst, I'm sure," he commented.

"I thought you had work to do."

"It will keep. I'm supposed to apologize to you. If I say I'm sorry will you stop frowning?"

"Yes."

"All right, I'm sorry."

"I truly doubt it, but thank you for the gesture. Besides," she continued with a sly grin, "I wasn't angry at you any longer. In fact, I wasn't thinking of that at all."

"What were you thinking of?"

"I was wondering why, with Chimera bordering the Savannah River, you don't dock here instead. It is closer than going on to Savannah and riding back. Wouldn't it be easier to unload your cargo here, especially things needed for the plantation?"

"Well, kitten, let me explain. Most of the time we do unload what is needed here, but we have to do it by raft, as it is much too shallow for the *Kat-Ann* along this part of the shoreline. Before I had a ship of my own, it didn't bother me, but it would be more convenient now that I am in port more often. However, most of my cargo is for sale in the shops of Savannah, and I keep a warehouse there for unloading and dealing with merchants. Besides, my crewmen spend their time in Savannah, not here.''

"Yes, but what do you do when you have cotton or other goods you export?" she asked.

"Sometimes we load up here, using rafts again, as does Kate O'Reilly. Most often, however, we deal through the cotton exchangers in Savannah.''

"That seems a waste of effort, time, and money when the river is handy and the ship available,'' she observed.

"What would you suggest, my sweet?" he asked, taking her hand as they both sat on the grass.

Kathleen thought a moment, her teeth worrying her lower lip. "You could dredge out a channel, deepen this section of the shoreline a short distance out. You could even build a small dock like some of the other homes have. Then you could load and unload easily what you need here. The rest could go on to the warehouse in Savannah. You could go together with Kate and some of the nearest neighbors if you want and sell your exports directly at better prices and cut out the exchangers altogether.'' She stopped and looked at him questioningly.

"Yes, go on." He found himself impressed with her reasoning.

"The house is really fairly close to the river. You can see the river easily from the north and east sides of the house, and from the widow's walk you can see a fair distance downriver.''

"You've been up there?" he interrupted.

"Oh, yes. Quite often. It's quiet up there and no one thinks to look for me there,'' she explained. "Now, as I was saying, you could go on to Savannah and unload your cargo and most of the men and return here to dock. Or, if you have a trustworthy man or two, they could drop you off and go on to Savannah without you, unload, and await word when next you want to sail.

At that he laughed. "And just suppose I get tired of trudging up to the house in the dark of night, especially when I'm tired and hungry?''

Again she stopped to think, and her emerald eyes lit up

with delight. "It's quite simple, really. Simply tamp in a post at this end of the pier and hang a large bell, or use the *Kat-Ann's* fog bell. It can be heard from the house and one of the servants can ride down and pick you up or bring Titan down to you. Set up your own signal if you want."

He laughed and drew her close. Looking into her clear green eyes, he commented, "You are quite a minx, my love. Who would guess that beneath that exquisitely beautiful exterior lurks the brain of a wizard?"

"Had you fooled, didn't I?" she teased. "Besides that, Reed, you've heard of Robert Fulton's steamboat in New York?"

At his nod, she went on. "They are building more of them I hear, not only for the Hudson River, but the Mississippi too. If they catch on, we may have steamboat traffic along the Savannah in a few years. I've seen your maps and noticed that the Savannah River runs the entire length of Georgia's eastern state line. Just think, Reed! It would be convenient travel from here to Augusta and farther. People could travel easily and swiftly to visit friends and relations, and you would have your own landing convenient for your own use and needs."

"And just where did you learn about steamboats, wife of mine?"

"I keep my ears open. Lots of things are discussed in front of women when men don't think they are listening. Most of the time I suspect you think we are too stupid to understand what is said, but we absorb more than you realize."

"So I'm discovering," he chuckled. Then, before she knew what he was about, he drew her closer and lowered his lips to hers. Instinctively she stiffened, but his lips were so warm, and the kiss so tender, that she relaxed in his arms and opened her lips to his gently probing tongue. A sigh escaped her as she raised her arms to lock them about his neck. A tingling warmth crept through her as his kiss deepened with desire. His palm caressed her breast through the fabric of her blouse and she shivered and pressed closer to him, her head spinning as he awakened her sleeping passions.

When he drew back from her, it was with reluctance. "As you pointed out, kitten, we can be seen from the house, or else I would take you here and now." His blue eyes sparkled playfully the next moment. "Perhaps I shall anyway," he teased as he attempted to draw her near once more. "After all, it is my birthday."

"You will not, you randy rogue!" she squealed as she wriggled in his arms. "You've had your birthday kiss."

"You miserly woman! Is that all I get?" he exclaimed in mock despair.

"For now." Her emerald eyes danced. "I have a gift for you which you can unwrap later," she stated demurely.

"That is not what I meant, and you well know it, temptress. The gift I desire is wrapped in golden skin as smooth as satin, with coppery hair as soft as silk and accented with emerald eyes as green as the hills of Ireland." His voice was as soft as velvet and husky with feeling, and his eyes were the darker blue they always became when he desired her.

"Come to me again tonight, Kat."

Choked by her own yearning, she could only nod her assent.

Kathleen dressed with care for Reed's birthday party. From her closet she removed the latest of Mrs. Fitz's creations. Again the daring design was Kathleen's idea, but the beautiful handiwork was Mrs. Fitz's. For a moment she debated whether to don the gown, then she waved away her doubts and rang for Mammy.

With the old woman's help, her hair was pulled back from her face and into a high knot atop her head. From there fell a long, thick coil that draped itself across her shoulder. Next she slipped into the gown. It was in the Grecian style, coming up over one shoulder and leaving the other completely bare. The shoulder clasp was of gold and secured a length of the sheerest silk that fell to drape one arm. The material of the dress itself was layer upon layer of filmy gold-colored silk. It sported a deeper gold taffeta sash and fell from there in smooth folds that draped her hips and long legs and made the viewer wish the material were even more diaphanous. In the shimmering, flowing gold gown, combined with the hair style, she looked like a Greek goddess come to life.

From her plunder, she had also taken a plain gold arm band which she clasped firmly about the upper part of her bare arm. About her neck she wore the emerald and diamond choker Reed had presented to her.

Mammy was aghast. "You cain't go down der in dat! Land o' goshen gal, yo is next to nakid! Whut is yo' thinkin' ob?"

Kathleen viewed herself in her mirror. The dress was indeed most daring, but so was she. Facing Mammy, she said, "I most certainly do intend to wear it. All of Europe is patterning itself after the Empress Josephine's manner of dress. If it is good enough for her, it's good enough for me!"

Mammy walked out muttering, "It's heathen, dat's whut 'tis, jes' heathen!"

Kathleen looked after her with a smile.

She waited deliberately until she knew most of the guests had arrived. Just as she was about to leave her room, she heard a rap on the door.

"Kat, are you coming down?" Reed opened the door and stood gaping. Momentarily he was awestruck. His eyes shone with desire as they traveled the length of her. When he finally found his voice, he said reverently, "You are a vision, Kat, a golden love goddess. Every time I think I am accustomed to your beauty, you amaze me anew."

Kathleen pirouetted before him and said with an impish grin, "Mammy says I'm a heathen."

"Then I am addicted to heathens, my pet. Come." He offered her his arm. "Let's join the party."

Suddenly unsure, she hung back. At his questioning look, she said, "Perhaps Mammy was right. It is a bit daring and I wouldn't want to cause you embarrassment before your friends."

His arm encircled her waist, "I am positive, darling. I am proud to have the most exquisite creature in Savannah at my side. After all," he added, "if I had wanted a plain, docile, unimaginative woman I would have married long ago."

Kathleen was the talk of Savannah the day after the supper party, and within a week Mrs. Fitz was besieged with orders for gowns such as Lady Hanley might wear. Once again the great green bug of jealousy bit savagely at Amy.

Throughout the evening, Reed, although courteous to all, paid court only to Kathleen. When he unwrapped his gift from her he was astounded.

"Kat, this is the most remarkable gift. It looks exactly like the *Kat-Ann*."

"Well I hope so, as it is an exact replica," she informed him.

"But how did you come by it? I don't understand."

"A friend made it and I convinced him to sell it to me. Actually, he was a friend of my father's. He's very good, wouldn't you say?"

"Indeed! Does your friend have a name?" He eyed her cautiously.

"He doesn't care for fame and you really shouldn't look a gift horse in the mouth. Next you'll be wanting to know how much it cost."

"You probably did spend overmuch, but I can't fault you for it. It is a unique gift, and I thank you. I shall treasure it always."

"You may, but I'm quite sure whoever has to dust it will soon grow to loathe the thing," she laughed.

"Since you purchased it, you should have that job," Ted teased.

"No, thank you. I respectfully decline. I claim a delicate wrist," she bantered lightly.

"Ah, maimed since birth, I expect." Reed's eyes danced with brilliance as he joined the fun. "It's my opinion you have had a spoiled childhood, that being your sole malady."

Kathleen tapped his chest lightly with her fan. "You,

sir, are fortunate I am so delicate, or I would surely call you out for that remark." She smiled up at him through her lashes.

"Oh-no!" he exclaimed. "Are you denying then that you are spoiled?"

"Not at all." She fluttered her fan before her face. "It is merely improper for you to say so. Your good manners are slipping dreadfully, sir!"

"My apologies, madam!" he continued playfully. "Perhaps you could tutor me in your spare time."

Their eyes met, green and blue, full of mirth, and neither could hold back their laughter.

They were still laughing hours later as they romped playfully in his huge bed. "Shh!" Kathleen giggled. "You'll have the entire household awake!"

His chest shook beneath her cheek as he attempted to stifle his rumbling laughter. "I might be able to stop laughing if you would quit tickling me, my sweet."

"Dreadfully sorry, Captain," she answered in her best English accent as she saluted smartly.

Fresh chuckles broke forth before he silenced her with a kiss. "You have had too much champagne this evening, Kat. I believe you are intoxicated."

"What a dreadful thing to say!" She paused to stare haughtily at him. Then the corners of her mouth struggled with a grin. "Even if it is the truth!"

Much, much later he sighed contentedly as he stroked her disheveled hair. She lay asleep, curled up against his warm body, a slight smile still upon her lips.

"My beautiful kitten," he whispered, and closed his eyes in sleep.

Much to Reed's dismay and anger, the next day brought an unwelcome change in his life. The fall social season was starting, and the Bakers were moving back to their home in Savannah. That meant Kathleen would no longer be conveniently at Chimera.

At the news he charged up to Kathleen's rooms, bursting into her bedroom, not bothering to knock. A little housemaid was packing Kathleen's trunk. A quick look around told him Kathleen was not there.

"Where is she?" he demanded gruffly.

"Miss Kafleen go to Miss 'Lenore's room, massa." The girl stood quietly wondering why the master was always upset these days.

"Unpack these things and put them away," he ordered.

A swish of skirts behind him alerted him to her presence even before she spoke. "Just one minute, Reed Taylor! Let the poor girl get on with her work. We must be ready to leave after lunch."

He swung around to glare at her. "You are not going."

Sweeping past him into the room, she answered, "Oh, but I am!"

She stood with her back to him, clutching the edge of the desk as she fought for control. A thousand thoughts raced through her brain as they had since news of the move had been brought to her by Mammy with her coffee. Part of her wanted to stay near Reed, the other part warned against such softness. Her path was mapped, and she could not allow him to make her falter.

"It is for the best, Reed," she said softly.

He came to stand behind her, his hands upon her shoulders. "How can it be best to have you away from me? How long are you going to keep this up, Kat? I know I promised you, but when are you going to see the folly of all this? What good does it do either of us? Wouldn't it be better for both of us to get on with our lives—together?"

She turned toward him, pushing his arms away. Looking at the young maid, she said, "Please leave that for now. I'll call you when I need you again."

When the girl was gone, she looked at Reed. "I'm going to Savannah, Reed. I'll not stay here with you. I'll not be your chattel, your faithful lap dog, your obedient bedmate. I've told you that before."

"Since when has being my bedmate become distasteful to you, Kat?" He leered. "It must be since this morning, for last night you were moaning and thrashing beneath me and begging me for more."

At her belligerant face, he warned her, "If you deny it, you're a hypocritical little liar. You enjoyed it and we both know it! What is more, you'll miss it." He motioned toward the door. "Go ahead, Kat. Leave! Go to Savannah and think of me when you awaken in the lonely dark night and I'm not there to hold you and warm you. And when you are restless and can't sleep, recall how we have passed the hours together. When your body betrays your stubborn mind and aches for mine, remember the passion we've shared."

"Stop it, Reed!" Kathleen nearly shrieked the words at him. Her hands clasped over her ears, she fled the room.

She did not appear for lunch and no one knew where she had gone. Just as the final bags were loaded onto the carriage, she came bounding into the yard astride Zeus.

One look at the windblown girl with her flying hair, flushed face, and unusual outfit, and Barbara almost fell from the carriage. "My stars!" she exclaimed, clutching at her bosom. "You're wearing breeches!" Mammy's eyes were are large as silver dollars.

Eleanore was the only one not upset by Kathleen's appearance. Her brown eyes glimmered with suppressed laughter as she viewed the gaping onlookers. Ted looked as if the goddess Diana had just appeared in flesh and blood, and Reed was glowering in dark anger, his hands clenched into tight fists at his sides. Susan stood beside her mother and looked unable to decide whether to idolize Kathleen or to be embarrassed for her. William looked away, thoroughly scandalized.

Amy was the first to recover. "Ladies do *not* ride astride in men's breeches, cousin!" She threw her nose so high in the air that if it had chosen to rain at that moment she would have drowned. "How dare you flaunt yourself like that!"

"I dare because it pleases me to do so." Kathleen's voice held a cool, deadly tone that made Reed shiver involuntarily. He had heard such voices as that at the other end of a loaded pistol. She eyed them all with the haughty air of royalty.

Eleanore felt the inane urge to curtsy as Kathleen's gaze reached her. "Eleanore, have all my things been packed?"

"Everything."

"Fine. I'd invite you to join me, but you could not keep my pace today." Her cool green gaze took in the other travelers. "I'll see you later at the townhouse."

She whirled her horse about and charged down the drive. A few yards away, she reined in sharply, causing Zeus to rear up in an effort to meet her command. Without looking back, Kathleen called out, "Ted, you may join me if you wish."

The ever-eager Ted responded like a bassett after a bone. Always in awe of his beautiful cousin, he was newly amazed at this strange, defiant mood of hers today. They raced down the drive together, leaving the others dumb-founded.

"What has gotten into that girl?" William questioned, shaking his head.

"I'll be double damned!" Reed said half to himself. "Lady Haley in all her glory!"

"I must have a talk with that young woman, William," Barbara sighed. "She has more of her mother in her than I'd guessed. Edward must be spinning in his grave!"

"Not gwine to be able to speak to her foah long," Mammy muttered. "She gwine bust her neck if'n she ride lak dat of'n 'nuff. Ride lak de debil afta' her."

Eleanore gave Reed a long look as she seated herself in the carriage. "You are so right, Mammy, but can he catch her?"

No one knew quite how to deal with Kathleen after that. Barbara tried to talk to her and was politely ignored. Mammy tried to shame her and got nothing but a wry grin for her efforts. She listened to Mary's advice and Kate's warnings, but said nothing.

Kathleen did not go out of her way to embarrass her family. Indeed, she behaved most properly, but the vivacious young girl they had come to know was gone over-night. In her place was Savannah's new ice queen. She maintained a cool, aloof attitude and her bearing was positively regal. She let everyone know that she would do, say, come, and go as she pleased, and they could like it or not; it mattered little to her. She created such an imposing figure that soon the elite society of Savannah was deferring to her, rather than the other way around. What Kathleen wore became the fashion; her hairstyles were copies overnight, her likes and dislikes were noted and religiously adhered to. Eleanore laughingly told Kathleen that if she shaved off all her hair and strolled bald through the streets, the next day all of Savannah would reflect the sunlight from their hairless pates.

Wherever she went, her favorite foods and drinks were served, her favorite music played. Her name topped everyone's guest list, and each hostess tried to outdo the next. It was the most exciting social season Savannah had known in years.

Reed was conspicuously absent from her side for nearly two weeks. He had stayed at Chimera when Mary and Susan had moved into the Baker's for the season. When he finally showed up, he was amazed to find the changes that had occurred.

Once again Kathleen was surrounded by the eligible male population of Savannah. They hung on her every word, fought one another for the privilege of a dance or to escort her to dinner, and nearly groveled for a mere smile from her lips. If she made mention of a favorite author or poet, the next day found scores of books pouring in from ardent admirers. Once she mistakenly mentioned her preference for roses, and for a week the house resembled a funeral parlor. So much candy arrived that if she had eaten it all she would have outweighed Mammy and been indistinguishable beneath the blemishes.

Through it all Kathleen remained outwardly poised and aloof. She found it ridiculous and vaguely amusing. She was constantly attending parties and other functions, and smiled her way graciously from one to the other, being at once congenial and yet somehow above it all. Rarely did she laugh outright anymore, or tease her suitors in a girlish fashion. It was as if she were the crown jewels, beautiful, glowing, and coveted, but cold, unattainable, and untouched by human emotion.

Only Eleanore, who now shared Kathleen's bedroom, knew differently. Only she knew that once the door closed for the night, once the jewels were put away and the gowns hung up, the much sought after ice queen of Savannah cried bitter tears into her pillow.

After the first few days, Ted cornered her. "Kathy, I don't know what happened to make you change like this—"

"Change?" she cut in shortly.

"Yes. Oh, you still do the same things, but you don't laugh anymore. You don't seem to enjoy anything anymore. Your lips smile, but your eyes don't." He paused to look at her in concern. "Do you know what they call you now?"

"Lord only knows, Ted."

"The ice queen. Oh, you intrigue them. Your reserve draws them like flies. Each man wants to be the one to thaw you out." He gazed at her in dismay. "You don't need to explain to me if you don't want to, Kathy, but if I can help in any way—"

"You are adorable, Ted." She took his hands in hers and smiled sadly. "No one can help, but don't worry yourself about me."

"I do worry. You don't have fun anymore. If you were a man I'd drag you along with me when I go out with the fellows. I'd take you to some establishments down by the docks and we'd drink and gamble till dawn. But you're a lady and so I can't help you."

Kathleen's eyes suddenly lit up dangerously. "Oh, but you can! You can take me along anyway!" He'd finally sparked her interest.

Ted was appalled. "You can't be serious! Ladies aren't allowed in places like that. I'm talking about brothels, Kathy!" His face turned beet red.

"Do you always gamble in brothels? Isn't there someplace else just a wee bit more acceptable? I've no wish to rub elbows with ladies of the evening, but I would dearly love to try my luck. Why, in London they have some very discreet places where lords and ladies can enter to gamble. They usually require masks. Ladies must have escorts, and generally you have to know someone to get in, but they exist.''

"Well, there is one very exclusive place I'm thinking of. There is a back room where the gentlemen gather if they wish to drink and play cards. Sometimes gentlemen arrange meetings there with their mistresses.'' Ted looked as if he were in severe pain.

"Oh, for heaven's sake, Ted! I know what goes on in a brothel, or at least I have a fair idea, and the idea of a man having a mistress is not news to me either. Don't look so distressed! I know young ladies are not supposed to be aware of all that, but facts are facts! Do you suppose I don't know why my father went to town every so often after Mother's death? Even in that girl's school in London there were stories. Several of the girls, all from prominent families mind you, had fathers who supported mistresses on the side. They tried to be discreet, of course, but everyone knew.''

"Oh, dear.'' Ted sank into the nearest chair. "I've been raised to think of ladies as delicate, innocent creatures, and you are destroying all my illusions.''

Kathleen laughed at the confused picture he presented. "It is quite all right, Ted. Ladies usually are innocent and many times most ignorant of such things. Papa would have fainted if he realized I'd guessed where he went. I don't want you to get the wrong idea. I wasn't sure what went on once he got there, but I did know where he went. Most young women, for instance, realize they will share a bed with their husband when they marry, but haven't the foggiest idea what goes on in that bed until the wedding night. We are left to wallow in our own ignorance and wonder about it, and believe me, we do our share of wondering and whispering.''

"I'd never realized ladies were curious about such things,'' Ted mused.

"Perhaps we wouldn't be if we were more

enlightened.''

"I think I see your point.''

Kathleen sat down across from him. "Do you think you could take me to this place you mentioned?''

"Good God, no, Kathy!''

"Why not?''

"Everyone in town knows you. Why, we're liable to run into the mayor or a half dozen of Savannah's leading citizens!''

Kathleen's smile was that of the cat that stole the cream. "I won't tell on them if you don't squeal on me! Don't forget, Ted my lad, most of them wouldn't dare let their wives know where they've been!'' Her laughter was tinkling and somewhat evil, and her eyes gleamed with deviltry.

Ted couldn't help but smile. It was the first time since coming back from Chimera he had heard her really laugh. "All right, but we'll have to sneak out and back in. How are you going to do that with Eleanore sharing your room?''

"We'll take her with us, of course,'' Kathleen replied smugly. "She'll just love it!''

Reed arrived in time for Thanksgiving. Only the Taylors and Bakers were present at the family dinner. He had wrongly assumed that Kathleen had had enough time to calm her anger. Throughout the meal she ignored him with frosty distain. When he asked her a direct question she answered him politely but curtly. By the time the meal was over he felt he had been seated near an iceberg.

When the men took their brandy and cigars in the study, he was told that she behaved in the same fashion to everyone. He thought they were joking as they related the events of the past two weeks and told him she was referred to as the ice queen.

"You can't be serious.''

"Oh, but we are,'' William assured him.

"But Kat is so vibrant, so alive!''

"You'll see,'' was their answer.

For a solid week Reed stayed with the Bakers, and not once could he find a crack in Kathleen's frozen façade. He insisted on escorting her to the theater and opera and various parties. Kathleen did not resist him, neither did she encourage him. It was as if she didn't care one way or the other. She remained distant, cool, and reserved. No longer did she laugh and tease with him. She seemed impervious to his attempts to arouse either her sense of humor or her anger. She was poised, polite, elegant, and beautiful, but emotionally as empty as a china doll—or so it seemed.

After a week of this new, unresponsive Kathleen, Reed threw up his hands in temporary defeat and sailed for Grande Terre. "It will be a short trip," he told Kate O'Reilly. "I'll be back two weeks before Christmas and I hope to God she has thawed by then. She's driving me berserk! I'd rather have her swearing at me and throwing china at my head than the way she is now."

Chapter 23

THE late afternoon was hazy, promising a foggy evening. Kathleen had given up her search for the *Kat-Ann* and was heading *Emerald Enchantress* toward her glade when the lookout shouted down from the crow's nest.

"Ship ahoy, Cap'n! Off the starboard bow!"

"Can you make her out?" Kathleen called.

"Aye! 'Tis the *Kat-Ann*, headin' for Barataria I'd say."

"Do we give chase, Cap'n?" Finley inquired.

Kathleen grinned rakishly. "Of course, Mr. Finley. Give the orders. Full sail and prepare for combat."

It took a full three-quarters of an hour to overtake the *Kat-Ann*. They stayed to her stern, thus avoiding all but two of her guns. By the time they drew alongside, they were only a league from their secret hideaway. As the grappling lines were tossed, Kathleen caught a boarding line and swung herself across, nearly landing on Reed's toes.

Reed, hands on hips, stared down at her with stormy blue eyes. "Are we to fight or dance, Emerald?" he taunted.

"Fight, of course, you big buffoon," she answered saucily. Then she gave him a flirtatious look and asked, "Just the two of us again?"

"Why not?" he shrugged.

The crews watched as the two captains squared off and the duel began. Reed matched her blow for blow. He was at his best this day and more determined than ever. It showed in his eyes, and as the contest wore on, Kathleen felt her confidence waning as well as her strength. Steel rang on steel as they lunged and parried and counter-parried, and still Kathleen could find no hole in his defense.

A part of her mind registered the fact that the fog was becoming thicker, and the deck was becoming slippery with moisture. As the minutes passed, Kathleen found herself more and more on the defensive. For a split second Kathleen's blade faltered, and suddenly she felt the tip of Reed's blade at her slim neck.

Time stood still. Her wide green eyes held his in astonishment. Cautiously she backed up a step and allowed herself to swallow. A multitude of impressions implanted themselves on her consciousness at the same time. First she realized that even as she and Reed watched one another in stunned silence, the crew of the *Enchantress* had cut the grappling lines and were quietly slipping away. This did not anger her, as she had instructed them to save the ship and themselves if she were ever caught. She knew they would immediately hide the ship and wait to see if she could escape and join them or manage to send word.

She sensed, rather than saw, Dominique's presence to her left. She felt the expectant stares of Reed's crewmen as they awaited their captain's next move. She heard her heartbeats thundering in her ears as she stared unblinking into his cool blue eyes.

At last she saw his eyes flicker with the light of victory. "At last!" he said in a near whisper. "At long last I have defeated you. I've waited a long time for this moment, my raven-haired piratess!"

His eyes flew to the rapier still hanging limply in her hand. "I'll have your weapon now."

Silently she turned it over to him, amazed to find that her hand was still steady. Tossing her head, she dared to say, "It is an empty victory, Captain Taylor. What have you to show for your efforts?" With a wave of her hand she drew his attention to the missing frigate.

He shrugged his broad shoulders. "I should have liked to have had the ship, yes, but your hold was empty, was it not?"

At her nod, he added, "There is always another day for that. For now I have my cargo still safe, and I have defeated you."

Under his powerful gaze she almost shuddered. He gave a short laugh. "What a bunch of bilge rats you have for a crew, to go off and leave you to face the consequences alone! And you will face them, my lovely. You have much to pay for." He reached out to caress her cheek with his knuckles, laughing again as she drew away from him.

"No, no, my dear. You will have to accustom yourself to my touch. You have lost and I have won, and now we play the game my way. I shall call the tune now and you will dance to it, like it or not."

Kathleen's chin went up in defiance. "You'll keep your hands to yourself if you mean them to remain attached to your worthless body, you braying ass!"

"Feisty wench, aren't you! Let's have a look at the nasty lady behind the mask." Reed made a grab for her mask and Kathleen backed away, only to find her back to the sternrail and Reed's rapier still aimed at her neck. From the corner of her eye she saw the waves far below, almost invisible in the thick, swirling fog. As Reed reached for her again, she braced herself against the rail, and bringing her feet up, planted both boots firmly against his chest. Before he could react, she pushed him back from her and launched herself in an almost impossibly graceful backward dive. In the blink of an eye she was beneath the waves and gone.

"Sweet Jesus!" Dominique swore. "Who would have thought she would try a fool stunt like that!"

He and Reed were hanging over the rail, trying to glimpse the woman through the fog.

"Can you see her?" Reed asked.

"Nay."

"The fool wench probably broke her neck."

"A pity, too. She was a fine piece," one of the crew added.

"God, what a waste," another said.

"Damn!" Reed swore. "To think she'd risk death before she would show me her face! I simply don't believe it!" Turning to Dominique, he asked, "Do you suppose she has a chance? We'll never find her in this fog."

Dominique was fighting nausea. He felt as if his chest was caving in, his breath coming in short gasps. All he could do was pray for a miracle. How could he tell Reed? What could he say? Better yet, what should he say? By her own actions, Kathleen had said she would rather die than have Reed discover her identity. Swallowing hard, Dominique decided to keep her secret and pray she found her way to safety somehow. He decided he would talk Jean into a trip to Savannah for the holidays. Maybe by then something would have happened; some miracle, for that is what it would take.

Aloud he said, "I don't think the odds are very good, my friend, but I hope so."

If Reed had known the pirate Emerald was actually Kathleen, he would probably have had more faith in her survival, and he would have been right. She surfaced about twenty yards from the *Kat-Ann*'s stern. Instinct alone guided her, for the fog was too thick to allow her to sense direction. If not for the voices, she would not have known where the frigate was. She struck out with strong, even strokes, knowing only that she must keep far from the *Kat-Ann* and Reed. She paced herself. When she felt tired, she floated and rested, and then went on.

After a while Kathleen began to doubt she would make it. Her arms felt like dead weights, and her legs were numb with fatigue. She had removed her boots and tied them under her vest in order not to lose them, and now the weight was dragging her down. She considered releasing them, but hated to lose the last gift from her beloved father. She stopped to rest, treading water and trying to gather more energy from within. She peered through the fog, trying to determine where she might be. She was not so much frightened as she was tired—bone tired.

Suddenly, through an opening in the grey curtain about

her, she caught a fleeting glimpse of a large dark shape to her left. Then the swirling mist closed about her once more. Straining her ears, she could barely distinguish the sound of waves lapping against a wooden hull. Then the distinct creak of a ship afloat reached her hearing.

Strengthened by renewed hope, Kathleen struck out with sure, silent strokes in the direction of the ship. What she would find when she reached it was anyone's guess. Since she'd lost all sense of direction, it could be the *Kat-Ann*, or one of Jean's fleet. Perhaps it was an English vessel or an American trader, or another pirate ship. She could only pray it was her crew searching for her, but at this point, even Reed's face would be a welcome sight.

A long hundred yards later, Kathleen at last came alongside the ship. The familiar green hue of the hull sent a shudder of relief through her as her hand traced the familiar lines of the *Enchantress*.

All was dark and silent aboard, but she could make out the shadowy forms of her men. Dan was leaning forward against the rail, his head in his hands, and she could see Finley standing next to him.

Chuckling to herself and sincerely hoping she wouldn't give poor Dan heart failure, she called up, "Hey, Finley! Get a move on and throw me a line!"

She wished she was close enough to see their startled faces as their heads snapped up in surprise.

"Cap'n?" Dan queried softly, as if fearing a ghost.

"And who else were ye expectin'?" She aped his brogue perfectly.

In the end, they had to pull her aboard, for she hadn't the strength to climb the rope. Her crew rejoiced fervently over her escape, and Dan repeatedly praised every saint he could recall. When the initial shock was past, Kathleen let him order up hot bricks for her bed. Feeling she had endured enough excitement for a long while, she told Finley to take the helm and set sail for Savannah as soon as the fog and tide permitted. Then she dragged herself to her cabin, pulled on an old flannel nightgown, and snuggled down between warm sheets with a hot buttered rum for a nightcap.

By the time Kathleen got back to Savannah she had regained her normal sense of humor. She was more like the old Kathleen and less like the ice queen. In public, she remained aloof. Almost every evening, however, found her at the Golden Slipper with Ted and Eleanore.

Kathleen enjoyed gambling. She always set herself a loss limit and stuck to it, for she realized the dangers of gaming. To Ted's surprise and her delight, she usually came away a few coins ahead. The club members still grumbled among themselves, but were gradually accepting her and Eleanore. The two women played seriously, which drew approval. The men were vastly relieved that they were not flighty, giggly and harebrained.

Although Kathleen tried each of the games at least once, she didn't really care for fantan or baccarat, and an hour of dicing was her limit. She most enjoyed blackjack and faro, or a good game of euchre or poker.

Reed arrived three days ahead of schedule. Kathleen and Eleanore were Christmas shopping when Ted discovered them in the milliner's telling them that the *Kat-Ann* was at the docks. The crew were full of news of good trading this trip, but mostly regaling others with tales of the fate of the daring lady pirate.

"General consensus is she is probably dead and dining with the devil by now," Ted repeated.

Kathleen glanced ruefully at Eleanore and muttered, "She'll probably be dining with him this evening, at any rate."

"Undoubtedly," Eleanore agreed with a grin.

That evening, much to Kathleen and Eleanore's surprise, not only did they dine with Reed, but Jean and Dominique were there, too. A quick look passed between the five friends, and before Kathleen could collect her thoughts, Reed said, "I would like to introduce two friends of mine, ladies. This," he said, indicating Jean, "is Mr. John Lafferty, and the other gentlemen is Mr.

Donald Alexander.''

Kathleen's eyes twinkled with impish delight as she caught their imploring glances, especially Reed's. Jean was his usual impeccable self, but Dominique was a delight. He had exchanged his pirate garb for excellently tailored clothing. His dark hair had been neatly trimmed, as well as his beard and mustache. To Kathleen, he looked almost naked without his weapons and earring.

Kathleen choked back a giggle and curtsied politely. ''A pleasure to make your acquaintance, gentlemen.'' She noted that Dom seemed extremely relieved to see her alive and well.

Dinner went well, considering. Eleanore gravitated immeidately to Jean's side and Reed was besieged by Amy. Kathleen, much to Reed's dismay and anger, seated herself next to Dominique and played him an extraordinary amount of attention.

Dinner had just ended when the butler entered to announce Charles's arrival. Chaos reigned for a few minutes as everyone tried to determine who they were supposed to be acquainted with previously, or not, as the case demanded, and whom to call what. It was so laughable that Kathleen wore an unmanageable grin throughout, and nearly burst into fits and giggles each time she glimpsed the humor in Eleanore's huge brown eyes.

As large as the Bakers' house was, there still was not enough room for all the extra holiday guests—the Bakers and Taylors, Eleanore and Charles, and the two new arrivals too. For the evening Reed would return to the *Kat-Ann* with Jean and Dom. Better arrangements would be made on the morrow. As usual, after the house settled down, Kathleen made her nightly pilgrimage to the Golden Slipper.

The next day Eleanore moved to the hotel with her brother. She and Charles shared a suite next to Jean and Dominique. Eleanore promised she would see Kathleen often, but Kathleen knew most of her friend's time would be spent with Jean.

In the afternoon Kathleen sneaked away and met

Dominique just outside town and took him to meet Kate. They spent the day with her, and it was almost dark by the time Kathleen stabled Zeus once more.

As she passed the library door, it flew open and Reed stood glowering down at her.

"It's about time you got back. Where have you been all day?"

"I've been visiting Kate, not that it is any of your business!" she retorted, flipping her copper hair from her face.

"Oh, but it is my business, my love." Then his eyes narrowed dangerously. "I can easily check on your story, you know."

"Go right ahead. In the meantime, would you mind removing yourself from my way? I must dress for dinner."

Reed gave her a mocking bow. "By all means, dearest, and please wear something stunning. I would hate to have my reputation ruined by escorting a frump to the opera." He leered at her.

"You pompous, overbearing offspring of a skunk!" she snarled as she stamped away, trying to block out his laughter.

Reed chuckled loudly and thought, "Ah, yes, that is more like my old Kat! All fangs and claws. Welcome back, my little spitfire! Farewell, ice maiden!"

Jean and Eleanore joined Reed, Kathleen, and the others in the family box at the theater. Jean seemed highly amused at the tension between Kathleen and Reed. At intermission he leaned over and whispered, "Ah, cherie, I see you two have not patched up your differences yet. I should beg you to do so soon. Never have I seen my friend so befuddled over a woman. You are leading him a merry chase, little one. How can a man keep his mind on my business that way? You are costing me profits untold!" Jean chuckled richly.

For a frightening moment Kathleen thought Jean had learned of her pirating venture. Then she realized she must be wrong, for he was not angry.

"Jean, you are misled," she answered with a smile. "It

is not me who has him in a stew, but his pirate lady. She has needled him like a pesky mosquito for months. Now it appears she is no more, so perhaps things will return to normal for you."

He nodded. "Perhaps, but then, perhaps not."

When the evening was finally over, Kathleen scurried off to her room. She and Ted left by the veranda stairs some time later on their way to the gambling club.

Kathleen triumphantly lay down her hand, showing four jacks, and raked in her chips.

"Sorry, gentlemen. It seems to be my night." She added her gains to the growing stack of winnings before her.

As she watched the man across from her shuffle the cards, she felt a hand come down on her shoulder. She looked up, expecting to see Ted, just as a familiar deep voice said, "Deal the lady out."

Kathleen's stomach lurched as she looked straight up into Reed's bland face. His carefully controlled features looked calm enough, but Kathleen knew better. She could feel his cold anger just beneath the thin veneer of politeness. The hand across her shoulder tightened in silent warning, daring her to defy him.

Blue eyes challenged green as he said quietly, "Collect your winnings. I'm taking you home."

She stood, sweeping her chips into her shawl, and walked sedately toward the cashier's window. "Ted brought me. He can escort me home as well." Her eyes scanned the room, coming to rest on Ted two tables away involved in a game of blackjack.

"Don't start a scene, Kat. I'm sure Mamie would not object if I were to request a private room in which to continue our conversation." He gave her an evil smile, his white teeth flashing.

Biting back an exclamation of surprise, she commented bitterly, "I should have known. Do you come here frequently?"

"Not as often as you seem to of late, my pet."

Ted, having finally taken notice of what was happening, approached hesitantly. "Reed, what are you doing here?"

"Collecting young truants. It seems you need some instruction on acceptable places to take young ladies." Reed glared at him hotly. "I'm not at all sure after this that I approve of you as a suitor for Susan."

Kathleen placed a restraining hand on Reed's arm. "It wasn't Ted's fault. I coerced him into bringing me here. Except for this evening, Eleanore has always been with us. Please, Reed, don't blame Ted. He'd never dream of bringing Susan." She glanced at Ted's ashen face and back at Reed. "He adores her. Don't destroy their happiness in your anger at me."

Reed surveyed her with bitter contempt. "I'm surprised that you can spare a thought for them. Most of your concerns seem to center about yourself. There may be hope for you yet. Shall we go?"

"I should take her home, Reed. After all, I brought her here. It is my responsibility." Ted spoke up bravely.

Reed wheeled about, his glare freezing Ted where he stood. His voice was dangerously soft, like velvet over steel, when he spoke. "I will say this only once, Ted, so listen well. Kathleen is my responsibility, solely mine. I will see her home. You stay out of it. You, young pup, have aggravated the situation greatly. I will speak with you later. Count on that! You had better see to yourself and hope my anger cools by then, my friend."

Without further comment, Reed steered Kathleen forcefully toward the door, leaving the red-faced Ted to contemplate his words.

"You don't have to be such a mule about things!" Kathleen complained as he pushed her inside her room and closed the door.

"I find my wife in a brothel gambling and I'm supposed to be calm?" he said through clenched teeth. His face was a dark mask of thunderous anger.

"I was only gambling, after all, Reed. It's not as if I were—well, you know." Her face reddened at the thought.

She turned from him, removing her shawl. Stepping up behind her, he started unhooking her dress. Startled, she tried to pull away from him, and heard a loud rip as the cloth stayed in his huge hands.

She realed on him. "Now look what you've done! If you would just leave me alone, I'll manage fine on my own."

"On the contrary, madam. If I leave you alone for long, you find much mischief. Now turn back around so I can finish unhooking you."

She glared at him defiantly and made no move to comply.

"Very well, the dress is probably ruined anyway." With that he hooked his hand in the bodice and ripped downward. Her gown split to the waist. Another swift pull and the waistband snapped and the cloth parted to the hem. He shoved the gown from her slim shoulders and it fell to the floor with a soft rustle.

"Now," he said huskily as he eyed her heaving breasts, "are you going to be more cooperative or shall I continue with the petticoats and chemise and underdrawers?"

"You are a vile, loathsome creature!" she hissed. She removed the petticoats, eyeing him carefully all the while. Then she presented her back to him while he unlaced her corset. His arms came about her waist, and he reached up and untied her chemise. Her bare breasts were exposed to his exploring fingers. His breath was warm and tingling on her neck as he nibbled his way up to her ear.

Kathleen felt his touch as he released her underdrawers, and they joined the other garments on the floor. Next he pulled the pins from her hair and it tumbled in a shining red-gold mass down her back to her hips.

"Beautiful!" he murmured as he buried his face in her hair. He picked her up and carried her to the bed, then stepped back and began removing his own clothing.

Naked, except for her stockings and jewelry, Kathleen raised herself on one elbow, watching him through slanted green eyes. "Just what do you think you are doing?"

"My dear, not-too-bright woman, I am preparing to bed my wife."

"Perhaps your wife does not wish to be bedded," she suggested with an edge to her voice.

"Perhaps she should reconsider. If she is at all wise, she would rather be made love to than beaten, which is what any other husband would do under the circumstances." He leaned over her, the look on his face telling her he was still quite angry with her.

"You would, too, wouldn't you?" she pouted as he removed her stockings.

"Don't push your luck. I may yet," he answered as he pulled her into his arms. "It all depends on how persuasive you are." His eyes held hers in silent combat for a long moment.

Knowing she was beaten, she sighed in defeat and brought her lips to his as her hands sought his body.

"I should hate you," she whispered mostly to herself.

"I thought you did," he reflected softly.

Before she could answer, his lips sought hers in a savage kiss that chased all thought from her mind. Her tongue darted into his mouth to duel with his, and her slim fingers twined themselves in his raven hair. She felt herself melting beneath his hands. Everywhere he touched, his fingers seared her flesh. Little tongues of flame seemed to leap to the surface of her skin, and all the while a wilder fire raged deep within, building until she felt she would burst.

Her fingers danced across his back creating exotic patterns of their own. Her legs entwined with his, pulling him closer. Her breasts strained upward to brush his chest, and she felt his hot lips along her neck.

She shivered in delight. His hand found her, and as his fingers teased and tantalized, she thrust her hips toward him, begging him for more. Then she felt his lips trail downward to her breasts and the ache in her loins was intensified. She twisted and arched beneath him, crying out her needs. His mouth left her breasts to trail kisses across the plane of her stomach until his lips replaced his fingers.

When she thought she could bear no more, he brought

her to climax, and her mind spiraled heavenward as her body shuddered beneath him. Then he was on and in her, and his lovemaking was as savage and wild as his kisses. Oblivious to everything but her need for him, she clung to him and wound her legs about his hips, showering his face with fevered kisses. Several times he brought her near the summit, only to draw back and leave her panting for him. She shook her head from side to side in frustration and tightened her pelvic muscles in an effort to hold and hurry him.

She heard him chuckle as he whispered words of love in her ear. "Patience, kitten," she heard him say.

She was not a kitten, but a tigress in his arms. She nipped at his neck and bit his shoulder, and her nails raked at his back in an effort to achieve her release. She ignited a fire in him to match her own as he drove powerfully into her body. His size, the feel of him within her, his masterful strokes excited her beyond measure. They climaxed together, a staggering adventure, like flying over a rainbow. Thousands of bright, glittering objects seemed to surround them, as if a stained-glass window had shattered in their midst.

Wet with perspiration, sated, and for the moment exhausted, they clung to one another and waited for their breathing to return to normal. He lay atop her still, his heart thundering against hers, his lips teasing the corners of hers, his fingers wound in her unbound hair.

"Sweet, merciful heavens, but you are beautiful!" he sighed fervently.

At that moment the door to Kathleen's room flew open and William stood framed in the doorway. Kathleen let out a shriek and Reed, conscious of their nakedness, threw out an arm and drew up the coverlet. As Reed slid off Kathleen, he turned to face the intruder, a scowl on his handsome face.

Now it was William's turn to gasp. "Reed!"

"Who did you expect?" Reed growled crossly.

"I don't know," William answered absurdly. "Amy heard noises and woke me up to say she thought someone was in here with Kathleen. I was sure she was wrong.''

At that moment Amy poked her head into the door. Her sly smile immediately evaporated upon seeing Reed in Kathleen's bed. *"Reed!"* Unconsciously she echoed her father's previous exclamation. From the look on her face it was obvious she had thought Kathleen's lover had been anyone else. In fact, she had been so sure that she had awakened her father so that he could catch them in the act. She was sure that having exposed Kathleen, Reed would never have anything more to do with her. Then, of course, he would turn his affections toward Amy, as he should have from the first.

Amy burst into tears and rushed from the room, nearly bowling her mother over. Barbara, hearing all the commotion, came hurrying down the hall and into the room. She froze and her mouth fell open in mute astonishment.

Kathleen hugged the coverlet over her breasts and glanced pleadingly at Reed. Casting a glare at William and Barbara, Reed reached down for his breeches. The coverlet dragged to his waist and Barbara turned her head into her husband's chest with a squeal.

''If it is not too much to ask,'' Reed suggested with a growl, ''may we continue this family conference once everyone is clothed?'' At William's doubtful look, he added, ''I am not going to escape out the window, William. I would not dream of leaving Kathleen to face you alone.''

At this William sighed and said, ''Of course you realize if you do not agree to marry her, we will have to duel.'' Barbara let out another squeal, this time in fear.

Kathleen looked from Barbara's terrified face to William's tired one, and then to Reed's. She let out a tremulous sigh, tears rolling silently down her cheeks. ''You needn't do that, Uncle William,'' she said softly. Under Reed's quizzical gaze, she choked out, ''Reed and I are already married. We have been for many months.''

The meeting that took place in the library was nothing short of mass confusion. William, of course, required a full explanation. He and Barbara and Mary listened quietly while Reed and Kathleen told of Nanna's death and their subsequent marriage. In answer to why they had kept it hidden, Kathleen admitted they had not been getting along well together. In addition, she fabricated, they had thought it best that the Bakers get to know her and she them for a short while first. Neither did they wish to spring a stranger on Mary Taylor for a daughter-in-law. Reed agreed to let her get to know them all first, and for them to adjust to her. As time passed, Reed explained, their deceit compounded itself and they did not know how to resolve the situation. Kathleen in all good conscience could not, as it had been she who had at last announced their marriage, not Reed.

"I suppose you have the marriage certificate as proof," William inquired.

"It is in the safe at Chimera."

"How could you have kept this from us? What are we going to tell our friends? However shall we face everyone?" Barbara lamented, balefully eyeing the two.

"Now, Barbara, I'm sure it will all work itself out," Mary soothed. "After all, it's not as if they have commited a crime."

Walking over to where Kathleen sat, she put her arm protectively about her. She bent and kissed her on the cheek. "Welcome to the family, Kathleen. I really should have guessed. There was so many little things that make sense now."

"They'll simply have to get married again," Barbara exclaimed suddenly. Everyone gazed at her in amazement. "Yes! Don't you see? Everyone is expecting it sooner or later. We needn't go through all these embarrassing explanations. Reed and Kathleen need never admit to being married before, they'll simply get married *now*. I can arrange it by next month, I suppose, if we rush things."

"But Barbara, dearest, what about the house slaves? You know how they talk," William pointed out. "Why, by morning every family in town will know that Reed and

Kathleen were discovered together.''

"We'll shut them up somehow," she declared adamantly, "even if we have to threaten to sell each one of them south.''

"No," Kathleen stated quietly.

"Did you say something, Kathleen?" Barbara looked at her as if she had forgotten her presence.

"I said no! No, I will not get married again."

Reed stretched his long legs out in front of him, highly amused by what he knew was coming. "I thought it was an interesting idea, my love," he said.

Kathleen stood and faced them all. "Absolutely not! Reed, I wed you once, and that was once too often to please me! If it didn't take the first time, that's just too bad." To Mary she said, "I'm sorry, Mrs. Taylor, but your son and I do not see eye to eye. We tend to disagree on nearly everything."

"I gathered as much already, my dear, and you must call me Mary or Mother, whichever suits you.''

Reed folded his arms across his chest. "Why not go through the ceremony again? I rather like the part where you vow to love, honor, and obey—particularly obey.''

"If you think for one minute that I'll go through that again, you have less brains than I credited you with! You'd have to drag me to the altar in chains!''

"That could be arranged, too." He gave her a devilish grin.

Kathleen tossed her long copper hair over her shoulder. "And how, pray tell, do you intend to force me to repeat my vows?" Her emerald eyes glittered with certain victory. "You can drag me to the altar, but what do you do when the minister says 'Will you,' and I say no! I swear to you that before hundreds of people I would yell no at the top of my lungs, and what would you do then?''

"You wouldn't, Kathleen!" Barbara was aghast. "Tell me you are joking. Surely you wouldn't!''

"Oh, but she would," Reed answered. He rose lazily. "You may as well tell everyone the best way you know how and forget about any wedding ceremony. I was just goading Kathleen at any rate. I would not agree to it either. It would be too confusing. Which wedding date would we celebrate? What if the truth leaked out anyway? What would we tell our children? Speaking of which, what if there is a child on the way already?''

"I hadn't thought of that," Barbara muttered.

"I had," William added.

"Come, Kat. Let's go back to bed." Kathleen felt Reed's hand at the small of her back. At her look of amazement he threw back his dark head and laughed heartily. "Yes, my pet. At long last we need not worry who may discover us. Neither of us need sneak back to a cold bed at dawn.''

Reed was on top of the world the next day. Finally he could openly lay claim to Kathleen as his wife. There would be no more Gerard Ainsleys to deal with, and no more sneaking down dark corridors at dawn.

He awoke to find Kathleen nestled in his arms, as content as a kitten. Her bright head lay upon his shoulder, and one long leg was thrown across his. He smiled to himself. This was as it should be. He placed a warm kiss on her bare shoulder and she stirred, snuggling closer. Then her sooty lashes fluttered open to reveal sleep-misted sea-green eyes. She smiled languidly and stretched like a cat.

"Good morning, kitten." His voice rumbled through his chest, tickling her ear.

"Good morning, Reed. How long have you been watching me sleep?''

"Just long enough to know you snore like a lumberjack." He smiled lazily.

"I do not!" She swatted at him playfully. Then she grabbed a handful of black, curling chest hair and tugged lightly. "Take it back.''

"Ouch! All right, you win. You don't snore like a lumberjack. You have a very delicate and ladylike snore. *Ouch!* Stop that! I was just teasing.''

She settled back into the curve of his shoulder with a sigh. "I hate to admit it, Reed, but this feels so right.''

"What does?''

"Being here with you. Waking up with your arms around me." Suddenly all sleepiness vanished and she sat up on her heels, her hair in glorious disarray about her.

"What do you suppose people are going to say?"

He propped his pillow behind him and sat up. "They'll be shocked, of course, but they'll accept it. You have been setting Savannah on its ear for some time now. They are getting to expect it of you. Besides, in two weeks something else will grab their attention and we'll be old news."

"I suppose so."

"Now, wife," he said with an evil glint in his eye, "greet your husband good morning properly."

"You randy old sea-dog!" she laughed as she fell into his waiting arms.

Chapter 24

EVERYONE'S reaction was fairly predictable. The news was announced at breakfast. Barbara, over her initial shock by now, was gearing up for the defense. William was back to business as usual, vastly relieved of not having to face Reed at ten paces. Mary was her calm, serene self.

Susan's eyes opened wide at the announcement, and then she smiled shyly at them both. "I'm so happy for you! I could not hope for a more beautiful sister-in-law. You belong together. I saw it from the start."

"You did?" Ted was genuinely surprised. "I say, Reed. I'm sorry. I really didn't have any idea. No wonder you were so upset last night."

"I'm still debating whether or not to take a willow switch to the seat of your britches." Reed gave him a mock glower, then grinned. "Let that be a warning to you. At it is, I'm feeling benevolent this morning, so you are forgiven."

"How touching!" Amy grumbled irritably. Her mother had broken the news to her earlier and her eyes were still red and swollen from crying. Even Mammy's remedies could not remove all traces. It was all she could do to pull her tattered pride about her and hold her head up. Her grand revenge had exploded in her face, and she was not taking it graciously.

"I suppose your grandmother knew all along," she said now.

"Amy!" Barbara gasped, glancing hurriedly at William.

Amy's comment may have passed unnoticed were it not for Barbara's reaction.

Reed's eyes narrowed as he glanced from one to the other. "Grandmother?" he queried.

"Yes," Amy retorted spitefully, "or hasn't she told you? It is the second-best-kept secret in Georgia—next to your marriage, of course. Kate O'Reilly is Kathleen's grandmother. Now I know what people mean when they say blood will tell!"

Reed shot Kathleen a quick look that clearly said to keep silent. "Of course she told me, Amy. Why wouldn't she? It is no sin to be Irish, you know. Kate is a great friend of mine, has been for years."

When they once more gained the privacy of their room, Reed exploded, "Damn it, Kat! Why didn't you tell me? What other little secrets do you have up your sleeve?"

"Reed, the Bakers are still very English and have never liked the fact that my father married an Irish lady. They never saw the need to acknowledge any relationship with Kate in all these years. Now suddenly I show up and it would have been awkward for them to all of a sudden claim Kate as one of their family. Kate understood all this. She and I agreed that when the time was right we would acknowledge one another. In the meantime, I have spent a great deal of time with her."

"But why couldn't you tell me? I'm your husband. Did you at least tell Kate that?"

"Yes, I did. After all, I had a miscarriage in her home."

"I should have seen the resemblance; the hair, the eyes."

"Everyone says I take after her. My parents named me for her."

"Are there any of your other relatives waiting to pop out of the closet?"

"No, Reed. I promise. I have a few shirttail cousins in Ireland, but no one close. I'm really rather an orphan." She gave him a sheepish grin and wrinkled her nose at him.

"How can I stay angry at you when you look at me like that?" he grumbled, pulling her into his arms. "Come here, orphan. I've adopted you now. You'll never lack for attention around me, my pet."

The general populace of Savannah was stunned by the news. In every parlor in town the topic was Reed and Kathleen. Kate had arrived in town right after breakfast and spent the remainder of the day visiting from home to home. As an interested bystander, she offered the explanations the couple had suggested and collected opinions of the leading families. After dinner that evening she returned to the Bakers' with her report.

"It appears to me that after the initial shock wears off, most o' the folks are takin' it fairly well. O' course, Mabel Ainsley isn't too sympathetic, since her son had his sights set on Kathleen here. Most people thought ye'd end up gettin' married soon anyway, so they just have to readjust their thinkin' a wee bit. Some o' them can see the humor of it and seem to think 'tis quite a joke. I tell ye, rather than being shut out, ye're a curiosity, and they're knockin' one another down to see ye're invited to their holiday festivities.

"They were sort of gettin' used to Kathleen lordin' it over them and having things her way, so ye haven't disappointed them, lass. O' course, all the lassies are woebegone over Reed here being wed, and most o' the lads are cryin' in their beer over Kathleen, but 'tis to be expected with two people as handsome as ye two are."

The gossip in the parlors the next day was a little more difficult to obtain, but the servants passed on what they had heard. The entire town was abuzz for the second day in a row as word leaked out that Kate O'Reilly was Kathleen's maternal grandmother. That, of course, only served to raise Kathleen's status in the community, since Kate was so well thought of. All in all, what could have been a social disaster turned out much better than anyone could have hoped.

On Sunday the entire family attended church. Kathleen

dressed carefully in a cream-colored lace gown trimmed in brown and sat next to Reed, who held her hand possessively throughout the service. By the time church let out, anyone would have thought Savannah had elected them king and queen. It took them three-quarters of an hour to get to their carriage, so many people stopped them to congratulate them and wish them well.

On the way home, Mary suggested that Barbara have a party for the newlyweds. "It is the only way to calm down the furor and pacify everyone. They are all so anxious to show their approval."

The ball Barbara and William prepared for the couple was an unqualified success. Everyone accepted their invitations, and Barbara had to rent the ballroom at St. Theresa's. There were so many gifts of silver, china, and linen that Reed laughingly declared he would have to add three more rooms at Chimera just to hold it all.

The gigantic white wedding cake that Barbara insisted upon was a mastery of the confectioner's art. It was seven layers high and covered with roses. At Kate's suggestion, which Kathleen enjoyed with subtle irony, instead of a bride and groom, a three-masted frigate rode atop the cake. Eleanore, upon seeing it, collapsed with laughter, eliciting quizzical looks from Jean and Reed. Kathleen smothered her giggles behind her fan.

As Reed escorted Kathleen down the long curved staircase, everyone sighed over the beautiful young bride. Kathleen had refused to wear her bridal gown. Instead, she donned a gown of white velvet. Its simplicity of line was stunning, with a low sweetheart neckline and long fitted sleeves cuffed in white fur. It clung to her youthful figure, draping itself gently along her body. The skirt, for once, was full.

Reed had insisted she wear the emerald collar and earrings. In remembrance of her wedding day, Kathleen drew her hair back on both sides and let the heavy, satiny length of it tumble down her back. The long, wavy tresses swayed like a copper curtain with each step she took. Reed was entranced.

The next weeks were so filled with activities that many evenings found Mr. and Mrs. Reed Taylor attending three galas in one evening. With the holidays upon them, everyone found the excuse they needed for inviting the young couple to their festivities. As Christmas approached, Kathleen particularly enjoyed the evenings of caroling. Groups of young people would bundle up and go from house to house serenading the inhabitants with carols, and nearly always were invited in for a cup of Christmas cheer. After one such evening, when Reed nearly had to pour Kathleen into bed in a fit of giggles, she learned to limit herself to a few sips per stop.

Kathleen chose Kate's Christmas ball to stun Savannah with another of her original gowns. Reed thought she was so ravishing he nearly refused to let her be seen in it. To be sure, it was fascinating. The gown was a lustrous dark blue silk sparkling with paillettes. The shocking thing was that it had no sleeves, merely molding itself across her breasts, exposing a generous amount of cleavage. How it stayed up was a miracle only Mrs. Fitz knew. The material clung to her waist and hips. Over it flared a skirt of sheer blue tulle. The tulle rose from the bodice to a jeweled collar at the neck and also formed long, loose sleeves with jeweled cuffs. The sheer tulle shadowed the cleft between her breasts, and only served to whet the imaginations of her dance partners. Reed stayed particularly attentive that evening.

Finally Christmas Eve arrived. Jean, Dominique, and Eleanore joined the family at the Bakers', and other guests came and went all evening. The house was beautifully decorated with holly and berries and scented candles. The yule log blazed brightly, adding to the warmth of the season. Laughter and cheer filled the rooms as they all trimmed the Christmas tree. The Laffites took Kate and Eleanore to midnight mass while the others attended a Christmas Eve service of their own.

Kathleen had never been so filled with the spirit of the season, not since many years before when Mama and Papa and Nanna were all alive. It had been so long since she had felt so loved and secure, surrounded by good friends and family, with the possible exception of Amy.

In the wee hours of Christmas morning, she sat snuggled next to Reed in the sheltered curve of his arm. The fire was crackling gayly, and all were feeling mellow and content in the presence of God and friends.

As Reed turned her face to his, her eyes were shimmering with unshed tears. "What's wrong?" he whispered.

She shook her head slightly and smiled. "Nothing, absolutely nothing. Everything is so perfect at this moment. I am so content and at peace that I would like to freeze time and stay right here. I want to savor the feeling as long as I can."

"I know. I feel it too. It's beautiful, isn't it?" He pulled her closer to him and they silently watched the fire, her head resting on his shoulder.

On Christmas morning everyone gathered to exchange gifts. Amid all the holiday rush, Kathleen had managed to sneak away long enough to purchase her gifts for Reed. She had wrapped each individually and handed them to him in sequence, in the privacy of their bedroom.

First there was a pair of snug black trousers, followed by a brilliant red silk shirt with loose sleeves and tight cuffs. Then came a fancy tooled black leather vest, and finally a beautiful pair of black leather boots of the finest Cordovan leather.

His blue eyes danced with pleasure and merriment. As he bent to kiss her, she whispered, "All for the best-dressed pirate this season."

Kathleen was speechless as she opened her present from him. Before her eyes lay jewels fit for a queen. First a five strand necklace alternating tiny pearls and rubies, with a huge teardrop ruby dangling from the center, the size of which could have graced a sultan's turban. There were matching drop earrings, a bracelet, and an oval ruby ring surrounded by tiny pears. All were cushioned in the fur of a magnificent white ermine cloak.

She gaped at Reed and swallowed twice before she

found her faltering voice. "Reed," she croaked. Finding no further words to express herself, she flung herself into his arms.

"If I'd known you would react this way," he chuckled, "I would have given them to you months ago. Now I know the key to your heart—fur and jewels."

She bestowed a kiss on him. Then she pulled away and gave him a wicked wink. "That, sir, is only part of the secret, but it is a good start, a tremendously good start!"

When she wore the jewels and fur that evening with a clinging red velvet gown, she seemed to very representation of Christmas. With her elegantly clad and extremely handsome husband, she created a picture of perfection.

"They are stealing the show," Dominique told Kate. "St. Nicholas must be green with envy. I know I am."

"Look at them," Eleanore sighed to Jean. "What beautiful children they will produce!"

Jean nodded. "If they stop fighting long enough, *cherie*."

The day after Christmas, Reed announced his decision to move back to Chimera with Kathleen. "I thought it would be nice to hold our New Year's Eve ball at Chimera this year instead of renting a hall in town," he explained. "Besides, it is time Kat started assuming her responsibilities as mistress of Chimera."

Kathleen gave him a startled look. "But, Reed, your mother is still mistress of Chimera. I wouldn't dream of usurping her position there!"

"Now, don't upset yourself, Kathleen," Mary put in. "Reed is right. You must learn how to manage the plantation. I'll be there to help you, but believe me, it will be a relief to have you taking charge. I've done it alone for too long. We'll start out gradually, of course, but it is time for the reins to change hands. It will leave me free for more enjoyments of my own. However, the season is not yet over and I'm afraid it will be dull for you to leave Savannah now." She gave her son a level look.

"Chimera is not that far, Mother. Kate stays at Emerald Hill nearly year round and still manages to attend almost every function you'd care to name. The weather is still fine and we'll be in town often, but we need a place of our own just now. It is kind of William and Barbara to open their home to us each winter, but there just isn't as much room here as at Chimera, or as much privacy. If truth be told, I'm not sure how much longer I can put up with Amy's attitude without exploding in anger.''

"She is being rather nasty," Mary agreed ruefully.

"She is being impossible," Reed corrected. "She is hateful and rude to Kathleen, and I won't stand for it."

"Not only that," Kathleen added, "but she is sickeningly sweet to Reed, especially in my presence. She nearly drools over him. She hangs all over him and makes a spectacle of herself. It is thoroughly disgusting!" Kathleen made a face.

"Why sweets, I believe you are jealous," Reed goaded.

"Jealous, my foot! As your wife I refuse to sit passively by while another woman plies my husband with her wiles, that's all. It is within my right to object, and object I will. If that flirtatious little baggage doesn't cease soon, she will need a wig to cover her bald head. I swear I'll pull every hair from her scalp one by one!''

Reed's dark head flew back as he let out a roar of laughter. Mary joined him, despite herself.

Kathleen stamped her foot at them. "It isn't funny! I refuse to be made to look the fool, and Amy had better watch her step. And you, Reed . . . '' She shook a slim finger at him.

"Yes?" He arched a dark brow at her.

"You had best not play into her hands, for you'll rue the day you do.''

"Would you snatch me bald too, my pet?" he challenged with an amused look still on his face.

"No," she answered cooly. "There are other means to use on you, dear husband, and I would not hesitate to use them.''

"That door swings both ways, Kat, so don't get caught behind it yourself," he warned.

Mary watched carefully the exchange between her son

and his wife. No wonder he had been drawn to her. She was not merely beautiful, but smart and spirited as well. Amy, too, was beautiful, but she had a vindictive nature. Behind that pretty face lay a shallowness of character that would have bored Reed to tears. One thing Mary felt for sure. Regardless of all else, Reed would never become bored with Kathleen as his wife. She was as changeable as a chameleon and as unpredictable as a leprechaun.

Mary smiled and shook her head. Her son was also strong willed and used to issuing orders. There would be storms ahead, she was sure. Reed and Kathleen would clash often, but if their love was strong enough, all would end well.

Kathleen had less than a week to prepare for the New Year's Eve ball, and she threw herself into it energetically. Jean, Eleanore and Charles had come along, so Eleanore helped as well. Dominique had also tagged along, but wisely spent most of his time with Reed and Jean, or visiting with Kate. Mary had stayed in town, allowing Kathleen her first challenge as mistress of Chimera.

As Kathleen busied herself supervising the cleaning and cooking and decorating, Reed attended to plantation affairs and business meetings with Dominique and Jean. The time sped by quickly and was well spent.

The evening of the ball, Ted and Mary arrived early, bringing the costumes. Reed's tailor had done a magnificent job on his Spanish outfit. The tight black breeches with the gold braid down the sides molded themselves to his muscular thighs. The white silk shirt was layered in ruffles beneath the gold-braided black vest. With the flat black hat, he made a dashing Spanish don, especially with the higher heeled black boots. The shirt he left open at the neck, tempting Kathleen with the view of his broad, hairy chest.

Admiration leaped into her eyes, and he chuckled. "We'll never go downstairs if you keep looking at me that way." He turned her around to hook the back of her dress. In the mirror he saw her watching him and smiled. Then the smile faded into a thoughtful look.

"Something wrong?" she asked.

"Isn't there supposed to be more bodice than that? My God, Kat, it is nearly indecent!"

She had chosen a flamenco dancer's gown to complement his outfit, of a flaming red-orange, with rows and rows of layered black ruffles. The bodice was extremely low and tightly fitted to the waist. From there the skirt flared out, with a generous portion pulled up and hooked near the waist, exposing a tantalizing glimpse of leg as she walked. Her hair had been parted down the center and wound into shining coils above her ears. A lacy mantilla hung from an ebony Spanish comb and was draped across her shoulders. A single red rose nestled behind one ear.

"Mrs. Fitz is going to have to pay more attention to her work," Reed went on. "She's gotten your measurements confused with someone else's."

"I doubt that. She is very competent and talented."

"And nearsighted."

"Just put your mask on and let's get downstairs. Our guests will be arriving at any moment."

Mary was already waiting near the door. She made a charming Betsy Ross, a colonial flag draped across her shoulders for a shawl. Eleanore appeared in the black wig and clinging gown of Cleopatra, with Jean as the adoring Antony and, of all things, Dominique as Julius Ceasar. He had powdered his dark hair, and sported a Roman toga and head wreath with much aplomb. Kathleen successfully stifled her laughter until she caught a glimpse of his hairy knees, and then she exploded in mirth. It took all her efforts to compose herself, only to collapse into giggles when Ted came down in kilt, sash, and bagpipes.

"Will you hide that silly smile behind your fan and behave?" Reed grinned.

She raised the fan from her bosom to her face.

"On second thought, smile all you want, but lower that fan again."

Their guests started arriving in droves. Barbara and William came as Napoleon and Josephine. Amy had powdered her blond hair and came as Marie Antoinette, as did seven other ladies. There were twelve Bo Peeps, eight dashing cavaliers, three court jesters, a doctor, a judge, a backwoodsman, an Indian and his squaw, three queens, and one king. Four knights complete with armor, five damsels in distress, and a devil or two showed up along with one Father Time, a six-foot leprechaun, two Puritans fresh off the Mayflower, and one Medusa, who turned out to be Kate in a green wig. Leif Ericson arrived in the form of Gerard Ainsley accompanied by Attila the Hun and a pirate. Susan made an exquisite and very believable angel.

"That wouldn't have worked for me," Kathleen commented.

"No," Reed agreed, "but you could have come as the devil's granddaughter."

"And you could have come as a snake and been right at home."

"Come to think of it, we could have come as Adam and Eve and really set Savannah on its ear."

"I could lower my decolletage a bit more if you wish, darling."

"You do and I'll cart you back upstairs and lock you in!"

"My, you are touchy tonight!"

"Hmph!"

The gala was going well and everyone was having a good time. About halfway through the evening, Kathleen slipped upstairs to the old schoolroom. There she changed into her pirate costume, which Kate had sneaked in under her cape, along with a black wig. She furtively made her way to the second story veranda and scurried down the veranda stairs to the outside patio. From the shadows she watched until she spotted Reed. Quietly she slipped inside the ballroom and walked quickly to where he stood.

A hush fell over the masked crowd, and then a buzz of whispers started and grew. Reed turned from his conversation with Jean to stare straight into emerald eyes behind a green mask. She stood before him, a mysterious little smile lurking at the corners of her mouth. A servant walked by with a tray of drinks and she calmly reached out and took one from the startled fellow.

"Nice place you have here, Captain Taylor." Her voice was as he remembered it, low and husky.

Reed could not believe his eyes. "You're alive," he said stupidly.

"I assume so," she smiled, her eyes dancing with mischief. "I was when last I checked. It was so thoughtful of you to have a masquerade ball so that I could feel comfortable this evening."

Reed glanced around uncomfortably, looking for Kathleen. Guests were gathered in huddled little groups, whispering in astonishment and casting curious stares their way.

"You are taking a great risk coming here. Aren't you afraid I'll call on some of my friends, tell them you are truly the pirate Emerald, and have you arrested?"

She held his gaze cooly. "I think not. First, I am not stupid enough to have come alone. Second, how could you prove I am not just some innocent young lady in masquerade? Third, you would not want your wife to learn of our activities the night of the hurricane."

At his raised eyebrow, she said, "Yes, I know of your wife. I have been watching for some time. I saw her go upstairs a few minutes ago, if that is who you are watching for."

"How did you know where I live?"

"You told me the night of the hurricane. At least, you mentioned Savannah. The rest was easy. Everyone in town knows of you." She set down her glass and glanced at Jean, who was eyeing her intently. "Aren't you going to introduce me to this handsome gentleman, Captain Taylor?"

"Emerald, may I present Jean Laffite. Jean, my troublesome lady pirate."

Jean bowed and kissed her hand. "I have been longing to meet with you, mademoiselle. You are even lovelier than I had been told."

"Merci. You are very gallant. You make me wish I could reconsider and join your crew."

"You still can, Emerald. Why don't you?"

"For personal reasons, I assure you I cannot. I met your brother Pierre at Matanzas. I do not care for him. He is definitely not the gentleman you seem to be. I did, however, enjoy Dominique You's company. Is that not he over there?" She gestured with a gloved hand.

"Oui."

"Why have you come here, Emerald?" Reed asked impatiently.

"Dance with me and I shall tell you."

"Do you dance?" he asked doubtfully.

"Take a chance and find out. That is a waltz they are playing, is it not?"

He took her gingerly into his arms, praying that Kathleen would stay upstairs for a few more minutes, until he could figure out what to do about Emerald. He had no doubt she would hear all about this, but he did not want her to meet Emerald. He feared what Emerald might say and what Kathleen might do.

They danced for a few minutes without speaking. Kathleen knew she was taking an immense risk. Both Eleanore and Kate had told her she was crazy to pull such an idiot stunt, but once the idea had taken root, she could not shake it.

She made quite a startling picture dancing in Reed's arms. Her long black hair was loose about her shoulders, and her equally long legs were a sight to behold. Every man under eighty was either openly or secretly lusting after her. In their minds they imagined those exquisite limbs wound with theirs, that tiny waist in their hands, those half exposed breasts fully exposed to their touch. The women, of course, were outraged, and if they would admit it, jealous.

"Have you ever seen such a vulgar display!" one whispered vehemently.

"Who is it, do you suppose?"

"Who would dare such a thing?"

"Do you see that heathen bracelet she has on her arm?"

"Why, she even has a knife in her boot, and a sword!"

410

"Do you suppose her gun is real?"

"Surely that isn't the real pirate lady we've heard about!"

"It couldn't be. She's supposed to be dead. Drowned, I heard. It has to be someone in costume, but who?"

"God forbid that it is the daughter of any of my friends!" one old crone exclaimed.

"But we are all friends here," her neighbor reminded her.

One young man leaned over to his best friend. "I'd give my best stallion to know who she is. I've never seen such legs!"

"Of course not. All the nice girls are in skirts to the tip of their dainty slippered toes. Who owns those legs? That is my question, and how will we ever know?"

"Well, it can't be Mary Jane. She came as a lady of King Arthur's court."

"It can't be Elizabeth Drury either. She's a shepherdess."

"It couldn't be the real piratess—could it?"

"Could it?"

"I don't care if it is. One night with her and I'd die a happy man!"

Reed spoke first. "You could at least have worn a shirt under your vest."

"What? And spoil the view?" she laughed. "No. You might have had trouble recognizing me then." She tossed back an inky length of hair.

"That's odd," Reed mused as he leaned closer. "You are wearing the same perfume as my wife. I couldn't mistake it."

Kathleen almost missed her step. "What is so odd about that, Captain? I often wear perfume. Just because I am a pirate doesn't mean I cease to be a woman."

"But the same scent?" he questioned.

Kathleen shrugged. "We have the same taste in men. Why not perfume?"

Now Reed shrugged. "You haven't told me yet why you came, Emerald."

She gave him a tantalizing smile. "Perhaps I was curious about your life, or maybe it was just to harass you. I've heard you thought me dead. Perhaps I came to see if you grieve for me. Or perhaps I decided to try a bit of piracy on land and relieve your home of a few items—or it could be I came only for my rapier." She patted the blade at her side.

Reed glanced down and immediately stopped dancing. Emerald was wearing the sword he had taken from her that foggy day when they had last fought. Or was it a twin to the rapier he had locked in his study?

Kathleen laughed at his stunned look. She had taken the rapier earlier in the day and hidden it. Now that she was in charge of the house, she held in her possession keys to all the rooms. It had been so easy.

"Captain Taylor, why are you so surprised? You know me for a thief already. Yes, it is the same rapier. I took it from your study. After all, I consider it still mine. I assure you I took nothing else, only what was mine."

Kate chose that precise moment to approach Reed. "Excuse me, Reed, but shouldn't you circulate among your guests? Kathleen will be down any time now."

"You are quite right, Kate." He turned to speak to Emerald, but she was gone. They had stopped dancing near the patio doors.

"Damnation!" he swore as he stormed out the door. A quick look into the dark night gave him no sight of her. He charged up the veranda stairs, Jean close behind.

"Perhaps she came up here instead. We'll check all the rooms."

It was a fruitless search. By the time they returned to the ball, Kathleen was there in her flamenco outfit, standing near the bottom of the stairs, and talking quietly with Eleanore. She glanced at him over her champagne glass, and her look nearly froze him in place. Someone had already told her, he knew.

He hurried to her side. "Kathleen, darling, I can explain."

"I can hardly wait," she snapped in a low voice. "However, now is neither the time nor the place, dear husband.

Right now it would be better if you could convince our charming guests that you adore me. We will discuss the latest gossip later, *if* I am still taking to you then. Now we must present the perfect, loving couple. Smile, Reed. Be charming and sweet, and rest assured I could claw your eyes out this very minute!"

Eleanore, who had heard Kathleen's words, choked into her champagne glass, trying to hide her laughter. Kathleen gave her what passed for an indigant look.

"Bubbles!" Eleanore gasped a lame excuse. "Champagne always tickles my nose."

"Then try another wine, Eleanore, for pity's sake."

Fixing a serene smile on her face and trying not to burst into laughter herself, Kathleen put her hand on Reed's arm and let him lead her onto the dance floor. Several times he tried to explain as they were dancing, but she put him off. She let him hold her closely, and when he chose to kiss her in view of their guests, she responded.

"That was for appearances only, you know," she hissed in his ear, and giggled to herself.

Much later she sat calmly brushing her hair as Reed explained as best he could. She hid her smile behind her cascading locks as he stumbled over his words.

Finally he stopped talking and walked to where she sat. "Aren't you going to say anything?" he demanded. "Damn it, Kat! Put down that brush and talk to me!"

She set the brush on her dressing table and faced him. "What would you have me say, Reed?"

"I'd like to hear you say you believe me, that you understand."

"That I absolve you?" she suggested.

"Of what?"

"Of your fascination with this creature, of course. You are smitten with her, aren't you?"

"Absolutely not!" he denied hotly.

"Not even a little bit?" she goaded. "I hear she is quite stunning, even exotic."

"Yes, she is a beauty. I would be lying to say differently. I do have eyes, Kat, and I am a man, but you are my wife.''

"What does that mean?"

"You have no cause to judge me or behave jealously."

"Oh?"

"I married you. You will share my life and bear my children. Besides, you are every bit as beautiful as she, and you are a lady. She may be a woman, but she is not a lady. She is a thieving pirate! Now can we drop this subject and go to bed?"

"You are the one who insisted on explaining, Reed," she reminded him. "All evening you have acted as guilty as a child caught with his hand in the cookie jar."

"Are you going to stay angry over this, Kat? None of this has anything to do with you and me."

She decided to let him off the hook. "Then how can I be angry with you?" she countered seriously.

"You aren't?" he asked incredulously.

"Let's compromise. I'll be just a bit angry, and you talk me out of it." She wrapped her arms about his neck and raised her lips to his. "Without words," she whispered.

Chapter 25

THEY spent the next couple of weeks in relative quiet. Kathleen diligently applied herself to learning how Chimera operated. It was not so very different from the estate in Ireland, except that instead of free white servants there were slaves to manage, and the crops were different from home. She found it relatively easy, incurring no real difficulties.

Some evenings they rode into Savannah to the theater or opera, but usually they spent cozy evenings alone with their friends. Kate visited often, and life became routine.

Halfway through January, Jean decided he must return to Grande Terre, and Charles and Eleanore departed from New Orleans. Reed wanted Kathleen to return to the Bakers' when he sailed, but she chose to visit Kate instead.

"That way I won't be tempted to scalp Amy," she reasoned.

"I'm going to miss Eleanore."

"I'm going to miss you," Reed countered. "I've grown used to having you near, kitten."

"You could always take me along."

"Let's not go through all that again, Kat. I'm not up to another six-month war with you, and I'll tell you right now I wouldn't put up with anymore of your nonsense. Your place is here at Chimera now. There is plenty to keep you busy here during the times when I am gone, and perhaps soon there will be children to occupy you."

She have him a disgusted look.

"I thought you wanted children, Kat."

"I do. It is just your attitude I object to. Why is it all men think alike? And why on earth do women stand for it?"

"I don't understand you sometimes," he commented.

"That is because you don't really try. I'm not so hard to understand. You are trying to fit me into a mold not made for me."

"It is you who won't try, honey. If you give it a chance, you'll settle into it."

She threw up her hands in dismay. "I give up! It's like talking to a brick wall. When are men going to realize there is more to a woman than being a wife and mother? You think all you have to do is give a woman a home and children and she'll worship at your feet! Well, I have news for you! Behind a pretty face and trim figure is often a fine mind that works as well or better than yours, and I want to use my mind. My body, too, has other uses than pleasing you and bearing children. My hands can do more than sew and pour tea and flutter a fan."

"Then paint or sculpt. I'll even agree to having Kate teach you horse breeding, though heaven knows that is not exactly an acceptable, ladylike occupation either. Take up gardening, ride that unpredictable creature you call a horse, redecorate the house. Read if your mind needs exercise, do charity work, plan parties. Do what all the other women do."

"Exactly!" she pointed out indignantly. "But above all, do *not* step into the man's world! God forbid I should want to learn the more earthy workings of a plantation, eh? *Don't* learn about crops and export taxes or the price of rice in China! *Don't* bother about the field hands, the planting, the harvesting, or when to spread manure. *Don't* pester the overseer to ask to see his books. Turn a blind eye when they castrate the bulls or turn one loose to service a cow! Delicacy forbids I notice the sow suckling her piglets or the hog being slaughtered, or that the bitch hound is in heat! I'm not supposed to know that the kitchen maid is five months pregnant by that big, muscle-bound buck who works the horses, or that you deliberately bred her to him, am I?''

"Enough, Kat!" he roared angrily.

"You had no idea I noticed all that, did you, Reed darling?" she cooed sarcastically. "Well, I have eyes and ears and brains!" She was past caring how angry he became. "I'm not a marble statue!"

"No. What you are is a very stubborn, spoiled brat! You obviously were let to do as you pleased as a child. It is a wonder you had any upbringing at all, and I'll guess Mrs. Dunley drilled that into you by force!" Reed grabbed her arms so she couldn't turn away from him. "Let me tell you right now, my willful little wife, that I am fed up with your headstrong ways, and you are going to do things my way from now on, if I have to resort to force. I'll beat you if I have to, Kat, so don't push me too far!" His hands on her arms tightened.

She stared at his hand, ignoring the pain, then raised narrowed cat eyes at his. Her voice was even, but filled with loathing. "Take your filthy hands off me."

"When I'm ready."

"Now."

"No."

"I despise you, you barbaric brute!"

"I realize that, and it doesn't really matter."

"It does to me, you fiend, you moronic lunatic! Your heart is as black as the devil's—if you have one! You're a monster, a snake, a cad, a cur, a swine, and a viper!"

"You are the one who has the tongue of a viper, and it is high time we cured that," he said as he started pulling her across the room. His jaw was set at a determinedly angry angle.

"Let go of my, you damned beast! You lousy Yankee, you paltry pirate, you . . ." her voice trailed off as she saw him reach for a bar of soap on the washstand. Then she shrieked, "You wouldn't dare!"

"Hah!"

She fought to pull loose, but his grip was like iron.

417

Finally he maneuvered both arms behind her back, and pulled her back across one brawny arm. "You've been asking for this for a long time now. Hasn't anyone ever told you ladies do not say such words, especially to their husbands? I'm getting tired of that nasty, dirty little mouth of yours. It's time we cleaned up your vocabulary!"

He dunked the soap into the bowl of water and brought it to her mouth. Kathleen clamped her lips shut and tried to turn her head away. She squirmed and kicked, but he caught her chin and held it firmly as he wedged the soap between her teeth and rubbed vigoriously.

Kathleen's eyes smarted and her tongue burned. It felt as if he had shoved the entire bar all the way to her tonsils, and she struggled not to swallow. Finally he took the soap away, and grabbing a towel, wiped her mouth and chin. Then, before she could recover, he carried her to the bed. There he sat down, tossed her across his lap, and flung up her skirts.

Guessing what was coming, she squirmed, desperate to escape his hold. He yanked down the lacy underdrawers. Kathleen drew in her breath and waited for the first blow. When it didn't come, she let out her breath, and then the blow came down hard.

"Yeouch!" she yelped involuntarily. He didn't let her catch her breath, but kept spanking, smack after resounding smack.

"Stop it, Reed! Stop!" she panted between swats.

"Are you ready to apologize and behave properly?"

"No!"

Another smart slap, followed by a twin. "Let me know when you are ready and I'll stop, but not until then," he advised.

A few more hurtful swats and her backside felt on fire. It hurt dreadfully. Finally, in tears of defeat, she cried, "All right, enough!"

He stopped swinging momentarily. "Did you say something?"

"I apologize," she said in a tiny voice.

"I beg your pardon, but I didn't quite catch that," he taunted.

"I said I'm sorry," she repeated more loudly.

"That's half of it," he prodded.

"You've got to have it all, don't you?" she cried out resentfully.

He delivered three more swift, painful smacks in succession.

"I'll behave!" she yelped with a sob.

"Oh, but you've taken so long to say so that I can scarcely believe you, my love. You'll have to be more convincing."

"What more can I say? What more do you want?" she sobbed.

"Let's see," he mused. "Say, 'I shall obey my husband.' That should do it."

"Reed, really!"

"I'm waiting."

"I shall obey my husband," she muttered through clenched teeth.

"You don't sound very meek, Kat. Try it once more."

She groaned aloud. Then with a small sigh she repeated quietly, but audibly, "I shall obey my husband."

He let her go then and she tumbled to the floor at his feet. She caught the edge of her skirt and scrubbed vigorously at her soap-filled mouth, eyeing him balefully. How maddening to see him sitting there, arms crossed over his chest, smiling like the cat that ate the cream!

"I despite you!"

"That is your privilege."

"I didn't know I had any left!" she retorted spitefully.

"A few," came the calm reply. "Now I hope you have a clear picture of the way things stand and what I expect of you. I regret having to resort to these measures, Kathleen, but you seem to have no respect for me, and I can't have that, you know. A man has the right to be master in his own home."

"Who died and made you king?" she muttered half under her breath.

He heard her and smiled broadly, his white teeth flashing. "My father did, sugar, when he left me Chimera."

An hour later he was packed and gone. Kathleen took

the carriage to Kate's.

"Why didn't ye ride Zeus?" Kate asked.

"Don't ask. You don't want to know." Kathleen followed Kate into the house and seated herself gingerly onto a chair.

"My stars, lass! What's wrong. Did ye hurt yerself?"

"Mostly my pride, Kate."

"From here it appears to be yer posterior."

"That, too."

"How are ye goin' to get to yer ship, gal? Ye can't sit a horse, can ye?"

"I'll have to I suppose, but it won't be comfortable."

"I'll tell ye what. Get yer things together and I'll take ye in the carriage. No one will think anything of two ladies out fer a breath of fresh air."

"But you'll have to drive back alone."

"So? I may be old, but I'm not dead! I'll be back in time for dinner and none the wiser. Besides, 'twill give ye time to tell me why ye seat yerself with such care."

Kathleen spent the remainder of the month at sea, pirating Reed with a vengeance. She attacked him at every turn, soundly defeating him each time. She wielded her blade with a righteous fury that astounded even herself.

The first three days at sea, every time she sat down she was reminded of her humiliation, and each time she faced him on deck she extracted her ounce of flesh in payment.

Finally, the hold full to bursting, they headed back. Kathleen was tired and dirty, her hair sticky with sprindrift. Her fury spent, she longed for a hot bath and a rest before Reed crossed the threshold of Chimera.

They had just turned north toward Savannah when Kathleen awoke one morning feeling dreadfully nauseous. She barely made it to the chamber pot in time, and when young Timmy brought her breakfast tray, the smell of gravy and biscuits promptly sent her stomach into revolt again.

Finally arriving on deck pale and shaken, she cornered Dan. "Tell the cook to have a care about the fish, Dan. It must have been tainted last night."

Dan eyed her critically. "Aye, Cap'n. Hope ye don't take offense, but ye're as green as a landlubber first time afloat."

"Kindly keep your droll observations to yourself and do as you're ordered," she grumped. "Better check and see if any of the others feel poorly."

The next morning was a repeat of the one previous, with Kathleen barely able to keep down a dry biscuit and a cup of coffee. She checked with Dan and he assured her he had found no one else who was suffering mal de mer, and the cook was sure all his supplies were still fresh.

By early afternoon she felt fit, and when a loose canvas on one of the masts was discovered, she climbed up to repair it herself. The day was bright and clear, and the sea calm beneath a fair breeze. Kathleen had cut away the frayed section of line and replaced it with fresh. All she had left to do was secure it to the yard, when suddenly a wave of dizziness washed over her, nearly causing her to fall from her precarious perch. She gasped and clutched tightly to the spar. Beads of perspiration popped out on her face and arms, and she closed her eyes against the grey curtain that threatened to engulf her. It seemed her fingers had gone numb, and the voices of the men below seemed to ebb and flow, first loud, then far away. Her knees had turned to water and it took all her willpower to keep them from folding under her. Forcing her mind to cling to conscious thought, she realized almost belatedly that she was holding her breath. Purposefully she began to breathe deeply and evenly, all the while praying she could hold on until the faintness passed.

"Oh, God!" her mind screamed. "I've got to hold on! If I pass out, I'll fall to my death! I'm too far up to even hope to survive a fall to the deck!"

An eternity seemed to pass before her stomach stopped quivering and her breath came more evenly. The fogginess behind her closed eyes slowly receded, and she sensed her equilibrium restoring itself. After a few more minutes Kathleen finally dared open her eyes cautiously. Her stomach lurched dreadfully and she snapped them closed again and waited. At long last sh e could open her eyes and not feel as if the ship were rolling in a gale. She stood quietly, not daring to move yet, and finally, when she felt her legs would hold her, she inched her way down the mast.

Never had she been more grateful to feel the solid deck beneath her feet. She stood for a few seconds clutching the mast as her breath came in great gulps that sounded oddly like sobs. When she gained sufficient control of herself, she walked wobbily to the wheel where Finley stood eyeing her worriedly. Luckily, no one else seemed to have noticed her strange behavior.

"Blast it, Finley! I haven't just grown three heads! You needn't look at me that way," she reprimanded shakily.

"Aye, Cap'n," he answered in a patient yet perplexed voice.

"I'm going below. Have one of the men climb up there and finish tying off that line, and send Dan to my quarters for further orders." She turned and started to walk away.

"Right away, Cap'n."

"Have someone else spell you at the wheel later if you need," she added, "and there's no need to go blabbing everything you know, or don't know."

"I didn't see anything unusual and I know even less, Cap'n. I know when to keep my mouth shut," he assured her.

She nodded her approval and left him wondering to himself.

"I must be coming down with something, Dan," she said later. "Don't say anything to the crew, but keep an eye peeled for anyone who seems sick. I'll stay in my cabin so if I'm contagious maybe it won't spread. Tell Timmy to leave my meals outside the door. Just say I'm not to be disturbed, and until we know what's wrong I'll give you all the orders to pass along. Don't come in. I'll talk to you through the door."

Two days passed and Dan reported all hands still healthy. Each morning Kathleen arose nauseated, but by late morning or early afternoon she was fine. She found if she ordered dry biscuits and tea for breakfast, she could get her stomach to settle itself sufficiently that she could avoid running for the chamber pot. It helped too, if she eased herself into her daily activities instead of her usual headlong rush. Still, the smell of fried bacon, fish, or strong cheese made her stomach immediately queasy.

By the third day of her self-enforced quarantine, Kathleen had finally convinced herself that her illness was minor, and probably nothing more than nerves and fatigue, since she seemed to require more sleep than normal. It was Dan, who came for the day's orders, who finally opened her eyes.

"Beggin' yer pardon, Cap'n K, but have ye considered, uh . . ." He stopped.

"Considered what, Dan?"

"Well, lass, it bein' none o' me business and all, I don't know quite how to ask ye, but . . ."

"Will you just say what you have to say and get on with it."

"Maybe ye're not sick at all, Cap'n."

"Are you saying it is all in my head, Dan?"

"Nay, lass, but maybe 'tis a natural thing. Could be ye're . . ." the poor man's voice trailed off again on the other side of the door.

"I'm what, Dan? Daft maybe?" Kathleen's temper was rising.

Dan straightened his shoulders, took a deep breath, and blurted, "What I mean to say is, when was yer last woman's time? Ye could be with child agin."

The silence from Kathleen's cabin was long and ominous, and Dan almost wished he'd kept his thoughts to himself.

Kathleen felt the blood drain from her face. She clutched at the back of her chair and eased herself into it. Could it be? Mentally she calculated, and suddenly all her symptoms fit, even her tender breasts. She was about six weeks pregnant, if she figured right. Just about the time she and Reed had admitted they were married and started sharing their bed nightly.

"Oh, damn that randy, rutting beast!" she swore silently, and then had to admit she had enjoyed it every bit as much as he.

"A baby." She breathed in wonder, and her hands went automatically to cover her stomach. Tenderness for her unborn child welled up in her, causing tears to blur her vision. "Oh, baby, how I will love you!" she thought dreamily. Then rational thought took over with a jolt.

"Child, while I long for you, and I know it isn't your fault, your timing isn't quite right," Kathleen sighed. She was so near to her goal. Before long, Reed would be forced to sell the *Kat-Ann*, but also before long her condition would start showing itself. In the time left to her, could she accomplish her task?

Kathleen squared her shoulders and unconsciously threw up her chin. "I'll have to," she decided. "I'll just have to hit him harder and make the time count. I'm so close! I can't quit now."

She rose and started pacing as her mind worked furiously. "I suppose Dan must know, since he's already guessed as much, and perhaps Finley in case something happens, but the rest of the crew must not find out. I'm sure they would refuse to serve under me if they knew."

Kathleen giggled silently as the ludicrous thought crossed her mind. "Who has ever heard of a pregnant pirate!" The vision of herself in her brief pirating costume, rapier in hand and stomach protruding, made her wince even as she laughed.

"And Reed must not find out, at least not now." Therefore she must hide it from everyone except perhaps Kate.

"How I wish Eleanore was still here!"

Kathleen had completely forgotten poor Dan until he called out nervously, "Cap'n?"

She strode to the door and jerked it open. "Come in, Dan, and close the door. We've a few things to discuss and plans to make."

As she had planned, Kathleen spent a few days with Kate. It helped her to have something different to occupy her mind and time as she attempted to learn the fundamentals of horse breeding. At times, though, she found her mind drawing ironic parallels to her own situation.

Kate was appalled that Kathleen meant to continue her raid on Reed. " 'Tis too dangerous, lass. Supposin' ye're caught, or killed, or wounded? Do ye want to lose this child, too?"

"I want the child, Kate. I want it desperately. It broke my heart to lose the last one, but I'm so close to breaking him! I can't stop now. All I can do is try to be careful. I wouldn't hurt this baby for the world. You must know that. A month and a half, two at the most, and it will be over with. Then I can rest easy. Then I'll hand up my sword and retire to Chimera and try to put up with all the boredom and Reed's tantrums and everything that goes with it. I promise."

Having finally wormed a reluctant promise out of Kate not to reveal her latest secret, Kathleen went home to Chimera. She needed a little time to herself to sort out her thoughts and her feelings before Reed arrived. After their latest encounter, which she recalled with a grimace, she wasn't sure what mood he would arrive in. He would be in a fit over his losses to his piratess she was sure, but could not determine how he would react to his wife.

"He'll probably expect me to be very contrite!" she fumed, recalling how he had soaped her mouth and then spanked her.

She didn't have much time to wonder. Reed returned the next morning at breakfast. Kathleen was sitting alone on the veranda munching her dry toast and reading the paper. She jumped as she suddenly felt his warm lips on the nape of her neck.

"Good morning, Mrs. Taylor." He laughed as she whirled about to face him.

"Land sakes, Reed!" she gasped. "Do you have to sneak up on a person? You are as quiet as a blamed cat!"

"And you are as jumpy as one," he countered with a grin. He seated himself at the small table. "Pour me a cup of coffee, will you pet?"

As she did so, she noticed the tiredness in his face. "Would you like me to order up some breakfast for you or have you already had yours?"

"Coffee is fine. I ate earlier. You could have Milly fill a tub for me, though."

"How was your trip, or dare I ask?"

"It was lousy, thanks, and I don't care to discuss it."

"Fine."

"What have you been doing while I was away? Mother says you haven't been to town at all."

"Oh, you stopped in at Barbara's then?"

"Just for a minute. I thought you might be there."

"No. I spent most of my time at Kate's. She is determined to teach me horse breeding. In fact, I've left Zeus there. She wants to try putting him in with one of the mares and see what happens." The thought made the color rise in Kathleen's face.

Reed grinned wryly. "You look very lovely when you blush like that, Kat."

"Do hush up, Reed," she whispered as she colored even more.

"Speaking of breeding," he went on as he caught her eye, "why don't you come up and scrub my back, and we'll see how we can do?" He grinned rakishly, his blue eyes alight with deviltry.

"Oh, dear!" she thought. "We do too well! If only you knew!"

Both Kathleen and Reed seemed determined to ignore their last spat and go on as if nothing had happened. She did, however, make an extra effort to curb her temper. She had said nothing to Reed about the Sweetheart's Ball to be held Valentine's Day, and was surprised when he insisted they go.

"I'd really rather not, Reed," she told him over dinner.

He cocked one black brow at her quizzically. "Why not? You've been hiding away out here long enough. Is the social butterfly suddenly content to be a country mouse? I can scarce believe it after all your protests to the contrary."

426

Her temper flared momentarily, making her green eyes flash, but she said quietly, "The ball is for sweethearts, dearest, not combatants."

He gave a short laugh. "Your point is well put, pet, but surely you can pull in your claws for one evening. You are such a convincing little actress when you set your mind to it. We are going. You'll find your gown on the chair in your dressing room."

She glared at him, but said nothing. Obviously he considered himself lord of the manor now and was putting her to the test. Was that his game? She wondered.

When Reed saw her coming toward him in the gown, his blue eyes darkened in obvious appreciation. Only Kat could have done this gown justice. It was a lustrous gold satin styled in a Greek fashion, leaving one tanned shoulder bare. It clung to her curves. With the heavy gold filigree necklace spread out on her chest, emeralds sparkling like defiant green eyes, she looked regal. She reminded him of some ancient Aztec princess, especially with the wide gold arm band that matched the necklace, and the dangling filigree earrings that fell nearly to her shoulders. Her coppery hair had been gathered at the crown of her head in a gold clip, and fell from there to the center of her back in an artfully tangled mass of gleaming curls.

For a moment Reed considered not going to town. He longed instead to carry this fiery, half-tamed creature to his bed and feel her silken limbs entangled with his, her thick lemon-scented tresses wound about him, her resistance melting and turning to burning passion as she cried out to him.

"God, but she's beautiful!" he thought as he offered her his arm. "If I don't watch myself, she'll have me stammering like a green schoolboy."

Once there, Kathleen enjoyed herself thoroughly. Mindful of the possessive gleam in Reed's brilliant blue eyes, she danced but once with each partner and often with Reed. As he whirled her about the room, she was glad of the strength of his arms, for she could easily have become dizzy. She said nothing about it to him, but took care to sip her champagne slowly. She often declined to dance and seated herself to talk with friends.

She glowed with good health and vitality, and if she curbed her natural vivaciousness a bit, she held instead a sort of secretive look that added a new appeal. She seemed unaware of the lustful looks cast her way, as well as the envy of most of the women present. If Kathleen was unware, Reed certainly was not. With a sigh he wondered if he would forever need to guard this beautiful jewel from covetous men who would seek to steal her away from him. How many duels would this vivacious vixen draw him into in the years to come?

"Perhaps I should lock her away in a room and keep her only to myself until age begins to dull her beauty," he mused. Then he shook his head and laughed at himself. "And I'd spend a fortune replacing the crockery she'd level at my head each day!"

With a start it came to him that he would not change Kathleen in any way. She intrigued him as no other woman ever had. He loved her beauty, her liveliness, even her temper. The only thing missing was her love. If it was the last thing he ever did, he would have her heart to call his own.

As he watched, he saw her unconsciously moisten her sensuous lips with her tongue, and his own smoldering desires flared up. "That does it," he decided.

A few minutes later he guided her down the steps and into their waiting carriage.

"Whatever will people think, Reed? Such a hasty departure! I barely managed to say goodnight to your mother and Kate, and you were whisking me out the door."

"You talk too much sometimes, sweet," he growled as he pulled her into his arms and lowered his lips over hers in a long, drugging kiss. His hands found the clasp at the shoulder of her gown, and before her mind had time to register this fact, her breasts lay bare to his exploring fingers. His lips traced a line down her neck and shoulders, making her shiver, and then they replaced his hands. His tongue played at her breasts and rising nipples while his nimble fingers sought the fastenings of her gown.

"Not in the carriage, surely!" she protested softly, already weakening under his forceful onslaught.

She heard him chuckle softly in response as he silenced her with a kiss. The next thing she knew, they were both undressed and he was lowering himself over her as she lay on the couch seat with her cloak beneath her.

"Oh, Reed! I've got the morals and restraint of an alley cat when it comes to you." She sighed to herself.

Only when he laughed wickedly did she realize she'd spoken aloud. "I've noticed that, you green-eyed baggage. A few well-placed kisses and you are practically begging me to bed you."

"It isn't fair," she murmured against his lips.

"Fair or not, and all other things aside, this is at least not a passionless marriage we find ourselves in, kitten."

His hands seemed to burn her already fevered flesh, and she clutched at his back to draw him nearer. "Hurry, darling. Please hurry!" she cried as she thrust her hips up to meet his.

Not by so much as the flicker of an eyelash did the footman or the butler let on that anything unusual was happening, as Reed alit from the carriage with Kathleen in his arms. Their blank faces showed no amazement that their mistress was wrapped only in her fur cape with her shoes in her hand and her golden dress draped casually over her husband's arm along with his cape, waistcoat, and vest. If they were blind to that, there was always the flushed, lazy look on a well-sated lover on her glowing face, and her passion-bruised lips that still drew the lustful blue gaze of her handsome husband, a look that foretold of more pleasures to come behind closed bedroom doors.

Chapter 26

DURING the next week and a half at home with Reed, only sheer willpoweer kept Kathleen from alerting him of her pregnancy. By keeping crackers in her bedside drawer, she held the nausea at bay, and if morning found her hair caught beneath Reed's shoulder or his arm about her waist, she nibbled quietly until it passed. More times than not, Reed was in a playful mood when he awoke, and she would have been obviously sick without her crackers. As it was, he noticed nothing unusual except for the crumbs in the bed.

"Why is it you find it necessary to eat in bed, Kat? It's damned uncomfortable, I tell you. Can't you wait until breakfast?"

Patiently she explained to him how often she awoke only to find herself longing for breakfast and trapped in bed with him. "Surely you don't begrudge me a few crackers to stave off my hunger while I wait for you to wake. Besides, you are usually so amorous, and I need the energy to keep up with you," she giggled as he glowered down at her.

"All right, keep your blamed crackers, you willful witch!" he growled as he began to shower her body with kisses.

Reed worked at plantation business and Kathleen kept the house running smoothly for him each day. They had breakfast and lunch at home, but evenings usually found them in Savannah. They dined a couple of evenings with Kate, and a few times with Mary Taylor and the Bakers, but more often Reed took Kathleen out to a fancy restaurant. Her favorite was near the riverfront, and built high enough along the bluff to provide a gorgeous view. They would dine on lobster, shrimp, and crab, and Kathleen swore she came away more intoxicated from the view than the wine.

The restaurant was actually part of a grand hotel, and whenever they dined there they would stay to dance in the ballroom on the third level. Once Reed even rented a suite, and they stayed the night in town.

Often they attended the theater, the opera, or the ballet. Reed even surprised her one evening by escorting her to the Golden Slipper for an evening of gambling, and shocked everyone when he rented a room upstairs so Kathleen could satisfy her curiosity and see what one of "those" rooms looked like. When they reappeared sometime later, Kathleen's carefully coiffed hair was somewhat less than perfect, and her color high, while Reed's blue eyes glinted with deviltry and self-satisfaction.

One young gentleman asked his older and more sophisticated friend, "You don't suppose they actually did?"

His friend gave him a disgusted look and said, "Now what do you suppose? They weren't up there playing chess all this time, you fool!"

"But to take your wife upstairs in a brothel!" The younger man gaped.

"Listen. If my wife looked like that, I'd take her on the dining room table while the servants served the soup! We should all have his luck!"

To Kathleen's relief, Gerard Ainsley had gone to Europe right after the New Year on banking business for his father. It was rumored he was keeping steady company with the pretty young daughter of a wealthy fur importer in Holland. Kathleen hoped the rumors were true.

There was still the sullen Amy to put up with, but things were looking up there as well. A certain dark-haired young man by the name of Martin Harper was taking up much of her time. He was the eldest son of a plantation owner from Augusta. He had come to spend the holidays with relatives in Savannah, and he had met Amy at Christmas and pursued her ever since. Unlike some of her other fawning swains, Martin did not cosset the pampered Amy, and perhaps this was why he was making headway. If she started to pout or argue with him, he merely walked away. If she displeased him in some way, he told her so, and warned her against repeating her misdemeanor. He was quiet, mannerly, and gentlemanly, but at the same time forceful. He told William and Barbara that he wanted Amy, yes, and intended to marry her, but she must change some of her petty ways. He would not have a spoiled child as a bride to meet his family in Augusta. Against her will, Amy was more than a little intrigued.

By the time Reed went off again, Kathleen had been having such a marvelous time and enjoying his company so much that she found herself reluctant to follow and engage him in combat. After all, he was the father of her unborn baby! She argued with herself all day, but by the time she left Kate's for her ship, she was resolved again. Besides, she must go to sea. It was her passion in life, her renewal. Heaven only knew when she would sail again once she stopped to have the child.

Kathleen's pregnancy did make her more cautious. For one thing, she no longer climbed in the rigging. They avoided other vessels, concentrating only on the *Kat-Ann*. Before they would have engaged another pirate ship or a tempting Spanish or English trader.

Dan hovered over her like a mother hen until Kathleen was at her wits end. Even Finley made sure she was not at the wheel too long or trying to lift anything heavier than her rapier. It did her no good to order them to stop.

"Mutiny! That's what it is! Pure mutiny!" she muttered darkly.

In the course of the next fortnight they hit Reed four times, and only when his hold was heavily laden. Kathleen could not afford to play games just for the fun of it now. When she and Reed squared off, Kathleen schooled herself not to overreact and become too cautious at the wrong times. She was just as wicked as ever with her blade, just as quick and vicious. She never took her eyes from his, and never hesitated to move in when she found the opportunity.

A couple of times, merely because of Reed's own expertise with a sword, he had her on the defensive for a while, but most of the time she fought the offensive, preferring to call the shots herself.

Their last meeting left her shaken and ready to head for home. Reed had fought particularly well, and the contest had dragged on until Kathleen's rapier felt as if it weighed a ton. Finally she left herself open a fraction of a second too long, and if not for a fast pivot to her right, he would have had her. She parried his blow from beneath, and as their rapiers swung upward hers glanced off his, the tip of her sword skimming across his cheek to leave a surface cut from just left of his mouth to the edge of his cheekbone. A thin line of blood, just two inches long, appeared diagonally across his cheek, ending just below the outer corner of his eye. It was far from a serious wound, but Kathleen realized it would leave a scar, a small one, but a constant reminder each time she looked at him. A few months ago she would have been glad, but now it distressed her.

In addition to everything else, somehow Dan or Finley had pulled Dominique aside and told him of her impending motherhood. As the doctor tended to Reed's face, Dominique cornered her.

"Cherie, you must give up this madness now. You could harm the child. Do you think I could live with myself if something happened to you or your little one? Think, Kathleen. What if you should slip? Reed could kill you accidentally, even while he only meant to wound, just as I know you did not intend to mark his face today. What could I say to him then? How could I face him and tell him I knew it was you all along?"

Kathleen eyed him balefully. "Dom, I am sickened by what I have done today, but I cannot stop while I am so close to the end. Only for a short while longer, I promise you. If I cannot force him to sell the *Kat-Ann* in two months time, I will give up gracefully. The Emerald Enchantress and her ship will disappear and never be heard from again. I swear it!''

He sighed and shook his shaggy dark head. "I will have to be content with that much, I suppose, but I will not breathe easily until then, I tell you. I will help you all I can until then, but it is against my better judgement.''

Kathleen had to stand on tiptoe to kiss his rough, bearded cheek. "Thank you, my adorable big brother."

Kathleen was on pins and needles awaiting Reed's arrival at Chimera. She alternately dreaded seeing him and viewing his latest scar, and wished he would hurry home so she could have it over with. When he did arrive, she was relieved to see that it was so small and healing well. The scab was off, and only a thin red line remained. When he smiled it nearly disappeared completely. In time the vividness would fade to leave only a thin white line. Rather than detract from his appearance, it gave him a rather devilish, dashing look, actually adding to his handsome character. Where the ladies had been captivated by him before, now they nearly swooned whenever he appeared.

Kathleen feigned surprise, of course, but the concern she revealed for him was real enough.

"Why, Kat," he said lightly, searching her face with curious blue eyes, "if I didn't know better, I'd swear you really cared. Don't you long anymore for widowhood?''

Anger covered her confusion. "I've never wished you dead, Reed. I've told you that before," she stormed.

"And pray tell, sweetheart," he teased with a mocking smile, "how was I to hang for piracy and not be dead?''

"Oh!" she exlaimed angrily, stamping her foot at him. "Oh, darn! You know what I mean!''

"Do I?" he queried. "How am I supposed to second guess you when you are not altogether sure of yourself half the time?''

"Let's just drop the subject, shall we?''

"Just like a woman! Just when the skillet gets hot, she doesn't want to cook.''

"You are not being fair, Reed!"

"That wasn't in the contract, sugar," he said, pulling her onto his lap and loosening her hair. "Not even in the small print."

Once a year Savannah kicked up its heels, let its hair down, and had a wickedly grand time. On St. Patrick's Day, they all turned out in costume for their own version of New Orlean's Mardi Gras. From early morning until long after midnight, the streets of this cultured, sophisticated city rocked with frollicking party goers.

Kathleen was amazed and delighted. She'd never seen anything like it before.

"This is nothing, kitten," Reed told her. "Someday I'll take you to New Orleans during their Mardi Gras. It is a sight to behold! It's even wilder than this."

"That's hard to believe," Kathleen marveled.

Reed had brought her to town the previous evening and they'd stayed in the hotel on the bluff, where Reed had rented a room overlooking the town instead of the river. Now Kathleen understood why. As they breakfasted on the veranda, they could watch the revelry on the streets below.

Already people were dancing and cavorting in costume, preparing for the parade through Savannah. The parade itself was scheduled for noon. The Taylors and Bakers were all going to meet in the park on Oglethorpe Square, where they would have a leisurely picnic lunch while they viewed the parade. Afterward, they would don their own costumes and join the melee in the streets until dinner. Later they would attend the masque ball in the hotel ballroom. It promised to be an exciting day, and Kathleen had plans to make it even more so. Her wide green eyes sparkled with suppressed anticipation.

The open carriage ride to Oglethorpe Park was slow, but entertaining. Kathleen felt sorry for anyone who would have a need to hurry through Savannah today. The streets were crowded and noisy, even frightening to the timid. The carriage crept forward in starts and lurches.

"It's a good thing we left so early," Kathleen admitted.

"We could walk faster, I'd swear."

"Yes, but then you'd get pushed and jostled about. You have a better view and the benefit of the breeze up here."

"I hope Martin is along today. Amy is almost likeable when he's around."

Reed chuckled. "She does seem to cheer up, thank heaven. I was beginning to wonder if her lower lip had grown into a permanent pout."

"She seems to be recovering from her heartbreak nicely. She's quite taken with Martin, from all accounts," Kathleen needled with a wry smile.

"My ego would be severely bruised if I didn't have you to soothe me, pet," he countered. "Especially since I expect wedding bells are in the offing for those two before long." He glanced at Kathleen, flashing his white teeth in a broad grin. "Are you sure you don't regret not having a big church wedding with the flowers and candles?"

"Don't rock the boat, Reed," Kathleen warned, narrowing her slanted green eyes at him. "Things have been going very smoothly lately. Don't rattle the bear's cage unless you're prepared to fight the bear."

He eyed her with amusement. "You certainly have a way with words, my sweet. It must be the Irish in you."

Since she'd dealt him the scar on his cheek, Reed had not shaved, and was now sporting a fair mustache and beard. It made him look even more like a pirate, Kathleen thought. She assumed it was to help cover the scar, but she discovered a hidden motive when he donned his costume that afternoon. He had chosen the dress of a desert sheik, and in loose robes, burnoose, and beard, he looked so convincing it was scary, especially with that wicked scimitar at his side. Kathleen had known his choice of costumes, but was still shocked at the transformation. He didn't even need a mask! Only those startling blue eyes staring out of his deeply tanned face gave lie to the fact.

"Saints, but you look so—so real!" she gaped.

He gave a devilish laugh. "Honestly, Kat! You look as if you expect me to carry you off to my desert retreat at any moment!''

"That is exactly the impression you create in that outfit!"

"And would you enjoy that, my dove?" His eyes gleamed wickedly.

Kathleen blushed furiously. "Don't be silly, Reed. Now get out of here so I can dress.''

"I still don't see why I can't see your costume. You are certainly being secretive.''

Reed was even more curious when twenty minutes later Kathleen joined him. She was covered from head to toe in the black street garb of a proper Moslem woman. Even the lower half of her face was hidden.

"That's it? That is your costume? I don't believe it!" He shook his head in bewilderment.

A soft chuckle came from behind the black folds. "This is just for this afternoon, my lord. This evening you shall see the rest of it, I assure you. Have patience, husband.''

"My, how meek you have become all of a sudden. This is certainly a novelty. It almost makes me wonder how it would be if I truly were a sheik and you my most devoted slave.'' He couldn't be positive, but he felt sure she was smiling behind her disguise.

At dinner the family gathered in the hotel dining room. Tears of mirth sparkled like emeralds in Kathleen's eyes as she viewed her relations. Uncle William arrived as a fat monk, with Barbara in tow as a nun.

"Don't ask me how I ever allowed him to talk me into this," she grumbled as he seated her at their table. "I feel like a short penguin! However do the good sisters stand these outfits! I swear I'll melt before the evening is done!''

"My dear, you are absolutely habit forming,'' William quipped.

When the petite geisha and her Oriental potentate arrived, it took quite a bit of guesswork before anyone realized it was Amy and Martin in black wigs. Amy was delighted that their deception worked so well, and doubly delighted to see Kathleen dressed so somberly in black.

"Why, cousin," she chirped. "You look as if you are in

mourning!''

"Yes, Kat," Reed spoke up. "Can't you unveil now? I'll admit I'm eaten alive with curiosity."

"Not until the dance," she said firmly.

Susan and Ted were charming in the native costumes of a Dutch girl and boy, though Ted admitted he felt ridiculous in knickers again. Resplendant was the only description for Kate, who came dressed as a gypsy fortuneteller, complete with jangling bracelets and big hoop earrings.

Looking about the dining room they saw many novel disguises. Over in a corner sat Galileo, globe and all, with Sir Walter Raleigh and lady. Next to them sat a magician and a witch with a farmer and his rosy-cheeked wife. As usual there was an abundance of damsels and knights, Romeos and Juliets, kings, princesses, queens and pirates. It was quite disconcerting, however, to see grown men dressed in bears, lions, rabbits and ghouls. Kathleen noted a court jester and a giraffe dining with two peasant girls. A fur trader and an Indian dined with a gypsy and Joan of Arc.

Excusing herself, Kathleen whispered to Reed that she was going to their room to freshen up. "Don't come up and spoil my surprise," she warned. "I'll meet you later in the ballroom."

It was perhaps fifteen minutes later that Reed looked up from the table to stare in mute amazement at the raven-haired beauty entering the dining room. Another part of his mind registered the fact that she was flanked by three men dressed as samurai swordsmen. The woman herself was magnificent as an Oriental. She stood proudly, her long black tresses hanging straight to her hips. Her dress was a sleeveless, lustrous, emerald silk sheath that clung to her figure as if painted on. It had a standing collar which melted into a plunging neckline, and that parted to reveal an indecent portion of breast. The dress was narrow, necessitating slits up each side, and clung to her hips. As she walked to her table, the skirt of her gown parted on each side, and those who watched gasped to see her shapely long legs revealed nearly to her hips. Tucked into a garter on one leg was a bejeweled dagger.

Kate, too, had noticed the young woman. "Oh, my stars!" she gasped faintly as one hand fluttered up to her neck. Quickly she glanced at Reed, who was still staring after the long-limbed beauty.

The woman was seated with her escorts before she glanced their way, and Reed got a quick look at emerald green eyes slanted upward at the corners behind a small green mask.

"It's her!" he murmured in amazement, not realizing he spoke aloud.

"It's who?" Ted prompted at his elbow.

Reed was too disoriented to reply. Instead he stood, quickly excused himself, and strode to her table, oblivious to the confused looks of the others.

The woman looked up as he approached. "Hello, Emerald," Reed unconsciously traced the small scar on his cheek.

"If it isn't Captain Taylor." Emerald inclined her head, indicating the remaining seat at her table. "Please join us."

Reed accepted the offer.

"We have been enjoying ourselves in your fair city today, Captain Taylor. It is almost like the Mardi Gras, don't you think?" she went on conversationally.

Reed noted, however, that as she spoke she leaned back in her chair and fingered the hilt of her dagger.

"It is similar, yes."

"You remember Mr. Finley and my bosun don't you? This," she said as she waved a slim hand at the remaining man, "is Mr. Kenigan."

Reed ignored the introduction, but registered the fact that all three men were armed. They obviously intended to protect their captain. The way she was dressed, she would probably have need of their protection.

"You have created quite a stir in that gown, Emerald. You seem to have a habit of displaying your wares."

His barb hit home as her eyes flashed dangerously. "I may be many things," she said in her low, husky voice,

439

"but I have never sold my body. I find it easier to gain my fortune with my talents as a pirate, especially when you have been so generous of late," she shot back.

She saw him stiffen in anger, and his jaw gave a definite twitch. She taunted him with a smile and, reaching out, she traced a tapered finger along his scar. "You are growing a beard, I see. I almost didn't recognize you. That is a very handsome costume. It suits you, love."

Reed caught her wrist in his large hand. "You enjoy taking risks, don't you, Emerald? You thrive on danger."

He released her when the waiter arrived with their drinks.

"Can I get you anything, Mr. Taylor?" the man inquired politely.

"No thank you, I'll be rejoining my table in a moment."

Emerald sighed heavily. "It is a shame you cannot spend the evening with us, Captain. It would be such fun! It's not often we can afford the luxury of an evening like this." Then she gave an elegant shrug. "But I forget. You have a wife to attend to."

She finished her drink and rose. "We must be going. Have a nice evening, Captain." With that she left him, and he walked back to his table with mixed emotions. Though he longed to best her, he had to admire her spit-fire courage, her absolute disregard for danger. Also, in spite of his anger, he recognized a lustful desire for her.

Reed had managed to get control of himself by the time he entered the ballroom. His composure was fleeting, however, and cracked completely when Kathleen entered the room. To complement his sheik's attire, she had dressed in a harem outfit. With her long copper hair loose and floating about her hips in lustrous waves, she walked sedately toward him. A sheer, gauzy peach blouse was overlaid with a short orange vest, from which dangled gold coins, forming a fringe. Several short vertical bands of beaded brain joined the vest to the slave-chain belt which rested on the curve of her hips. A large pearl lay nestled in her navel. The belt itself was attached to a panelled orange skirt, which when she walked, parted to give a glimpse of filmy harem pants. These were gathered about her ankles by ankle chains above small orange slippers. To complete the costume, a band of gold coins lay across her forehead, and the most translucent of veils covered the lower half of her face.

To the utter amazement of all present, she walked straight to her husband, bowed before him, making her obeisance as if he truly were a lord of the desert. She arose, and her laughing green eyes met his from above her veil.

"Well, master, does your humble slave please you better now?"

"Humble slave, my foot!" he muttered to himself, and then huskily, "Yes, Kathleen, you know damned well you do. In fact, if not for these loose, flowing robes of mine, the entire population of Savannah would know how well you please me, you green-eyed witch."

Savannah would long remember that night and talk about it for years to come. They would recall not only Kathleen's startling outfit and actions, but the fact that her fierce young sheik kept her all to himself that evening, neither of them dancing with anyone else. Indeed, it seemed as if they had forgotten anyone else existed, so wrapped up in each other were they that evening.

The weeks that followed were filled with activities. First upon them was Palm Sunday. Kathleen organized an Easter egg hunt for the children of the slaves. Much to Reed's amusement, his young wife seemed constantly to have a small child or two following her about, and often clinging to her skirts. Kathleen did not seem to mind, and hardly ever had to scold one of them. They adored their new mistress. He teased her unbearably about her little flock.

"Go ahead and laugh, Reed. Have your fun. They are children and I love them, but Lord help you if any of them ever bear any resemblance to you! I know what goes on under the very noses of many plantation wives. So does everyone else. Just because those wives are too well brought up to mention it and turn a blind eye, doesn't mean I will. I can not condone the practice of the master having free access to any and all the female slaves, and I won't stand by quietly and pretend not to notice, so consider yourself warned."

"I really haven't any interest in bedding any of them, but what would you do if I did?" he wanted to know.

"I'd probably leave you," she told him bluntly.

While Kathleen busied herself with overseeing the job of spring cleaning, Reed was seeing to the task of clearing and plowing the fields for spring planting. Kathleen started to worry that Reed was never going back to sea in time for her to follow. She might have time to launch a final foray against him if he didn't dally. So far he hadn't noticed anything different, but she'd had to let out the waistline and bust of a few of her gowns already. Her secret wouldn't keep forever.

Also, one day she looked in the posting tray and found a letter he had written and left to be mailed. It was addressed to Mr. Kirby. Kathleen tucked the letter in the pocket of her gown, and when she gained the privacy of her rooms, she tore it open and read it. Reed had written telling of their marriage and asking for information concerning the estate and land holdings, which by the laws of marriage now belonged to him. Kathleen had known he would get around to it sooner or later, and considered herself fortune to have intercepted the letter. Now was not the time for Reed to learn that he owned seven more frigates and a shipping firm! She shuddered to think what he would do when he eventually did find out.

"At least he can't beat me while I'm pregnant. He wouldn't risk it then," she reasoned.

The last week in March, Barbara threw a ball announcing the formal engagement of Amy to Martin Harper. It was an elaborate affair, and they all got to meet Martin's family at last. This was followed immediately by the annual peach festival, at which Susan reigned supreme as Savannah's Peach Queen, with Ted at her side. There was a grand parade through the countryside past acres and acres of blooming peach trees. The fragrant white blossoms filled the air with their beautiful scent, making everyone heady on their perfume.

April brought with it balmy temperatures and Kathleen's eighteenth birthday. Kathleen, who was busy trying to coordinate a ball the next week to announce the betrothal of Susan and Ted, didn't give her own birthday a thought. Kate remembered, however, and with Reed's help, contrived to throw her a surprise party.

The morning of her birthday dawned bright and clear. The rain had stopped sometime during the night, and the sun was drying up the last puddles. Reed came down to breakfast whistling merrily, and suggested she go in to town with him that day.

"Oh, Reed, I have a million things to do. I really shouldn't."

"Come on, Kat. You've been working too hard. Take a rest. Make out a list of things you need for Susan's ball next week and you can pick them up in town and do some shopping while I see if my seed order is in at Gavin's, among other things. We'll make a day of it, and I'll take you out to lunch. Besides, Mrs. Fitz cornered me last week and said you have a new gown ready for the final fitting. Maybe you can pick it up today."

Kathleen grinned at him. "You surely can sweet-talk a girl, Mr. Taylor," she said, aping his Southern drawl.

He dropped her off first at the dressmaker's. The gown she had ready was sure to be one of Kathleen's favorites. It was a pale aqua of fine silk netting so sheer and soft the material seemed to float. The merest movement set the full folds to billowing. The top had two inch straps that came down and melted into a crossed bodice that accented the fullness of Kathleen's breasts and whittled her waist down to nothing. This alone cheered Kathleen, even though Mrs. Fitz had to let it out a bit. The skirt fell in soft, whispering folds to her toes.

The dear lady told Kathleen to stop by after lunch, and she could take the dress home. Then she fitted Kathleen for another dress due to be done in time for Susan's betrothal ball. It was to be of shiny black satin with slim pearl-trimmed straps and bodice, with another belt of pearls about the empire waistline. Kathleen's hair and coloring would set the dress off perfectly, Mrs. Fitz assured her. She would look stunning.

Kathleen spent the rest of the morning shopping for the items on her list, and finally met Reed for lunch at the hotel on the bluff. She was famished, and Reed teased her terribly about the amount she managed to eat.

"If you don't watch it, you'll end up fat, and then I won't take you anywhere with me. I'll hide you away and spend my time with several mistresses."

Kathleen stuck her tongue out at him and scooped up the last of her whipped cream. All the same, she wondered how Reed would react a few months from now when she would indeed be fat and awkward. She tried to tell herself it wouldn't matter if he didn't desire her, but she knew she was lying to herself.

They picked up the gown from Mrs. Fitz's, then went by to visit with Mary and Barbara. They were out, and Mammy didn't know when they would return. Susan and Amy were gone, too. Even Ted was nowhere to be found.

"My goodness! This is unusual. I hope they didn't decide to ride out to Chimera and find us gone."

Reed merely shrugged. "It's too pretty a day to go back and work. Let's go for a drive along the river and take the long way home."

It was nearly dinner time by the time they finally got back. Reed steered her upstairs as soon as they were inside. "You go freshen up and have Milly do your hair for you. Try on your new dress for me, all right, kitten? I want to see what it looks like."

"What are you going to do? Aren't you going to change? Dinner will be soon. Whatever Cook has been fixing sure smells good."

"I'm going to have a drink before dinner, and I'll be right up to change."

Between them, Reed and Milly kept Kathleen occupied in the bedroom until dinner. When Reed finally escorted her downstairs, he led her out into the courtyard.

Kathleen was agape. Colored lanterns hung all around, lighting the courtyard and fountain. Tables had been set up and gaily decorated, and a band of musicians played softly from a dim corner. At her entrance, all her friends and family stood up and began singing, wishing her a happy birthday. Two servants wheeled out an elaborate cake sparkling with eighteen lighted candles.

Tears blurred Kathleen's vision, and she stumbled and grasped onto Reed's arm as he led her to her seat. She was touched by his thoughtfulness, and could not find the words to thank him. He kissed her tenderly and laughingly told her to stop stammering or they would all think she'd gone daft.

Among her gifts, she found a beautiful aquamarine pendant and ring from Kate. "These were me mother's and mine. I would have given them to yer mother, had she lived longer. Now they're yers, and will go to yer own daughter some day," Kate said with feeling.

Ted presented her with art supplies, and Uncle William and Barbara had bought her an elegant yellow brocade lounging gown with backless heeled slippers to match. Even Amy and Martin had brought a lovely porcelain musical figurine that somehow resembled Reed and Kathleen waltzing.

"I'll treasure it, Amy," Kathleen said truthfully. "Thank you." Perhaps someday they would actually be friends.

Susan and Mary approached her together, and Susan carefully deposited a tiny fluffy white kitten in Kathleen's lap. "Mother and I figured since Reed is always calling you Kat or kitten, you should have one of your own."

"Oh, he's adorable!" Kathleen crooned.

"She's adorable," Mary corrected with a laugh.

Reed waited until last to present her with his gift. It was the most magnificent saddle and bridle Kathleen had ever seen. Of the finest leather, hand-tooled and inlaid with silver.

"Of course, you'll have to wait until Zeus is back from his little vacation to use them," he joked, his diamond blue eyes twinkling.

Kathleen could never remember having such a splendid birthday. She was asleep almost before her head hit the pillow, but it was a contented exhaustion.

Somehow Kathleen managed over the next week to accomplish the impossible, and the ball for Ted and Susan was grand and elegant. Finally even that was behind her, and at long last Reed was preparing to sail again. Kathleen breathed a silent sigh of relief. She had exactly two and a half weeks to bring Reed to his knees. By May first the summer season would be starting, and Mary and Susan would be returning to Chimera. The only reason they were not already in residence was that Mary thought Reed and Kathleen needed the time to themselves. That and the fact that she and Susan were having their summer wardrobes fitted.

Kathleen was no longer bothered by morning sickness. By her calculations, she was nearly three and a half months along in her term. This was absolutely the last opportunity she would have to defeat Reed. It was now or never, and she sailed after him.

Chapter 27

HIDDEN away in her secret bower, the *Emerald Enchantress* waited, and so did Kathleen. She paced anxiously up and down the deck. It wasn't just in anticipation of the coming dawn, or the expectation of battle with Reed. Something else was niggling at her nerves, and she couldn't quite put her finger on it. It was an odd feeling, almost a premonition, a feeling that something was not quite right. It was similar to a slight change in the air or the sea, sensed rather than seen, something out of place. This sixth sense, or whatever it was, gnawed away at her, making her restless and on edge.

Kathleen was not the only one who felt it. Perhaps because sailors spent so much time at sea, so much of their lives alone, they became in tune with their feelings, and when something disturbed the subtle rhythm of their routine they knew it immediately. Dan was near the bow, his jaws working overtime on his wad of tobacco, his eyes constantly searching for something just out of sight. Finley was pacing an alternate route to Kathleen's and cracking his knuckles at regular intervals. The noise of his joints cracking in the quiet predawn sounded like light artillery, and that in itself was adding to Kathleen's irritation.

Finally she could stand it no more. "Haul anchor," she told Finley. "Set the sails. Tack her sou'east and we'll try to intercept the *Kat-Ann*. There is no sense in sitting around her waiting. Frankly, it's getting on my nerves."

Dan nodded his agreement. "Glad ye said thet, Cap'n. I been feelin' like a hen in a fox den."

"I don't know what's got into all of us, but I'll feel better with the wind in my face and the sails unfurled. I guess I'm anxious to have this done with," Kathleen thought to herself.

Had Kathleen known, Reed was having similar feelings. Through deliberate calculations, an educated guess, and his own intuition, he had chosen these islands as the most likely area for Emerald's hideaway. He had already meticulously searched the northernmost of those in the chain, and was slowly and methodically working his way from east to west along the few remaining isles.

He was becoming impatient and beginning to doubt his own wisdom. Perhaps he'd been wrong after all. The wily sea-witch could be miles from here.

Just as he was ready to call off the search and go in search of more profitable quarry, he caught a glimpse of green sails that seemed to materialize from nowhere. The sleek frigate was just nosing cautiously from a perfectly camouflaged little inlet not a hundred yards from the *Kat-Ann*. They had not spotted him slipping quietly along the shoreline of the islands, possibly because the sun was just rising in a huge golden ball almost directly behind him, blinding them to the oncoming danger.

Reed recognized his advantage. Under full sail, with the added momentum of the early inland breeze, his maneuverability was at its maximum, whereas the *Enchantress* had yet to unfurl all her sails.

Just as Kathleen recognized the danger, she also realized the hopelessness of her situation. Damn! She should have listened to her intuition!

The *Kat-Ann* swooped down upon them like a bird of prey. Within seconds the grappling hooks had latched onto the *Enchantress* like giant claws, and Reed was swinging aboard to meet her on her own deck for once.

He landed as lightly as a cat, executed a dashing bow, and gave her a roguish grin. "Well, well. Imagine meeting you here. Lovely morning, isn't it?"

Emerald glared at him from behind her mask. "I really hadn't noticed, Captain Taylor."

"Oh, believe me, Emerald, it is indeed. It is a perfect morning for a sword match."

Emerald gave a forced groan. "Pray, not before my morning cup of coffee. I'm such a bear if I'm not properly awake. Care to join me, Captain?"

One dark brow arched upward. "Are you stalling, Emerald? It will do you no good, my dear piratess. I'm going to defeat you this morning. I can feel it in my bones." His blue eyes sparkled in the sunlight.

Hands on hips, Emerald looked him over head to toe. "Are you always so full of good news so early in the day? It's disgusting!" Then she smiled up at him and asked, "Do those ancient bones of yours also tell you when it is going to rain?"

He threw back his head in a deep, rumbling laugh. "Ha! Perhaps you do need your coffee in the morning. I'm really sorry to have to make you forfeit it, wicked one, but I'm in a bit of a rush this morning, so let's have at it. Just the two of us once more?" he queried.

"Why not?" she shrugged. Giving a quick look around, she noticed Reed was not taking any chances this time. A number of his men had crossed over and were standing on the deck of the *Enchantress*. If she lost, there would be no chance of rescuing her ship this time.

In a short space of time, she realized that they were evenly matched this morning. Reed's confidence communicated itself through his sword. Kathleen, perhaps because of her pregnancy, felt sluggish and slightly awkward, not a very desirable circumstance at the moment. They fought silently for some time, neither gaining the advantage, and then somehow, Kathleen never quite figured out how, Reed maneuvered her so that the sun was in her eyes. Just for an instant she was blinded by the glare, and the next moment she felt his swordtip rake ever so slightly across her exposed midriff. Instinctively she brought both hands across her stomach in an unconscious effort to protect her newly conceived child.

Within seconds Red had disarmed her of both rapier and pistol. He swung her about with her back to him, his arm a band of steel about her waist. Recovering from her shock, Kathleen wriggled and kicked wildly, screeching at the top of her lungs as she tried in vain to reach the knife in her boot. Failing that, she grabbed for Reed's pistol, but he was too quick for her. Readjusting his hold on her, he managed to restrain both of her arms tightly to her sides, her hands in front of her.

Kathleen was like a wild young tigress, frantic to escape before Reed discovered her identity and fearful of what would happen to her faithful crew. Twisting and squirming against his relentless hold, she raged at him in her anger, calling him every vile name she could think of. After one long-winded stream of vindictives, she paused to gather her breath for more, and felt his arms tighten with vicious pressure under her breasts. Try as she might, she could not catch her breath. Her world began to tilt crazily. Her view became distorted, and from the edges of her sight, everything began to darken ominously. She felt as if she were spinning into an ever darkening void, and then . . . nothing.

As soon as Reed felt her go limp in his arms, he threw her across his broad shoulder, grabbed a line, and swung across to the *Kat-Ann*. Issuing terse orders to his crew, he carried Emerald to his cabin, where he tossed her onto his bunk. Already she was coming to. Quickly he removed her knife from her boot, and stood back to lock the cabin door.

Kathleen's first view as her eyes fluttered open was of Reed standing a few feet from the bed, a smug look of victory on his face. She scrambled off the bed and made a mad lunge for the door, knowing all the while it was hopeless. Finding it locked, she wheeled about to face him.

"Where are you taking me?" she demanded. Already she had noticed that they were aboard the *Kat-Ann*. "What have you done with my ship and my crew? I demand to know!"

Reed laughed wickedly. "You are in no position to

demand anything, Emerald. You and your crew are my prisoners, and your ship is my booty."

A quick look of relief slashed across her face. "You didn't fire her, then? Or kill my crew?"

"I'm not stupid, Emerald." He stood relaxed but ready for any move she might make, like a sleek dark panther eyeing a rabbit. "The *Enchantress* is too fine a vessel to sink."

She was not ready for him when his hand shot out and ripped the mask from her face. Immediately she lowered her face, and her long black hair fell about it as she turned away from him. Just for a second did he glimpse her face, but it was enough. Stunned, disbelieving what he had seen, he swung her about to face him again. One large tan hand gathered a handful of hair at the nape of her neck, and he forced her face up.

His breath left his lungs in a whoosh. "Damn! I don't believe it!" Then, as if he needed further proof that this was indeed Kathleen and not some dark-haired mirror image, he undid her belt and yanked at her brief breeches. There, high on her left hip, was the odd-shaped mole he'd seen a thousand times before.

Anger boiled up in Reed, choking off all reason. His world seemed fringed in red as a searing anger raged up in him. Without thinking, he raised his arm, and with all his strength, he backhanded her hard across the face. Under the force of his blow, Kathleen's head snapped back, and she fell across the bunk face down. She had given one sharp cry as she fell, and now she hid her face in the coverlet, trying to hold back the huge sobs that threatened to choke her. The entire left side of her face felt on fire.

Above her Reed raged. "All these months! All these months it was you! Damn you for the vicious, convincing little witch you are! Look at me, damn you!" He reached down and flipped her over on her back, his strong hands biting into her upper arm.

She bit her lip to keep from crying out. and stared up at him through tear-filled eyes that glittered like fresh-cut emeralds.

"Why?" he bellowed, shaking her roughly. "I want to know why! You've chased me over half the Caribbean how many times now? You've robbed, cheated, and pirated me. You've pulled every trick in the book. You've fought me, bested me, scarred me, and damned near defeated me altogether. You humiliate me at every turn, both here and at home. You," he paused to utter an ugly laugh, "you even had the nerve to blackmail me for being a pirate! Ha! I'm a privateer. You, my dear barracuda, are the true pirate of the family! Is there nothing too low for you?" He gave her a final shove, and released her. "You make me sick!"

Kathleen cringed under his furious glare. With that, he promptly removed all the weapons from the cabin, walked out, slammed the door, and locked it.

Reed stored the weapons in the arsenal and started for the bridge at a lope. "If I had stayed in that room one more second I'd have killed her!" he thought. He knew he needed time to cool down and think this out rationally. Too many questions were racing through his brain. Where had she gotten her frigate? How had she assembled a crew? Who had covered her absenses from Savannah?

Suddenly one of the answers clicked in . . . Kate! Of course! Naturally she would help her only granddaughter. Now he remembered the countless times Kathleen had visited with Kate while he was gone. His anger flared anew, his face twisting in a pained grimace. He had always liked Kate, confided in her, trusted her, and she'd turned against him. Why? Kate was another who would have a great deal of explaining to do. Sweet, kindly old aristocratic lady! Ha!

Halfway up the ladder to the bridge, Reed suddenly halted as another thought jolted through him. Dominique! A kaleidoscope of scenes flitted across his brain. Kat and Dominique on Grande Terre, developing the close relationship that had riled Reed's jealousy. Dominique protecting Kathleen, protecting her chiefly from Reed. Dominique telling him how he and Pierre had come across the lady pirate at Matanzas, and how she had given them the slip. Dominique, so anxious to join Reed as his gunner on the *Kat-Ann*, and all the artillery problems and missed shots since he had. Dominique looking so pale and stunned after Emerald had plunged into the foggy, churning sea. Dominique showing up with Jean at Christmas, and looking so relieved to see Kathleen.

Almost as an afterthought, he realized that not only Dominique, but Eleanore knew as well. Eleanore, who could easily keep tabs on activities at Grande Terre, and cover up for Kathleen's absences while she was at Chimera. Jean, at least, he felt sure knew nothing.

Reed spun about, shielding his eyes from the sun as he scanned the decks of both frigates. Having spied the man he was looking for, he leaped down and strode briskly across the deck.

"Dominique!" he roared.

Dominique turned from his duties to meet Reed's blazing blue gaze. A kind of sad resignation reflected itself in Dom's black eyes.

They were nearly nose to nose, Reed's right hand resting on the hilt of his rapier. "You knew, you damned cutthroat! You knew all along, didn't you?" he demanded.

Dominique answered with a calm he really didn't feel. "I didn't know until Matanzas. I tried to talk her out of it, but failing that, I've tried to see at least in part to her safety." His dark eyes clouded slightly, and he frowned. "I heard her scream. You didn't hurt her did you?"

"I could have killed her, but I didn't. She'll be fine, at least until we reach Grande Terre. Then it depends on what kind of answers I get."

Reed glared at him. "What about you? How much do you know about this?" He sneered hatefully. "I'm sure she's told you everything. Can you tell me why she's been doing this?"

Dominique looked pained. "That is something Kathleen will have to tell you. It is between the two of you."

Reed's blue eyes narrowed dangerously. "That is something you should have remembered months ago, old friend," he spat out. "I'll talk to you again later. We've a lot of scores to settle, but right now I can't stand the sight of either you or her. Get your rotten carcass off my ship before I kill you where you stand! Take what men you need to sail the *Enchantress* to Grande Terre, and get out of my sight!"

Reed turned to leave, and then faced Dominique again. "I might warn you, Dominique. Don't let anything happen to any of our hapless prisoners. I want them all accounted for when we get there."

Before Reed could stalk off, Dominique stopped him. "Don't harm her, Reed," he warned with quiet dignity. "For all our sakes, please don't harm her. I've sworn to protect her, and as much as I'd hate to, I'd see you pay in hell for it if you do."

Reed laughed hatefully. "Don't fear, faithful watchdog! I wouldn't trust myself anywhere near that backstabbing viper right now. By the time we reach Grande Terre I just might have calmed down enough not to kill her, or you if you are lucky. There will be a day of reckoning, however, rest assured of that!"

When Kathleen was assured that Reed was not going to return immediately to the cabin, she arose from the bunk. Even though she knew Reed had locked the cabin door, she tried it anyway. Then she crossed to the porthole. It was open, and though she could not see the *Enchantress* from this angle, she could tell from the sounds of activity that they were making ready to sail. She strained her ears to hear a familiar voice of any of her crewmen, but could not identify even one. She hoped none had put up resistance and gotten wounded.

At the thought of wounds, she crossed to the small mirror above the basin. The entire left side of her face was a brilliant red. Her cheek bone was already swollen and turning purplish. She touched it cautiously and worked her jaw tentatively. As far as she could tell, it was not broken, but she was going to have one dandy bruise, and it hurt like the devil! In fact, she was going to sport several bruises where Reed had grabbed her so roughly.

Kathleen's eyes flashed in danger, and then she sighed heavily as she turned away from the mirror. For a few moments there, she had thought he was going to kill her. Lord knew he still might if he didn't calm down! She should have been more prepared for the chance of getting caught, and Reed's resultant anger.

"You didn't expect him to smile and serve you tea, did you?" she berated herself. "What did you want? Congratulations?"

Still, she hadn't really considered being bested by him. She'd beaten him so consistently, perhaps she'd become overconfident. He'd appeared so quickly this morning—out of nowhere. Well, now she'd certainly landed them in a pretty pickle!

Kathleen paced the cabin as she mused. Then she gathered her wits and methodically searched each nook and cranny. She went through the wardrobe, his desk, the cupboards, under the bed, everywhere, but could not turn up one weapon. "Blast him for being so efficient!"

They were underway now. She had felt the lurch as the sails caught the wind. Now she could see the *Enchantress* from the porthole. She guessed Reed would sail for Barataria. Kathleen flushed as she thought of having to face Jean. Facing Reed was one thing, but explaining to Jean, who had befriended her so unhesitantly, was quite another. Now she was glad she'd had the forethought to store what would have been Jean's share of the booty in the Savannah warehouse.

Kathleen braced herself for another encounter with Reed, but he never ventured into the cabin. Neither did he send anyone else, and as the day wore on Kathleen became aware of her thirst and empty, rumbling stomach. The cabin heated up quickly, and soon Kathleen felt very grimy and sweaty. She had no extra clothes left here to change into anymore, but she did find water in the pitcher. Some she drank, and some she used to damper a cloth and wash, and then she wet the cloth again and held it to her aching cheek. She used Reed's brush to straighten her tangled hair.

She had already been through her old cabin in search of weapons and clothes. Now she sat at the desk before the open porthole and watched the *Enchantress* as it followed beside the *Kat-Ann*. Lunchtime passed and no one brought her food. She wondered if Reed meant to starve her to make her more malleable. She didn't wonder why Dominique hadn't sneaked down to speak with her. She thought she recognized him aboard the *Enchantress*. She was on her own now.

Kathleen knew if the opportunity arose, she would still attempt to escape, even though Reed now knew her identity. She had seen the contempt in his eyes, heard the anger in his voice, experienced his rage and humiliation. Whatever might have been between them was now lost. Not even for the sake of her child would she abide his scorn.

She scoured her brain for a plan that would allow her and her crew to escape, and she came up empty. Perhaps once they got to the island Dominique could aid her. She would simply have to wait. One thing was for sure. If she could get away, she would return to Ireland, to her friends and her homeland, to await the birth of her child. She wondered to herself, if the opportunity arose, would she tell Reed about the baby? Would he care? Would he ask if it were his as he had done before?

Reed stood at the helm of his ship and steered for Grande Terre completely by reflex, for his mind was occupied with his problems.

"Why? Why? Why?" The question repeated itself in his brain admist a hundred others. Bit by bit, some of the answers started to sift through the muddle. He recalled the morning of his duel with Pierre, and how Kat had bested the pirate. Reed hadn't really thought it was a complete fluke that she had handled herself so well, but he had since forgotten the episode. Now he remembered Kat saying her father was an expert swordsman. Perhaps he had taught her to fence, but had he also taught her to sail? He knew she ran the *Enchantress* herself and gave all the orders. She wasn't just a figurehead. If she and

Dominique would not tell him, perhaps Finley would. He made a mental note to interrogate Finley, then Kate.

That thought prompted another, and his temper rekindled. Kat seemed so at home and familiar with some of her crewmen. She sailed alone with them for weeks at a stretch, dressed in the briefest of costumes, her long, tanned limbs exposed to the view of each and every randy sailor. To top it all, she had gone at least once to Matanzas, where some of the roughest, most notorious pirates congregated. They were all immoral, fiendish jackals who would slit your throat and your pockets without blinking an eye. The thought of Kat parading through Matanzas, being ogled, and perhaps mingling in the commonest sense with that filthy scum, turned Reed's stomach. And all the while, she had been holding him at bay with her blackmail! It made his blood boil just to think of it!

How could he have been so stupid—so blind! How had he failed to recognize his own wife as he faced her over crossed swords time after time, and gazed into those immense emerald eyes? If Dominique was telling the truth, he had recognized her at Matanzas months ago, but Reed had not in all this time. Now little things came to mind that he should have noticed before. He recalled Kat's penchant for climbing into the rigging on the trip over from Ireland, and her familiarity with the ship. He had simply assumed she had sailed often with her father. Now he knew she had actually captained the *Kat-Ann* before. Those spectacular green eyes should have been a dead giveaway, but with the ebony hair and the mask to camouflage her features, he hadn't connected the two, especially since Kat was supposedly safely in Savannah. He recalled asking her once about her tan lines, and how easily he had accepted her excuses that she had been sunning herself in her camisole. How stupid could one man get!

Then there was the New Year's Eve ball when Emerald had appeared, then disappeared into thin air. Reed laughed ruefully to himself. How easy it had been for her! Then and at the Mardi Gras Festival. He'd recognized her perfume, but had shrugged that off too.

Now Reed wondered how many people had known all along and laughed at him. Dom knew for sure, and Kate and Eleanore he was certain. Jean and Pierre could not have guessed, but what about Barbara, William, and especially Ted? Ted had certainly been easy for Kat to manipulate at times. Susan and his mother couldn't know, or they would have said something by now, and Amy would have been the first to squeal.

For some reason, the day of the hurricane and their wild run on the *Enchantress* came to mind. He could still recall the thrill of it through his rancor. And after the storm had passed, they had created a storm of their own in her cabin. She had been so wild and uninhibited that night! Reed felt himself grow rigid with desire at the remembrance of it all. Now it came back to him how familiar Emerald's voice had seemed at times, that husky timbre Kat got in the heat of passion.

What a fool he had been! Here she had practically ruined his privateer's business, and all the while he had been lavishing clothes and jewels on her! The nerve of that green-eyed baggage! Numerous clashes, each incident of personal humiliation, curious comments with previously vague nuances now suddenly came clear, and rankled as nothing else could have.

Yet under it all, Reed had to admit a grudging admiration for her skill in handling both her rapier and her ship. He'd never known another woman with abilities to match hers. Now he understood why he'd been so drawn to Emerald—Kat, he corrected himself. The one time she'd fought at his side against the pirates, he'd admired her courage, her bravado, her unquestioned rule over her crew. He'd once thought Kat would faint at the sight of some of the things he had seen, and now he knew better. Undoubtedly, she'd killed and maimed with her slashing sword. She'd kept a cool head, and stomach as well. Truth be told, she probably outfought, outsailed, and outwitted him most of the time, and he applauded her skills even while he despised her. His male ego had been severely bruised, but he could have handled that. He might have been able to accept all these attributes in Kat if she'd only been honest with him, but not this way. Not behind his back, not at his expense, not with all her energies bent toward his destruction. For him, yes; but not against him.

Reed had timed their arrival at Grande Terre perfectly. They negotiated the shoals long after nightfall. Hence, the famous green frigate and her notorious lady captain arrived under cover of darkness. This suited Reed well, since it saved him the embarrassment of having the entire island aware of the fact that the infamous piratess was in fact his wayward wife.

Kathleen was relieved to be spared the fanfare also. Since early morning, she had sat alone in the cabin. No one had ventured in to relieve her hunger or her curiosity. She was tired and hungry, and not at all anxious to face either Reed or Jean.

As soon as they had docked, she heard Reed's footfall outside the door. He opened it and stood towering formidably in the doorway. "Let's go," he growled. "It's time to face the music, and I'm calling the tune this time. If you recall, this is quite a fortress, and if you are as smart as you seem to be, you won't even think of trying to escape. There is nowhere to go, and nobody to help you. Even your dear friend Dominique would not dare to go against his brother to aid you."

Kathleen stood up, flipped a long strand of inky hair back across her shoulder, and strode arrogantly to Reed's side. Inside she was shaking, but she'd be damned if she'd let him see it.

Reed took her firmly by the arm and marched her off the ship and up the hill to Jean's house. He nearly threw her through the doorway into Jean's study.

Jean was seated behind his immense desk when they burst through the doorway. For an instant surprise registered on his face, immediately replaced by a smug pleasure. "Ah! So you captured your piratess at last, Reed. And the ship?" He came around the desk toward them.

"Both," Reed answered, "but all is not as it seems, Jean. Here. Have a closer look." With that, he shoved

Kathleen toward Jean, grabbed her by her long hair, and tipped her face upward for Jean to see.

For a second time surprise flitted across Jean's features. "Kathleen!"

"Indeed." Reed was indignant. "Quite a revelation, eh, Jean? Dominique has known for months, and I suspect so has our dear Eleanore. They've been aiding her."

"I must say, this is the last thing I expected," Jean commented.

"You and me both," Reed added dryly. "What I need now is a safe place to confine her. Do you have an upstairs room that she can't escape from?"

"Come with me." Jean led the way upstairs. "I suppose for now you want her identity kept secret."

"If possible." Reed dragged Kathleen along beside him as he strode briskly down the hallway.

Jean opened a door near the end of the hall. "You'll want some privacy," he said as he handed Reed the key and left, not even glancing in Kathleen's direction.

Reed shoved Kathleen roughly into the room. She lost her footing and landed in a heap in the center of the carpeted floor. She scrambled to her feet as Reed lit the lamp and locked the door.

"Now." Reed towered over her. "I want some answers, and they'd better be good ones." He gestured toward a chair. "Where did you obtain your ship?"

Kathleen swallowed the lump in her throat, but refused to answer.

"Ok, let's try another question. Where and how did you collect your crew?"

Still Kathleen maintained her silence, staring down at her hands, clasped in her lap.

"Not very talkative now, are you, Kat? I must say, I've never seen you at a loss for words before. This must be a first."

She glared up at him.

"Good! Now that I have your attention, tell me why you did it." He leaned down to place his hands on the arms of the chair, effectively hemming her in.

"You stole my ship." Kathleen spoke quietly, but clearly. "The *Kat-Ann* was mine. You had no right to her."

Reed straightened up as if she'd struck him. "Do you mean to tell me that all of this dates back to the same old thing, after all this time?" He was stunned.

"Yes," she said simply.

Reed laughed harshly. "Oh, Kat, if you only realized! As soon as I'd made enough profits from my raids, I intended to buy another ship and return the *Kat-Ann* to you. But you don't believe that, do you?"

"No."

"Well, you managed to keep my profits fairly well trimmed. Tell me, how long had you intended to keep pirating me?"

"Until I'd forced you to have to sell the *Kat-Ann* to a buyer that I would have arranged to purchase her in my behalf," she admitted.

"And then?" he prodded.

"I don't know. I hadn't planned that far ahead."

"I know Dominique and Eleanore and Kate knew what you were about." At the look on her face, he waved a broad hand and shook his head. "No, don't bother to deny it. Does your Aunt Barbara and Uncle William know?"

Kathleen shook her head, again studying her fingers.

"What about Ted?"

"No."

"Fine. Now answer me this. Where did you learn to fence?"

"My father hired fencing masters for me in Ireland."

"You're good. Very nearly the best I've seen."

"Thank you."

"You are also very expert at captaining your vessel. Where did you learn that?"

"I've always loved the sea and ships. Papa always took me along, and I've been learning since I was a young child. It's second nature to me."

Reed nodded. "Where did you get the *Enchantress* and

461

her crew, Kat?''

"If I tell you, will you release them unharmed? They were only following my orders."

"Orders or no, they are pirates. No, I'll not deal with you, Kat."

"Then I'm afraid I can't tell you."

One dark eyebrow quirked upward. "Oh, you can't, can you?"

"No." Kathleen met his look squarely.

"It will go harder on you if I have to find out from other sources."

Kathleen only shrugged.

Reed was becoming angry at her again. "Suppose you tell me where you hide your ship up Savannah way. I know we surprised you coming out of your lair this morning."

Kathleen clammed up once more.

"What have you done with all your booty and your profits, then?"

She stared at him mutely.

Reed eyed her with distaste. "Look at you!" he grimaced. Grabbing a large handful of hair, he dragged her to the full-length mirror on the armoire in the corner. "Just look at yourself! You look like a harlot! No," he corrected, "more of them dress more tastefully."

Color rushed to Kathleen's face, put there by anger and embarrassment. "You didn't seem put off by my dress, or should I say lack of it, a few months ago, Reed. In fact, if memory serves me correctly, you jumped at the chance to bed your lady pirate!"

Now it was Reed's turn to register embarrassment.

"Technically, of course," Kathleen went on, "you are innocent of extramarital cheating, but you didn't realize that then, did you? In that light, you are as guilty as sin, and we both know it, you faithless fiend!"

"What do you expect when you parade around half dressed, your breasts half exposed, your legs entirely bereft of cover? Oh, and don't try to tell me you've been so pure either, my sweet! I'm not so much a fool that I'd believe I was the only one you've taken to your perfumed bunk!

You'll answer for that, too, before we're through! I don't take kindly to playing the cuckold."

Kathleen drew in her breath sharply. She felt as if he'd just dealt her a physical blow. "You despicable, double-dealing snake!" she hissed as she launched herself at him, her fingers curled into claws.

He caught her arms and easily held her at bay, the light of battle making twin stars of his eyes. "You deny it, then?" he taunted.

"Yes, damn your eyes, I do!" Her chest heaved with the effort to control the sobs that threatened to surface.

"Well, I believe you just about as much as you believe me, my dear little wildcat!" He pushed her away from him and walked to the windows. After testing the shutters to make sure they were securely barred from the outside, he stalked to the door. "I've got some other matters to attend to now. Jean will see to it that you get proper clothing and something to eat. He'll also make sure you stay safely and securely locked away until I return, so put away any thoughts you have of escaping me this time, Kat. I'll leave you now, but I'll return, and when I do I want some better answers than you've given me tonight."

He reached out and clamped his hand about her delicate jaw, drawing her toward him. She flinched at the pain she felt from her swollen cheek. "You should count yourself fortunate that I'm such a restrained man," he commented. "Anyone else would have beaten the day-lights out of you, if not worse."

He drew her lips up to his in a harsh, bitter kiss that ground her lips into her teeth. Abruptly he left her with the taste of his lips, her blood, and her now free-flowing tears on her mouth.

Chapter 28

REED went downstairs and closeted himself with Jean for over an hour. From there he went directly to the fortress compound where they had jailed Kathleen's crew.

The men were installed in one large cell in the fort itself. A door at one end led into a walled yard, in the corner of which stood an outhouse of sorts. The entire yard could be seen by the guards patrolling along the upper walls. The only furnishings inside the cell were small woven mats for sleeping, and these were old and dirty and smelled abominably.

The men were huddled together in a small group talking quietly in the dark, when Reed and the jailor approached. Looking them over, Reed quickly picked out Finley and motioned him forward. It took him a few seconds more to recognize Kenigan.

Dan, remembering what Dominique You had said, sat huddled against the wall. He still had on his adopted disguise, but he was taking no chances. He bowed his head upon his chest, as if asleep, and tried to make himself as inconspicuous as possible. He prayed that Reed would not shine the lantern his direction, or take much notice of him. He knew Finley and Kenigan would not give him away, nor would any other of the crew. If he could keep Reed from recognizing him, Dominique might still be able to help them all yet. If he knew Kathleen, she wasn't telling too many secrets. She'd play it pretty close to the chest for as long as she could.

Reed questioned Finley and Kenigan, but he could tell from the start that they wouldn't tell him much. He was tempted to drag them out into the compound and take the whip to them to try and loosen their tongues a little. He'd often had success that way in the past, but Reed got the impression none of these men would betray their captain. They'd lie before they'd be disloyal to Kathleen in any way.

The only thing he got out of Finley was repeated questioning as to Kathleen's welfare.

"You're mighty damned concerned with my wife's well-being," Reed sneered acidly. "Why would that be, Finley?"

Finley had it on the tip of his tongue to tell Reed that Kathleen was pregnant, but thought better of it. He didn't like Reed's attitude, and could tell by his tone what the man was getting at.

"You're mistaken in your views, Captain Taylor. She's our captain, and we respect her as such."

Reed laughed harshly. "Honor among thieves, eh? You're hardly convincing, Finley."

Finley's eyes hardened perceptively. "You're disgusting, Captain. No wonder your own wife revenges herself against you!"

Reed's control snapped. His barely leashed anger flared, and before anyone could move, he felt his fist explode in Finley's face.

Finley staggered backward, blood gushing from his broken nose.

Reed glowered at him, and then turned to the jailor. "Get him out of my sight before I kill him. Take him back to his friends and let them clean him up."

Reed walked back up to the house, when he found Dominique attempting to explain matters to an irate Jean. Between the two of them, they bombarded the hapless Dominique with questions, but after a time they both gave up. Dominique was being as stubbornly uncooperative as everyone else.

Reed was at a loss as to how to proceed. After a sleepless night and too much to drink, he hit upon an idea that might work. He would leave Kathleen under Jean's guard and sail to Savannah, where he would corner Kate O'Reilly. As he set sail out of the bay, he determined to use every threat, every underhanded means he had to, to get the information he sought.

Just before he left, he told Jean, "You tell that hard-headed hellcat that if she even thinks of escaping while I'm gone, I'll sell her to the first Turk I can find. At least I'd recoup some of my losses that way."

"Ye're a fool!" Kate's emerald eyes, identical to Kathleen's, spit fire at Reed. Kate was hopping mad. Reed had never seen her so angry.

"I'm a fool!" he roared. "What about your darling granddaughter?"

"Aye. Ye're right! Ye're both fools," Kate amended vehemently. "A pair o' blind, bloody fools! Too stupid to see the forest for the trees bein' in the way." Kate's Irish brogue was even more pronounced in her ire.

"You're not entirely blameless either, Kate. You aided her in her piratical endeavors. You are as big a fool for helping her as she was to try it to start with."

"Aye, but Kathleen thought she had a valid reason. Ye are the biggest fool o' us all because ye could ha' prevented all o' this simply by telling the lass that ye loved her beyond all else in the world. Ye never have swallowed that enormous pride o' yers and told her that, have ye?"

"Do you honestly think that would have changed things?" Reed scoffed.

"Ye bet yer boots I do! I'd stake my life on it. Ye're second huge flaw is yer immense jealousy. Ye even went so far as to suggest her baby was someone else's and not yer own. I suppose ye think the same o' the child ye've got her with now?"

"What!" Reed nearly came up off his chair. At the same time, he was recalling his last conversation with Kat. "Oh, Lord," he groaned. He didn't even have to wonder why Kat hadn't told him. He really had a knack of knowing how to stick his big foot into his even bigger mouth!

"So!" Kate nodded. "She didn't tell ye, did she? I only hope ye haven't done anything to cause her to lose this one."

Reed grimaced as he recalled how he had lost control and struck Kathleen. Her cheek had been swollen and already turning purple when he'd left Grande Terre.

"Ye haven't, have ye?" Kate sat the look of apprehension on his face.

"No. I did strike her," he admitted ruefully. "I lost my temper and struck her across the face. I regretted it immediately, but, my God, Kate, she made me so angry! Can you imagine how I felt finding out that my own wife had been making a fool of me?"

"I can sympathize with that. I tried to reason with her, but she's a stubborn one, nearly as stubborn as ye are. I just hope the two o' ye can straighten this mess out. Ye shouldn't have run off and left her. She loves ye, Reed, and she's carryin' yer child." At his look, she said, "Aye, 'tis yer baby all right. Kathleen may be spoiled, sassy, and a pirate to boot. She may be a red-headed Irish spitfire who sails and fences and rides astride, and an outrageous flirt, but she's not a hussy. She's not promiscuous. She's too honest for that."

"Honest! Kate, she's a pirate, a thief, and a liar. What makes you think infidelity isn't one of her vices as well?"

"I know that girl, Reed. She has a reason for pirating ye. That prompted her into lying to ye, but she had no reason to be unfaithful to ye. She couldn't live with herself if she did that. Besides, she loves ye too much."

"Somehow this is so crazy it's starting to make sense," Reed sighed and slouched in his chair.

He'd told Kate all that had happened, and she had told him a few things as well. Now he knew about Kathleen's father and his small shipping firm. He knew how Kat had commanded the *Enchantress*, and that there were six other vessels as well, still operating legitimately out of Ireland.

Kate had told him how Kathleen had assembled her crew, and where she had hidden the ship near the mouth of the Savannah River. She also told him of the leased warehouse where Jean's portion of the goods were stored, as well as all of Kathleen's personal profits. Once things were settled, neither he nor Jean would come up short, financially at least. Their booty would be recovered in full, and the profits would be even greater with the blockade still in effect and prices rising ever higher.

Reed finally understood why Kat had struck out at him so viciously. It had seemed petty to him before, but thanks to Kate, he now saw things through Kathleen's eyes. He saw now that if he'd told her, convinced her of his love, she would have ceased to believe he'd married her for the *Kat-Ann*.

Now that his anger had had a chance to cool, he realized that over and above all else, in spite of all that had happened, he still loved Kathleen. His pride was severely injured, but that would heal. If he lost Kathleen, the wound would never heal. Armed with his new understanding, he intended to go back to Grande Terre and face her. He had an impossible task ahead of him, but somehow he had to convince her of his love, and hope she didn't throw it in his face. He prayed it wasn't too late for them. They were so good together when they weren't at odds.

Somewhere along the line, he had to get Kathleen to tell him about the baby so that he could convince her that he believed without a doubt that it was his, and how pleased and proud he was. But that would come later. Kate had advised him on that score. If he didn't play his cards right, he'd have Kathleen thinking he wanted her only because of the baby; a possible son and heir. That would be a disastrous mistake.

He'd never realized how calm and serene his life had been before he'd met Kat.

Three days cooped up alone in that hot little room was enough to drive Kathleen to distraction. The shutters remained barred, and the room grew sweltering during the day, and little better at night.

Jean had located two dresses that fit her, but most of the time she wore her vest and short breeches. They were cooler. Besides, she hadn't forgiven Jean for keeping her locked up. He'd come only once, and she'd told him she'd saved back his part of the booty. When she tried to use it as a bribe for her freedom, he flatly refused, and she hadn't seen him since. She knew she had hurt him badly. He was severely disappointed in her, and it would take time to rebuild his faith. His look told her he still loved her, but the trust was broken.

It was Rosita who brought all her meals and supervised while a black slavegirl cleaned the room. Surprisingly, Kathleen and Rosita got along fairly well. She brought Kathleen news of Dominique and her crew, and tried to make her comfortable. She'd even helped Kathleen wash the black dye from her hair, so now it was a radiant copper once again. Rosita had not forgotten that Kathleen had saved her life, and she promised to help her if she found a way. Right now the island was a beehive of activity, unfortunately, and it complicated matters greatly. Besides, there was generally a guard outside Kathleen's door, with strict instructions.

Rosita had raided Jean's library and pilfered a couple of books for Kathleen's entertainment. This afternoon was dreadfully hot, and Kathleen had fallen asleep across the bed. Her book lay open beside her.

Kathleen never stirred until the hand clamped down tightly across her mouth. Then she came wide awake to stare up into the shifty hazel eyes of Pierre Lafitte. Her heart gave a mighty lurch, and her eyes widened in surprise.

"So. We meet again, and now it is I who have the advantage," Pierre said conversationally. "Your guard is new. He did not know of our previous disagreement and let me in," he went on to explain.

Pierre was half lying across her, effectively pinning her down. With his free hand he caressed her breast, chuckling when she squirmed in protest. "I have waited a long time for this, *cherié*. You owe me something for nearly costing me my sword arm. That was very bad of you, Kathleen, so if I am a little rough with you, you will know it is because I have not forgotten or forgiven.''

His hot, rancid breath fanned her cheek. The thought of being violated by this mad beast was terrifying, and Kathleen wondered if either she or her unborn child could survive it, or even if she would want to after he had sullied her.

Now he was fumbling with the ties to her vest as Kathleen renewed her struggles. She managed to wriggle free of his hand long enough to emit one long, healthy scream before his grimy fingers closed over her mouth again.

As he tried to tug her breeches down, she scored a connecting blow with her knee to his groin. It gained her nothing, as he cuffed her roughly across the breasts. Pain raced through her chest, and she lost precious time just trying not to pass out. Pierre's hand was cutting off most of her air, and she was fuzzily aware of him unfastening his breeches. With a strength born of desperation, she freed her arms and clawed at his face. As he tried to ward off her attack, she fought and screamed for all she was worth.

Suddenly the door to her room burst open, hanging crazily on its hinges, as Dominique flew into the room, Jean right behind him. They were followed by the red-faced guard, his pistol drawn.

Dominique was ready to forget the weapons and strangle his younger brother with his bare hands, but Jean intervened, sending Pierre from the room at gunpoint with the guard.

''This is what you call keeping her safe from harm?'' Dominique challenged Jean as he flipped the coverlet over Kathleen's near-naked body.

''It will not happen again, I assure you both.'' Jean's gaze went from one black set of accusing eyes to Kathleen's deep emerald glare of unspoken accusation.

''You're damned right it won't!'' Dominique answered. ''Because I'm taking her away from here as soon as I can arrange it.''

''You'll do no such thing.'' Jean was in command once more. ''I've given Reed my word as a gentleman.''

"Your word almost cost her life just now, brother."

Jean looked at Kathleen with dismay. *"Mon Dieu,* Kathleen, I am so very sorry! Are you all right? Did Pierre do you any harm?"

"I'll be fine, no thanks to you," she retorted. "I'm suffocating up here, I haven't seen the light of day in half a week, I've barely escaped rape, and there is no telling what Reed has in store for me whenever His Majesty deigns to return. Shall I go on?"

"Besides that," Dominique cut in, "all this is not good for the child she is carrying."

Now Kathleen turned her glare in Dominique's direction. "Why don't you take out an advertisement in the *Washington Gazette,* Dom?"

Dominique looked down at her with regret. "I'm sorry, little one, but I feel it is something he should know."

"Is that so? It is not a trick?" Jean questioned.

'It's true, Jean," she admitted. "The child is due in September."

"Yes, and look what that idiot of a husband does to her." Dominique waved a huge paw at her bruised and swollen face.

"Again I am sorry, but a husband has the right to discipline his wife. In this case, I think Reed had every justification."

Dominique looked thoughtfully at his brother for a moment. "Maybe so," he said at last, "but right now perhaps he is a little hotheaded, and his judgement is unfair. It is up to us to save him from committing a terrible mistake, and perhaps harming his unborn child. It is also our responsibility to remove Kathleen from Pierre's reach. You know how he feels about her. He will try again, and perhaps the next time he will succeed. We must help her off the island and away from here."

Jean contemplated his a moment. He knew Dominique would do anything to aid Kathleen's cause, yet his words held the ring of truth to them. "You are right," he said quietly, then to Kathleen, "Dominique will see that your ship is made ready."

"And my crew, Jean?"

He smiled, knowing she was pressing her advantage. "Your crew also."

"Thank you, Jean. I shall never forget you for this. You will always hold a special place in my heart." Only her eyes told him just how special.

"Where will you go?"

"Back home to Ireland."

"Write to me now and then, and to Dom, of course." She nodded. "And Eleanore, too."

He came close and leaned down to kiss her forehead. "Reed will have my skin for this. I will leave now so you can dress. Have Rosita bring you anything you need. Hurry now, before I think better of this and change my mind."

"Goodbye, Jean. I'm grateful. I may even name my child for you."

Dominique looked hurt. "What about me?"

"You, too. How does Jonathon Alexander sound?"

Jean smiled gently. "I like it."

"So do I," Dom said, "even if I do get second billing."

"Well," Kathleen mused. "If it is a girl, I could name her Alexandra Jean. We'll just have to wait and see, won't we?"

Kathleen and Jean parted as special friends once again. Since it was too late to load the ship before dark and still catch the tide, she would sail first thing in the morning.

Using every available man, they hauled the frigate out of the water that afternoon. Quickly scraping the hull as best they could in their rush, a necessity for the long voyage ahead, they then repainted the hull a gleaming white. While one crew was busy painting, another was changing the green sails for white, and yet another was loading food and water and other needs aboard. When they ran out of daylight, they worked by torchlight. Dan had even put together a crew to restore the figurehead to her original state, and paint out the name. All in all, over a hundred men accomplished in one day what would have normally taken three days.

By the time they sailed the next morning, the *Emerald*

Enchantress was no more. Kathleen had even left behind her pirate outfit, and wore a pair of Jean's black trousers and a white ruffled shirt, hastily altered by Rosita. Kathleen shrugged. Physically she was growing out of her costume anyway, as her waistline started to thicken.

Dominique had wanted to escort her safely to Ireland, but Kathleen vetoed that idea. "I don't need a protector, Dom. I'm going home. I have friends there. I'll be fine. Don't forget, I have a faithful crew, and I'm a skilled captain and swordswoman. Besides, there is a special feeling, a certain understanding, between the sea and me. I'm as comfortable on the sea as most women are in their own parlors."

He regretfully agreed, and they parted with the promise that one day they would again see one another.

Fireworks would have seemed quiet compared to Reed's explosive reaction when he sailed back into Barataria Bay three days later to find Kathleen gone. It took all of Jean's patience and persuasion to keep Reed from murdering all three of the Lafitte brothers. Dominique practically had to sit on Reed to get him to stay still long enough for Jean to explain. Once Jean's words began to penetrate through the red haze of Reed's anger, he calmed down a little and started to listen to what was said. At last he realized what they were telling him, and his common sense told him they had done what they thought was right, both for him and Kathleen.

He lost no time in replenishing his supplies. Once he had convinced Jean and Dominique that he meant Kathleen no harm, they assisted him in every way. Kathleen was three days ahead of him, and he had to travel light and fast.

As they waited for the tide to turn, Reed had a long talk with Dominique. His jealousy had allowed him to read more into Dominique's relationship with Kathleen than was actually there. He realized that now, even as he saw that Dominique truly loved Kathleen. If not for the respect and friendship between Reed and Dominique, Dom would have tried to win Kathleen for himself. As

Dominique explained, it would have done him no good, as Kathleen's heart belonged to Reed alone. Dominique's place in Kathleen's affections was that of a dearly loved brother, and he would be content with that.

"I hope we can remain friends, Reed. We go back a long way, and true friends are too hard to come by these days," Dominique said. "I'd never try to come between you and Kathleen. The happiness of both of you means too much to me."

Reed clasped the huge hand Dominique extended toward him. In a rare moment of humility, he admitted, "Kate was right. I've been a blind, jealous fool. I've almost lost two of my best friends, and I still have to see if I can win my wife back."

"Shall I go with you?" Dominique offered.

"Thank you, but no. This is one time when I have to do my own explaining."

It took Reed over a week to overtake the *Starbright*. He recognized the sleek lines of the infamous *Emerald Enchantress* even under her new sails and paint. It still came as a shock, as they came alongside, to see Kathleen at the wheel, her coppery hair flying out behind her, her long legs clad in well-fitted black trousers instead of her usual brief outfit. Somehow, he'd expected to see a black-haired temptress in green, a mask across her face. Here stood his wife, tall, proud, with her rapier at her side.

The grappling hooks found their mark, and Reed swung swiftly aboard to meet an irate Kathleen, rapier drawn and ready.

"Get off my ship!" she commanded, her eyes flashing emerald fire.

"I've come to take you home, Kat." He stood fast, eyeing her calmly.

"I'm headed for my home under my own power, thank you." Her voice was full of spite.

"Your home is not in Ireland any longer, Kat. You are my wife. Your home is in Savannah with me." He spoke firmly, but without rancor. His expression was almost tender.

Kathleen's eyebrows arched upward in surprise. At the very least she had expected his usual domineering attitude, with a tad of barbed sarcasm thrown in.

"What do you need with a full-fledged pirate for a wife?" she demanded. Then her brows knitted thoughtfully. "Or have you devised some especially demented, subtle torment for me as punishment?"

"Kat, I'm willing to forgive and forget if you are. Regardless, you are coming home with me. I'd hate to use force on you. I'd much rather you come willingly."

"I don't know what your game is this time, Reed, but I'm not playing. Now remove your carcass from my deck and let me go on my way, before I cut you down where you stand." She brandished her rapier at him.

He took a step backward, his hand going to the hilt of his own weapon. "I don't want to fight with you, Kat. I only want to take you home with me."

"Never!" She screamed the word at him. "I'll never leave with you willingly."

"And I won't leave without you," he stated determinedly.

"Then one of us will leave feet first, because I won't surrender to you. Arm yourself, Captain Taylor. It goes against my ethics to slay an unarmed man."

When he made no move to do so, she aimed a blow close enough to slice the ruffles from his shirt front. Thus prompted, he drew his weapon out of defense only. He had no other recourse until he could make her listen to reason.

"Kat, this is crazy!" he declared, as he parried her next two blows. "Let's talk about this!"

"There's nothing left to talk about. I'm not going back with you." She aimed a blow at his head, but he deflected it with ease.

He parried several more of her thrusts. "Why?" he pressed. "What is so terrible about it? Why shouldn't a wife live with her husband?"

Kathleen was tiring, and her pregnancy was obviously affecting her usually graceful movements. She wasn't exactly clumsy, but neither were her steps and movements as fluid as they should have been. To her own amazement, she started to cry. Through her tears, she continued to attack him.

"Damn it, Kat! Enough is enough!" His voice came through sounding tormented, even though she could no longer see his face clearly for her tears. "Will you only be satisfied when I am dead? Does widowhood appeal to you that greatly?"

She was sobbing desperately now, and his words cut her deeply. What was she thinking of? What was she doing, she wondered. How could she kill the father of her child, the man she loved more than life itself?

With an oath, she flung down her rapier, sinking swiftly to her knees, her hands brought up to cover her tear-stained face. "Oh, God! I can't do it!" she sobbed. "I love you too much!"

In an instant, he was down on the deck beside her, cradling her gently against his hard chest. "Oh, Kat. My beautiful, stubborn, Kat." He buried his face in her tumbled hair and rocked her back and forth until her sobs eased.

"Look at me, darling." He tilted her face up to his. "Look at the fool of a man who loves you more than anyone in all the world, and was too proud to tell you. I didn't want to give you that ultimate means to destroy me. I felt it would have made me too vulnerable, and maybe it does, but I can't bear to lose you." He held her tightly to him. "I need you, kitten. I love you so much!"

"Reed, Reed," she moaned against his shirt. "Don't say it if you don't really mean it. I've waited too long to hear those words, and I love you too much. If you don't love me back, I think that I shall die."

"I won't let you. I couldn't bear my life without you. You've brought chaos into my well-ordered life, played havoc with my nerves, and turned my world upside down, but I've never known more joy than waking up beside you. I've never felt such splendor as I do when I make love to you. Will you please come home with me?"

He was asking her, not demanding, Kathleen realized as her eyes searched his face. He was offering her not only a home, but his heart as well. Her heart sang with joy as she gazed up at him. He loved her! He had said so, and the truth of it was written on his face.

"Oh, Reed, I thought you'd never ask! I'd love to go home with you!" She melted against him, and held him as if she'd never let him go.

Somehow she found herself once again aboard the *Kat-Ann* in Reed's bed. She still wasn't too clear on how it had happened, but she vaguely recalled Reed taking charge and giving orders to both crews.

She lay back and watched as Reed removed his clothing. She watched as each garment fell away to reveal his gloriously muscled body, and she loved him with her eyes.

Then his long fingers were working on the fastenings of her own clothing, his loving gaze never leaving hers. Within moments she was luxuriating in the feel of his hands on her bare body, as they worked their special magic.

His warm lips came down to claims hers, branding her as his own, and she sighed contentedly, glorying in her surrender. All the barriers were down now, and she showed him her love in the movements of her hands and body. She drank in the male scent of him, headier than any perfume, and let her fingers experience the texture of his skin.

There was no urgency in their lovemaking. It was slow and sweet, and poignant with the sense of what they had almost lost. His hands and lips explored every inch of her body, as hers did his, as if to commit each touch to memory.

As his lips traveled the planes of her body, he whispered words of love to her, letting her know how precious she was to him. He told her how exquisitely she was formed, and of his delight in her. Kathleen felt prized and beloved above all else, and she was moved to tears at the depth of her emotions for this man.

She worshipped his body with her own, letting her

warm, salty tears wash away all the bitterness, leaving only the sweetness behind. As her tears fell upon him, she licked them away, and tasted the flavor of him on her tongue.

Every inch of her was sensitive to his touch, and she shivered with delight and anticipation as he deliberately sought out each erotic spot so familiar to him. His lips came back to capture hers in a drugging kiss, his tongue mating with hers, and her senses reeled. One hand found her breasts, his fingers teasing the sensitive tip into a rigid peak, and longing flared through her to the center of her body. His other hand searched out and lightly caressed the tender button of sensual delight between her legs that sent desire coursing through her. His fingers teased her even as his mouth demanded her complete surrender.

Her body felt on fire, and when she could endure no more, she tore her mouth from his and begged him to satisfy her. Obligingly he fitted his body to hers, entering her with a long, smooth stroke that took her breath away and increased her arousal. He moved slowly and tauntingly, savoring every glorious sensation. She met his thrust, twisting her hips beneath him in an exotic, sensual manner that drove him to increase the tempo of his thrusts. Their passions mounted until they were almost mindless with the exquisite torment of it. Then, like a sunburst, they climaxed together, pounded by crashing waves of release that resounded time after time until they finally receded to leave them replete.

Much later, still basking in the glow of their love-making, Kathleen asked him when he had first begun to love her.

"I didn't realize it at the time, but looking back, it must have been the first time I saw you aboard the *Kat-Ann*. There was something about you that disturbed me from the very start. I wasn't sure what it was, or that I liked it at all, but I believe I was smitten right then, and I sincerely hope there is no cure." He touched his finger to the tip of her nose. "You sort of grow on a person, you know."

"Yes, like a wart." Kathleen wrinkled her nose at him as he chuckled. "I fought my feelings for you for a long time, too," she confessed. "I knew I was lost the first time you kissed me and I felt flushed and feverish and fluttery in the pit of my stomach. Every time I tried to argue myself back to some semblance of sanity, you'd turn those brilliant, blue eyes on me and I'd melt inside."

Truths were coming out on both sides, and as he cradled her in his arms, she lay her head on his chest, comforted by the steady rhythm of his heart. As they lay listening to the slap of the waves against the hull, she told him that she carried his child. Careful not to arouse her suspicions, he did not let on that he already knew. He kissed her gently, his eyes glowing with pride and tenderness, as he told her how pleased and happy he was.

They wandered up on deck after a while, and he stood holding her as they gazed at the star-filled heavens. It was then that Kathleen told him of her idea to name their child after Jean and Dominique. "Jonathon Alexander if it's a boy, and Alexandra Jean if it's a girl. What do you think?"

"I think after all the trouble you've put me through, you owe me a son," he chuckled.

"All the trouble I've put you through!" she chided. "Really, Reed Taylor, you are incorrigible! If you want the truth, I let you off easy! Just for that, I'm going to present you with a daughter, just to spite you. You see if I don't!" She jabbed him playfully in the ribs.

"I'd love a daughter, of course, but I really could use a son or two first to help out around Chimera." He laughed, white teeth flashing, as she jabbed him again.

"Definitely not! A daughter! You'll just have to wait until next time for a son," she teased.

"But an older brother can protect a younger sister. Let's have a boy first, love," he argued with a charming smile.

"It's no use arguing, Reed, darling. I've made up my mind."

Reed tilted her face up to his and brought his lips to hers in a searing kiss that immediately rekindled the banked fires of her passion.

"That's unfair," Kathleen murmured dreamily in his arms, "but it's your most convincing argument yet."

"Shall we continue this discussion below deck?" he suggested as their eyes met in a passionate gaze.

"Let's," she answered softly.